The Refrigerator and the Universe

THE REFRIGERATOR
AND THE UNIVERSE
Understanding the Laws
of Energy

MARTIN GOLDSTEIN

INGE F. GOLDSTEIN

Harvard University Press

Cambridge, Massachusetts

London, England

1993

This book is printed on acid-free paper, and its binding
materials have been chosen for strength and durability.

Library of Congress Cataloging in Publication Data

Goldstein, Martin, 1919 Nov. 18– .
The refrigerator and the universe: understanding the laws
of energy / Martin Goldstein, Inge F. Goldstein.
p. cm.
Includes bibliographical references and index.
ISBN 0-674-75324-0 (alk. paper)
1. Entropy. 2. Force and energy. 3. Entropy (Information theory)
I. Goldstein, Inge F. II. Title.
QC318.E57G65   1993
530'.7—dc20

93-2084
CIP

To Theodore Saul Miller,
and any others
who come along

Contents

Acknowledgments

~~~~~~

This book might be said to have been started when one of the authors took a correspondence course in thermodynamics while serving in the U.S. Army. In the years since, we have both been inspired by the various professors from whom we have learned—and the various students to whom we have taught—thermodynamics as well as other branches of science.

We have profited particularly from discussions with Professor James L. Anderson of Stevens Institute of Technology on the irreversibility paradox and on relativity, with Professor George Schmidt of the same institution on irreversibility and on chaos, and with Professor Erick Weinberg of Columbia University on cosmology.

We thank our son Michael Z. Goldstein for making a number of line drawings. Computer graphics were prepared by Ms. Vaun Hatch of the Center for Biomedical Communications, Columbia University.

It was a pleasure to work with Kate Schmit of Harvard University Press, who gave the book a thorough and critical editing.

Our children have been kind enough to tell us that they did not feel neglected while we were writing it.

The Refrigerator and the Universe

1
Energy and Entropy in Everyday Terms

~~~~~~

Some people want to know how a refrigerator works. Others want to know the fate of the universe. The purpose of this book is to show that the answers to the two questions are related. The science that relates them is thermodynamics, and we can illustrate its laws with an even simpler device than a refrigerator: a spring-driven wristwatch.

We take the watch and wind it up. The hands begin to move and a ticking is heard, which continues for a little more than a day, and then the watch stops. The spring is unwound, and the hands no longer move. Except for one day's wear and tear, the watch is in the same condition as before we wound it. What has happened to the effort, small as it was, that we put into the winding of the spring? Has it disappeared with the sound of the ticking, or is it in some way still around?

Thermodynamics is the science that deals with energy; its transmission from one body to another and its transformations from one form to another. The bodies we will apply it to—watches, refrigerators, microbes, the earth, the stars—all can gain energy from other bodies or lose it to them. Energy comes in different forms—the energy of water stored behind a dam, of chemical or nuclear fuels, of electric currents, of those electromagnetic waves we call "light" or "microwaves," and the energy stored in the moving atoms of which matter is composed—and it can be changed from one form to another.

The first law of thermodynamics says that among all these transmissions and transformations, the net quantity of energy does not change. The second law says that while energy is never lost, it can be wasted, in the sense that it can become unavailable for further transformation, and thus unavailable for human use.

The spring-driven watch serves not only as an illustration of the laws of energy but as a metaphor for them as well. Invented in the fifteenth century, such watches had been developed to a state of high accuracy by the nineteenth century. In the twentieth century, though, they became obsolete: the new electric clocks and quartz watches are just as cheap to manufacture and more accurate than any spring-driven watch can be (although some of us, out of habit or nostalgia, still wear them). The laws of thermodynamics are among the nineteenth century's greatest scientific achievements. New scientific discoveries made in the twentieth century required that these laws be modified and identified limits on their applicability, but within those limits they have extraordinary power, generality, and accuracy.

We will describe these laws first in their nineteenth-century form. The reader may reasonably ask why we have chosen to take this approach. We have three reasons:

First, the laws of thermodynamics are more easily understood that way, because they are in accord with the common experiences and intuition of even twentieth-century individuals, who are more familiar with refrigerators, batteries, and automobile engines than with the interiors of stars.

Second, like spring-driven watches, the laws in their earlier form are sufficiently reliable and accurate for most purposes.

Third, one can get some sense of the significance of the revolutionary discoveries of the twentieth century—quantum mechanics and the theory of relativity—by comparing what we know now with what we knew then. In the last part of the book, we will discuss how the laws of thermodynamics and our understanding of what they mean have been changed by these discoveries. We will also describe how the laws apply to nuclear explosions and to an expanding universe.

The First Law

The first law, known as the law of the conservation of energy, states that energy is indestructible and uncreatable. The idea that there is a "something" that cannot be created and cannot be destroyed is not of itself difficult to comprehend, but we need to know more about what scientists mean by the term *energy* and how they established the impossibility of making some of it out of nothing. Scientists are hard-nosed about such matters. Energy for them is not the vague and mutable entity it is in ordinary speech—"vigor or power in action, vitality or intensity of expression, the capacity for action or accomplishment," as the *American Heritage Dictionary* (1969) defines it. They want precise definitions and quantitative measurements. Some crass bookkeeping is involved: the statement that energy is conserved means that it is conserved to the last penny, or whatever units energy is measured in.

Because the concept of energy is a difficult one, we will have to approach its meaning slowly. One problem is that energy comes in many different forms, and for most of history the different forms were defined in different ways and measured in different units. Let us give a short list of familiar examples of energy. First, energy is associated with matter in motion: an apple falling from a tree or a wrecker's ball swinging toward its target. This kind of energy is most easily visualized and is closest to the common meaning of the term. But there may be energy where there is no motion but only the potentiality of producing it—the energy of water stored behind the dam of a hydroelectric power plant, the chemical energy of gasoline, the nuclear energy of plutonium or hydrogen. From any of these we can produce the energy of motion as needed—the motion of automobiles, chain saws, or food processors. Sometimes we use the stored energy to heat our immediate environment rather than to produce motion, which tells us that "heat," whatever we mean by the term, is in some sense also a form of energy. Then there is electrical energy, often a medium by which the potential energy of fuels or water is transmitted to a new setting, where it is transformed into another form of energy: the motion of a fan or the heating of a stove.

We measure, and pay for, electrical energy in kilowatt-hours (costing approximately 10 cents per kwh in New Jersey in 1992). If energy is both convertible and indestructible, it must be possible to measure all forms of it in kilowatt-hours and show that, whatever changes in the form of energy have occurred, the total number of kilowatt-hours has not changed. As we have indicated, other units for measuring energy are also used, some quite familiar: Calories, for the energy value of foods, and British thermal units (BTU), as an indication of the power of air conditioners and refrigerators. Less familiar units are the foot-pounds of the engineer and the joule and electron-volt of the scientific laboratory. There are fixed factors for converting quantities of energy from one system of measurement to another: unlike currency conversion rates, however, they do not fluctuate from day to day. A dieter's booklet we consulted gave the energy content of a banana split as 1,165 Calories: this translates to 1.35 kwh.

Let us now apply the principle of the conservation of energy to the burning of a log of wood. A quantity of chemical energy is stored in the wood and the oxygen of the surrounding air, and this energy is released when the log is burned. Chemically, wood and oxygen are converted to carbon dioxide, water, and ashes. The chemical energy has been released and the surrounding environment has been heated. In more scientific language, the "thermal energy" of the environment has been increased, as a thermometer would show us. Kilowatt-hour for kilowatt-hour, the chemical energy released can be accounted for by the increase in the energy of the surroundings.

We can trace the chemical energy stored in the wood back through previous forms. The living tree converted carbon dioxide, water, and minerals into living cells with the aid of the energy of light from the sun. The sun emits light because it is hot: its surface temperature is about 6,000°C (about 10,000°F). The source of the sun's energy is the nuclear fusion reaction in which hydrogen atoms in the core of the sun are converted to helium.

Having traced the chemical energy of the wood this far back, we will change direction now and follow the sun's energy to other destinations. Not all plants are used for firewood. Some are grown by human beings for food and for fodder; some grow wild, die, and are converted by molds and bacteria back to carbon dioxide and water. Food and fodder give both humans and animals the chemical energy needed for their muscles and their internal temperature-regulating systems to function. Thus some of the food energy goes to motion and some to keeping the body warm. But not all the sun's energy falling on the earth goes to grow plants—only about 0.1 percent does. Of the rest, some 35 percent is reflected or re-radiated back into space. Some, most strongly the ultraviolet light, is absorbed in the atmosphere, increasing its temperature and thus its energy but also producing some chemical reactions. The remainder warms the surface of the earth, evaporating water and warming air next to the surface. The warmed air rises, winds follow, the atmosphere of the earth circulates, rain and snow fall, tradewinds blow, and sometimes hurricanes. All is powered by the energy radiated by the sun.

Throughout all the transformations that take place, the bookkeeping goes on: kilowatt-hour for kilowatt-hour, all the energy can be accounted for. The total amount remains the same regardless of the changes of form. According to the first law, the total energy of the universe will remain the same for all time, the same as it is today, and the same as it was in the distant past. Forward or backward, it makes no difference.

The Second Law

But forward or backward must make a difference somehow. Think of the log of wood burning: the flames start out and spread; slowly the wood blackens, glows, and then turns to ashes. The heated gases from the combustion rise and warm the surrounding air while they mix with it. Gradually it all dies down as the gases and the thermal energy spread out more and more in the atmosphere. Eventually nothing is left except a pile of ashes.

But can the clock be turned backward to reverse this process? We have mentioned that plants convert carbon dioxide, water, and minerals into wood. Are we not back at the starting point again? Not quite. The reconstitution of the log of wood from ashes and the other products of combustion—or, in other words, the growth of a new tree—requires an input of the sun's energy. Turn off the sun and the ashes remain ashes. Well, then, could we not reconstitute the wood by other means: powerful lamps to take the place of sunlight, for example? Yes, but we need a source of energy for the lamps. To reverse the burning, a price must be paid in energy. Note that there was no external energy cost to burn the wood. True, we needed a match to start the fire, but one match can burn down a forest; we can leave it out of the bookkeeping. There is a natural direction to processes in the universe, and no energy cost is required to make them go in this direction. To make them go in the opposite direction we must pay.

Imagine that we had made a film of the burning of the log and then projected it backward. First we see a pile of cold ashes. Gradually warm air containing carbon dioxide and moisture descends on the ashes, which begin to get warm. The rate of descent of warm air, now mixed with smoke, increases: the ashes begin to glow, then break into flame. As the flames rise, wood is formed. Soon the flames die down again, and the log has been recreated.

No one watching this film would believe for a moment that it represents a possible event in the real world. Yet nothing that has happened violates the first law: the energy can be accounted for, kilowatt-hour by kilowatt-hour, whichever direction the film is projected. We need a new law of nature to tell us which is the natural direction and which the impossible direction for things to happen. This is the second law of thermodynamics.

The second law can be stated in a number of different ways that reflect different perspectives on its meaning, implications, and consequences. We will use various statements as we go along. Let us give one here, one related to the question of which way to run the film through the projector.

In the nineteenth century, Rudolf Clausius, a German physicist and one of the discoverers of the second law, formulated that law in an almost innocuous-sounding statement that heat will flow spontaneously from hot bodies to cold ones, never the reverse (heat flows from cold bodies to hot ones in the operation of refrigerators, but not spontaneously: an expenditure of electrical energy is required to make it happen). From this starting point, he was able by logical analysis to prove that matter must have a previously unrecognized property, which he called *entropy,* that can be readily determined in the laboratory. He further showed that in all natural processes the total entropy of everything involved in the process can never decrease: it can remain unchanged in certain idealized processes, but in all real changes, the entropy will always increase. Note the contrast to the total energy in any real change, which neither increases nor decreases. It led Clausius to formulate the second law in a new form: *the entropy of the universe tends to a maximum.*

The universe of the twentieth century—now expanding, perhaps infinite in extent, and obeying the theory of relativity—raises questions about Clausius's formulation. We will discuss them only briefly here, but we will discuss them in more detail in a later chapter. Dealing with the whole universe at once is an awesome project anyway, and we more commonly deal with processes taking place in small pieces of it at a time, pieces the size of a laboratory. To the extent that we can imagine each process carried out in isolation from all others, the total entropy within such isolated small regions must necessarily increase. Let us consider a few examples:

1. The flow of heat (thermal energy) from hot bodies to cold ones, leading to an equalization of temperature.
2. Any process involving friction, in which the energy of motion is converted to a heating of the moving body and its environment.
3. The expansion of a compressed gas into ordinary air, as when a tire is punctured.

4. An ice cube melting in a glass of warm water.
5. A teaspoon of sugar dissolving in a cup of hot coffee.
6. The combustion of gasoline in an auto engine.

We immediately recognize a film of any of these processes run backward as representing an impossible event. In all of them it can be shown that the total entropy of everything involved in the process has increased. The total entropy thus tells us objectively whether the film is being run forward or backward. Entropy has been called "the arrow of time."

The Molecular View of Entropy

Just what is entropy, and what does it mean to say that it is a property of matter? Entropy was discovered by an analysis of the conversion of the chemical energy of fuels into the energy of motion by steam engines. But the concept is more easily understood from the viewpoint of the molecular theory of matter, so we will set aside our burning logs and melting ice cubes for a moment.

Matter is made up of atoms of various kinds, usually combined together in stable aggregates called *molecules* (for example, H_2O is a molecule). Atoms and molecules are very small, and even a small amount of matter—an ounce or a gram—contains an enormous number of them. Further, they are in constant motion, colliding with each other repeatedly and therefore undergoing frequent changes in speed and direction. We may imagine them to act something like billiard balls colliding elastically on a billiard table. It would be difficult to predict the path of one ball when there are as few as 10 others on the table for it to collide with, but nevertheless we are able to study the millions of molecules in a vial of gas, say, or a drop of liquid, if we ask different kinds of questions: we cannot know the path of a particular molecule but we can describe the *average* behavior of enormous numbers of them. This is just the kind of problem for which the theory of probability was designed, and while it may seem easier to apply it to the tosses of a coin or to the shuffling of a deck of cards than to the motion of atoms, the theory has allowed us to make successful predictions of molecular behavior. Let us first talk about cards, though.

A deck of fifty-two cards can be arranged in order of suit and value. When shuffled, the deck tends toward a more disordered arrangement. A disordered deck of cards can also be shuffled, and it is not impossible that it could be shuffled into a perfectly ordered arrangement. But it is highly unlikely; it is much more likely to go from one disordered arrangement to another. The tendency of systems to approach a state of maximum disorder, and once there to stay there, is an example of the operation of what is often called the *law of averages*. It takes no deep grasp of the theory of probability to see that if we repeat the experiment with decks of trillions or quadrillions of cards, the tendency to go from an ordered arrangement to a disordered one is even more overwhelming, and the likelihood of any large deck ending up in a neat order after shuffling is inconceivably small.

Just like decks of cards, large collections of atoms or molecules can be classed as being either in ordered or disordered arrangements, and the vast numbers of disordered arrangements involved ensure with virtual certainty that they will end up disordered.

To return to our example of the burning log of wood: it can be shown that there was more order in the arrangement of the molecules of the wood and the air than there is, after burning, in the molecules of the ashes, the combustion products, and the heated environment. Entropy is related in a very simple way to the degree of disorder. The law of increasing entropy is thus equivalent to the statement that ordered systems tend to disorder while disordered systems tend to stay that way. The outcome is not inevitable, only very, very probable, but the probability can be overwhelming. It is safer to gamble that the tendency to disorder will prevail.

Quantum Mechanics and Relativity

A quantitative calculation of how much disorder there is in a particular collection of molecules, and therefore of the entropy of the collection, requires us to know how molecules move under the influences of whatever forces they exert on each other. The assumption was made in the nineteenth century that molecules move according to the same Newtonian laws as falling apples do, but the results of such calculations were often in stark disagreement with experimental measurements. In the twentieth century, the new quantum mechanics was shown to describe the experimentally observed properties of molecules, including their entropies, with, as far as we can tell, complete accuracy.

The theory of relativity, also a product of twentieth-century science, provided a framework for studying the history of the universe, from its beginning to its ultimate end. One of its most remarkable implications is that the universe is not static: it must be either expanding or contracting. Astronomical observations show in fact that the universe is now expanding. The theory of relativity shows it will either continue to expand forever or at some point turn around and begin to collapse.

The most significant consequence of relativity for thermodynamics is that energy is not always conserved. When the universe expands, the total amount of energy we can account for decreases, and if the universe were ever to contract, the total amount would increase. This does not deprive energy of an important role in our unstable universe; the energy content of the universe determines its rate of expansion and whether it will ultimately contract or not.

The second law must also be modified to apply to the unstable and possibly infinite universe of the theory of relativity, but there is as yet no good reason to doubt the inexorable increase of entropy. It is still, as far as we know, time's arrow.

In what follows we have three goals: to tell how the concepts of energy and entropy were discovered; to explain how they apply in a variety of fields—for example, in the study of radiation (including the "greenhouse effect"), chemistry (the synthesis of diamonds), biology (how the muscles do work), and geology (the age of the earth); and to describe how the concepts of energy and entropy have been modified by quantum mechanics and the theory of relativity, and how they apply to the expanding universe.

2
Work and Force

~~~~~~

Major scientific discoveries convert a large number of "small" and unconnected discoveries, some made years or centuries earlier, into special cases of a single general statement. The laws of thermodynamics are an example: the first law combined several thousand years of discoveries in mechanics with several hundred of the study of heat. In this chapter we will review another example, Newton's laws and the science of classical mechanics, in which the concept of energy makes its first appearance.

Newton's laws of motion enable us to predict how the motions of bodies are changed by the forces acting on them. We need to know the strength of a force acting on a body and the direction in which it acts, but it doesn't matter whether the force arises from gravity, elasticity, friction, magnetism, or whatever. In the study of mechanics it is sometimes useful to imagine an idealized world from which frictional forces have been eliminated. Many processes that go on in the real world are only slightly affected by friction, and for these the idealized world is a close approximation. In this frictionless world, mechanical energy—the sum of the energy of motion and the energy stored as potential motion—always remains the same. Here we will describe what is meant by the energy of motion, or *kinetic energy,* and the energy stored as potential motion, or *potential energy;* why the knowledge that the total doesn't change is useful; and what happens when friction takes its toll.

To convey the concepts of kinetic and potential energy we will need to refer to more basic concepts first, some of which were discovered by the ancient Greeks and some by Newton's immediate predecessors or contemporaries, and all of which are combined in Newton's laws. The most difficult of these is Newton's concept of *mass* and how it is distinguished from *weight*.

Work

In science familiar words are often given an unfamiliar meaning: one such word is *work*. Work is defined as the product of a *force* acting on a body and the *distance* the body has moved under the action of the force. The common meaning of the word *force* is close enough to the scientific one for us to forgo a formal definition; we rely instead on the reader's intuitive understanding of it and some examples. One of the most familiar forces is *weight*, the downward force on any object at the earth's surface resulting from the gravitational attraction between the object and the earth. Another is the force of *friction*, which we must overcome to slide an object along the floor. The quantity of work done when we lift something is equal to the product of its weight and the height it is raised; when we slide an object along the floor the quantity of work done is the product of the horizontal force exerted by our muscles against friction and the distance the object is moved. This definition of work implies that no work is done when a weight is held in the air but not lifted, as when we stand holding a suitcase. Still we get tired when we do; why we get tired in spite of the fact that we are doing no work (in the scientific sense) is a question that will be discussed in Chapter 12.

We can clarify the concept of work further by an analysis of the "simple machines," such as the lever, the wedge, or the pulley. The lever, as exemplified by the crowbar, the bottle opener, and the seesaw, is the most familiar. The law of the lever was known to the Greeks of the fifth century B.C.E. It was clearly formulated in a book whose earliest extant version, a Latin translation, bears the name "Aristotle" as the author. Scholars are agreed that its style and approach are not that of the master and that the work is probably by one of his followers, name unknown, who is therefore referred to engagingly as "Pseudo-Aristotle." The lever, although simple and familiar, has a quality of magic about it: the ability to produce a large force at one end by the application of a small force at the other, with no limitation on the amount by which the small force is magnified. It led Archimedes to say, "Give me a lever, a fulcrum, and a place to stand, and I will move the earth." With the aid of a seesaw, for example, a 60-pound girl can lift her 150-pound father against the pull of the earth's gravity. If the relative distances of girl and father from the bar over which the seesaw turns (the fulcrum of this particular lever) are adjusted correctly, the two bodies can be held in perfect balance, and by moving a tiny bit further out the child can lift the adult.

To achieve the state of balance, the distances from the fulcrum must be in the right proportion to the weights: the lighter of the two must be further out. Expressed more precisely, the product of the force F (the weight) by the distance d from the fulcrum must be the same for both. If the father is 4 feet from the fulcrum, the daughter must be 10 feet from it ($150 \times 4 = 60 \times 10$). While the quantities on either side of this equation are the products of a force times a distance, they are not a measure of "work" because the distances are not distances that the bodies have moved in the direction of the force. (The direction of the force of gravity is, of course, downward, so the motion we are interested in is vertical.)

From this state of balance, the up-and-down motion that makes riding a seesaw fun can be produced if each rider "pushes off" each time he or she is down. (The additional force of the "push-off" need be only a little more than enough to overcome friction, and we leave it out of the following discussion.) We can express the law of the lever in terms of work if we introduce the vertical distances (which we symbolize by h) traveled by the girl and her father. Note that the father, closer to the fulcrum, rises a shorter distance than the daughter descends. Geometrical intuition suggests that the vertical distances traveled are proportional to the distances from the fulcrum. Suppose that the father travels a vertical distance of 2 feet as he goes up or down; the daughter then must travel 5 feet. Again, force times distance must be the same for both, but now instead of $F \times d$ our equation involves $F \times h$: $150 \times 2 = 60 \times 5$.

In other words, the work done *on* the father to raise him is equal to the work done *by* the daughter as she descends. We have introduced two innocuous-sounding prepositions, *on* and *by*, in the preceding sentence: work is done *on* a body when it is moved in the opposite direction from the force acting on it (up, if the force is gravity), and work is done *by* a body when it moves in the same direction (down, for gravity) as the force acting on it.

While the two riders on the seesaw travel on arcs of a circle, not on straight vertical lines, the direction of the force of gravity is strictly vertical, and the "distance" that we use in the definition of work is that part of the distance traveled that is strictly parallel to the direction of the force.

Obviously there is a difference between the scientist's definition of work and the ordinary use of the term to mean any activity that requires effort and is tiring.

Waterpower

Throughout human history we have relied mostly on our own muscles and those of animals to supply the force needed to do the work we wanted done. Prior to the Industrial Revolution there were two other important sources of force, wind and falling water. Waterpower will be of some interest to us. On the left side of Figure 2.1 we illustrate the

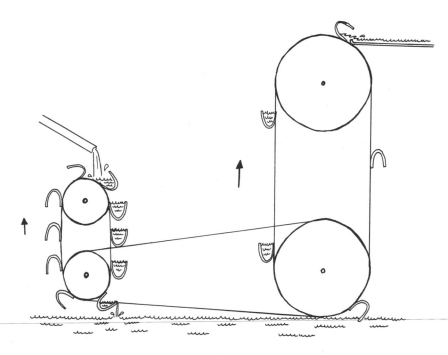

Figure 2.1   Waterwheels

Here one waterwheel is doing the work necessary to drive another and pump water uphill. Water can be pumped up to a higher elevation than that of the falling water driving the work-performing wheel, but the weight of water pumped up must be less in proportion.

basic principle of a waterwheel. The cups of the wheel fill with water and exert a downward force (weight) on one side of the wheel. When the filled cups reach the bottom they are emptied so that as they rise they do not counterbalance the filled cups. The wheel thus turns the shaft, and the shaft is connected to some machine that does work for us. We can arrange things so that the shaft drives millstones and grinds corn, or it can drive a lathe for woodworking. Or we might want to pump water out of a coal mine.

Suppose our waterwheel is driven by a waterfall 20 feet high. Can we use the power to pump water out of a mine 100 feet deep? The answer is that we can: the problem is not in principle different from the use of a lever that allows a small force to overcome a large one. A possible design is illustrated in Figure 2.1. A waterwheel with a radius of 10 feet, driven by a 20-foot waterfall, is connected by a belt to a wheel with a 50-foot radius, which can raise water 100 feet. The law of the lever requires only that the weight of water raised 100 feet must be much less than the weight of water falling down our 20-foot waterfall, in accord with the rule previously given for the lever: the weight of falling water × height of fall = the weight of water pumped up × height the water is raised. This can be put in another way: *the maximum work our waterwheel can do is equal to the work done by the falling water that drives it.*

The qualification *maximum* is important. Waterwheels and drive belts are affected by friction and other factors leading to inefficient operation: the actual work a wheel will do will in reality be less than the amount ideally possible. But the ideal case is an important special case, and no one designing a mechanism for accomplishing some practical purpose under real conditions can afford to ignore the upper limitation it imposes.

Friction

Turning now to the question of how much motion can be produced by a given quantity of work, we find ourselves at the beginning of the modern era of science, with Galileo, Newton, and their contemporaries. Before we go on we would like to suggest that the reader do a simple experiment requiring equipment easily available at home: to construct and observe the motion of a simple pendulum, which illustrates beautifully almost everything about classical mechanics that is needed here.

To prepare a simple pendulum, tie any object weighing a few ounces

Perpetual-Motion Machines

The relationship we have stated for the maximum amount of work a machine can do sounds reasonable enough to modern ears, but it took a long time for it to become accepted. Repeated attempts were made, continuing until the nineteenth century, to design waterpower systems in which the fall of water could somehow be used to pump enough water back up to the top of the waterfall to keep the waterwheel turning forever, and still have some capacity to do other work besides. They represent specific examples of a device called a *perpetual-motion machine,* which by definition is any device capable of performing more work than the work that must be supplied to keep it running. Illustrated here are a waterwheel that both drives a wheel to grind grain and pumps the water that drives the wheel itself, and a wheel that is kept by moving balls in a permanent state of unbalance, so that it keeps turning forever.

A perpetual-motion machine, once in operation, could do the work necessary to run itself and at the same time do an unlimited amount of external work as well, thus giving us something for nothing. Inventors

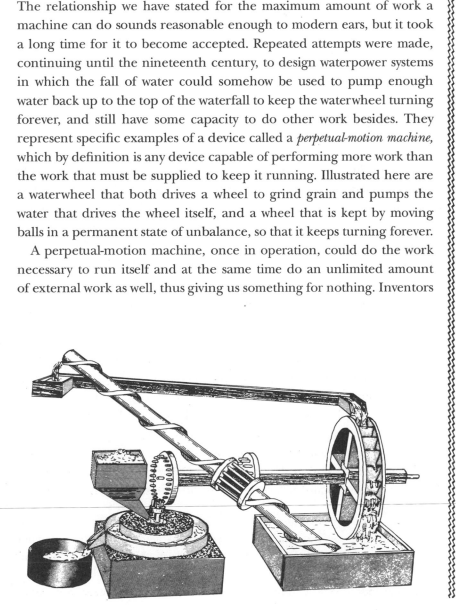

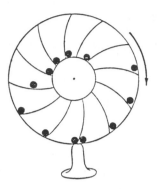

of such machines have not limited their efforts to falling water. Recently an electrical device claimed by its inventor to put out more power than the power input needed to run it was refused a patent by the U.S. Patent Office, on the grounds that it was a perpetual-motion machine and therefore couldn't operate as claimed. The case was taken to the U.S. courts, which backed the Patent Office.

Most inventors of perpetual-motion machines have not been scientists or trained engineers, who, both on philosophical grounds and on experimental ones, have taken for granted ever since the Middle Ages that no such device could ever be made. Galileo assumed its impossibility to make some predictions about the behavior of pendulums, which he then confirmed by an experiment. A less well-known Flemish scientist, Simon Stevin, preceded Galileo by a few years in this kind of reasoning, and both were anticipated by Leonardo da Vinci one hundred years earlier. In his notebooks, after describing a number of designs for perpetual-motion machines, da Vinci wrote: "O speculators about perpetual motion, how many vain chimeras have you created in the like quest? Go and take your place with the seekers after gold [the alchemists]."

to a thread or light piece of string about a meter long, and tie the other end of the thread to a hook or nail in such a way that the object can swing freely over a distance of 30–40 centimeters. Now move the weight to one side, keeping the string straight, and let it go. Watch it move.

The object will swing from side to side. The amplitude of its swing will gradually diminish, and after a long period of time and many swings back and forth it will come to rest in the vertical position. Because the thread is of fixed length, we know that the object is traveling on an arc of a circle, and therefore as the object swings back and forth it is successively rising and falling: at the ends of its swing it is higher above the floor than in the middle. It is less obvious but easily shown that the speed of motion of the object changes in the course of the swing: it is moving fastest at the middle of the swing (when it is in its lowest position) and actually comes to rest for an instant at the extremes. The eye can be fooled here—some people perceive the object as moving at a constant speed, with abrupt changes of direction at the ends of the swing. The speed can be estimated better by intercepting the object with the hand at various positions in its path and comparing the force of the different impacts. Stroboscopic photography shows the changes in speed more elegantly (see Figure 2.2).

The gradual decrease of the amplitude of the swing and the fact that the pendulum eventually comes to rest remind us that we live in a world of friction. Whatever moves, moves against the resistance of air, of water, of the ground with its pebbles and its mud. We take for granted that to keep things moving we must keep pushing. If we don't, they slow down and stop. Aristotle was led, by such common physical experiences, to formulate this behavior as a fundamental law of nature: the normal state of a body is a state of rest. To make a body move, a mover—an external force—was necessary. Galileo could go beyond Aristotle only by an imaginative act: he invented an ideal world in which friction was abolished. A moving body in this world remained in uniform motion unless some force acted on it to change its state of motion. Friction, in this view, is simply one of a number of forces capable of changing the motion of a body.

Note that both Aristotle and Galileo would have agreed that in the real world bodies do come to rest unless something or someone keeps pushing them. For Aristotle they did so because rest is their natural state. For Galileo they did so because the frictional forces that arise from their motion slow them down.

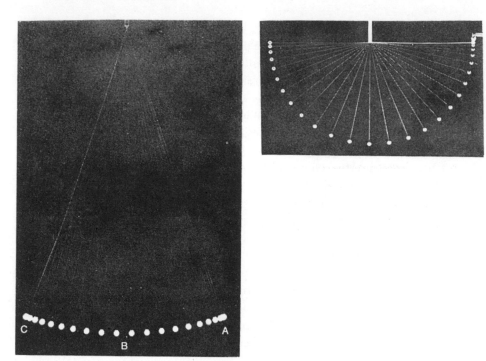

Figure 2.2   Pendulums
Stroboscopic photographs of moving pendulums, one swinging with a large amplitude and one with small, show the different speeds at different points in the pendulum's path. The stroboscopic camera superimposes a series of photographs taken at equal time intervals on a single film, so that two successive images taken when the pendulum is moving at a high speed will be farther apart on the film than when it is moving more slowly. It can thus be seen that the pendulum moves more rapidly at the bottom of its swing.

Newton's Synthesis

In the seventeenth century Isaac Newton formulated his laws of motion, incorporating and extending Galileo's insight in a form that could be applied to an enormous variety of problems. These laws relate changes in the speeds and directions of motion of bodies to the strength and directions of the forces acting on them. To calculate the orbits of the planets—their first dramatic experimental test—Newton needed to know in particular how the force of gravity depended on the distance between the sun and each planet, and how strong it was. He combined a few plausible assumptions in his famous law of gravitational attraction and then used this law together with his laws of motion to account for all that was then known about how bodies moved under the influence of gravity. These few laws described not only the planets in their orbits and the moon's path around the earth but also the fall of an apple, the flight of a cannon ball, and the swing of a pendulum.

Newton's work has been described as the greatest single advance in human knowledge ever made. The exploration of the consequences of his laws of motion and their applications to problems he never thought of remained one of the main activities of physicists for the next two hundred years.

The Distinction between Mass and Weight

We do not need to give Newton's laws in full here, but we do need to understand some of the concepts he used in stating them. In particular we need to know what *mass* is, and we need to distinguish it from *weight*.

Weight is familiar enough: we deal with it every day. We feel the weight of a body when we lift it, and our common experiences tell us that anything that is hard to lift is also hard to push. But there is a distinction between lifting and pushing that reflects the distinction between the weight of a body and its mass, a distinction recognized by Newton when he formulated both his law of gravity and his laws of motion. We know that weight is a force due to the gravitational attraction of the earth, and that at distances far enough away from the earth weight no longer exists. But there is in outer space something left of what we would have called "weight" on earth. If a body were standing still in a place free of gravity and we wanted to get it moving, we would still be required to exert a force on it. We can test this, even in the presence of gravity, with a child's wagon bearing a load of bricks and with wheels well-greased to reduce friction. We find it requires a considerable force to get the wagon moving at an appreciable speed, even on a horizontal surface; and once it is moving, an equally large force is required to stop it. If one is tempted to attribute the forces needed to the *weight* of the bricks, one should be reminded that horizontal motion involves no effort against gravity. If it were feasible to bring the loaded wagon onto a space vessel and try the experiment in gravity-free outer space, the forces to get it moving or slow it down will be found to be the same as for horizontal motion on earth.

What is this ghost of weight, present when weight is absent? Newton called it the *mass* of the body. He defined it, somewhat tautologically, as "the quantity of matter" in a body; it is a property which remains the same whether or not gravity is acting on it and which can be measured by the force necessary to produce a given change of the body's speed. Mass is equivalent to what is called the inertia of a body: the difficulty of setting it into motion when it is at rest, and the difficulty of stopping it if it is moving.

The mass of a body is thus a property that can be observed and measured in the absence of gravity. Measurement requires that we first choose an arbitrary standard of mass, just as we have chosen arbitrary standards of length (the meter, the foot) and time (the second). The standard of mass used internationally in scientific laboratories, the kilogram, is defined as the mass of a particular piece of platinum-iridium alloy kept at Sevres, near Paris, copies of which are in standards laboratories all over the world.

The mass of any other body can be determined in principle by comparison with this standard or any of its copies. Mass can be measured by what is called *centrifugal force:* tie a string to the body whose mass is to be determined and whirl it in a circle at constant speed. The force in the string is determined by the speed of rotation, the length of the string, and the body's mass. If the rotational speed and the length of the string are kept the same, the relative masses of any two bodies are proportional to the two centrifugal forces.

This procedure is based on a deduction, using some mathematical reasoning, from Newton's laws. We will need to give two of them here: (1) If no force is acting on a body, its speed and direction of motion remain the same. (2) To change either the speed of a body or the direction of its motion, a force must be exerted on the body. The greater the force, the greater the change; but the greater the mass of the body, the smaller the change—meaning only that it is harder to get heavier bodies moving or to change their speed or direction of motion if they are already moving.

Newton's Law of Gravity

To use these laws to predict how planets or apples move under the influence of gravity, Newton needed to know the strength of the gravitational force on a body at the earth's surface. He (and Galileo) were aware that, when air resistance is neglected, heavy bodies fall at the same accelerating speeds as light ones. Acceleration is defined as the rate of change of motion: the concept includes changes in either speed or direction. We feel the acceleration in a moving car when the gas is stepped on suddenly (change of speed) or when the car swerves suddenly (change of direction). If the force of gravity were the *same* on each body, the *less* massive bodies would be accelerated *more* and fall faster, but they don't. Since the downward acceleration was the same for all bodies, whatever their masses, it followed that the gravitational force on any one body must be greater the greater its mass. This greater gravitational force, then, just can overcome the greater resistance of a more massive body to being accelerated. In brief, the mass of a body, a property defined independently of gravity, is found to determine the strength of the gravitational force acting on the body.

In his law of gravitational attraction, Newton therefore assumed that the strength of the force of gravity between two bodies was proportional to the product of their masses. Although he had good reasons for making this assumption, there is something surprising about it: why should there be an attractive force between two bodies proportional to the difficulty of getting each of them moving? Despite this mystery, the assumption worked so extraordinarily well that for over two hundred years no one thought of questioning it, and even then it took an Einstein (see Chapter 15).

The attractive force exerted by the earth's gravity on bodies at its surface is therefore greater the greater their masses, and since this attractive force is what we mean by "weight," we can describe the relation between the weight w of a body and its mass m by the equation $w = g \times m$, in which g is the strength of the earth's gravitational field at its surface. The gravitational-field strength of a planet or a star depends, according to Newton's assumption, on two things, the mass of the planet or star and how far the observer is from its center. Any body weighs less on the moon and more on Jupiter than on earth, reflecting the differences in mass and size of the planets and the moon. The numerical value of g can be determined without knowing either the mass or size of the earth simply from the weight of the standard kilogram. The equation above provides us with a quick and easy way to measure the *mass* of any body: compare its *weight* to the *weight* of that arbitrarily chosen standard, an experiment we can do on the moon or on Jupiter as easily as on the earth (in principle, anyway). Although we can determine the mass of a body without using the force of gravity, weighing it is usually easier (Figure 2.3). One of the crucial properties of mass (as far as pre-Einstein physics is concerned) is its unchangeabil-

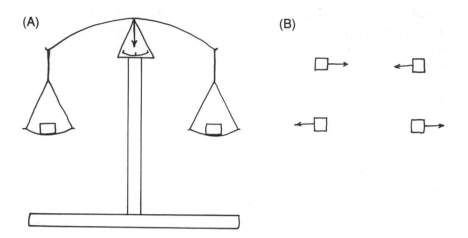

Figure 2.3 Mass vs. Weight
It is possible to compare the mass of a body with an arbitrarily chosen standard
mass. *A:* If there is a gravitational field present, compare the weights of the two
bodies on an equal-arm balance. *B:* If there is not, set the two bodies on a direct
collision course with equal speeds. If their masses are equal, their speeds after
the collision will also be equal.

ity. Like energy, it cannot be created or destroyed; the total amount of
it must remain the same for all time.

Kinetic and Potential Energy

Newton did not establish the laws of mechanics all by himself. He had
predecessors—Galileo died the year he was born—and great contemporaries as well: Descartes in France, Huygens in The Netherlands, and
his arch-rival in the invention of the calculus, the German Gottfried
Leibniz. These men made a number of discoveries in mechanics and
motion independently of Newton, but in the fullness of time everything
they discovered was shown by others to follow from Newton's laws.

Gottfried Wilhelm von Leibniz was a man of remarkably diverse
interests and accomplishments. He was active in mathematics, science,
engineering, theology, and metaphysics while he pursued a busy career
in public affairs and diplomacy. His generally optimistic philosophical
outlook led Voltaire to satirize him in *Candide* as the philosopher
Pangloss, for whom everything was for the best in this best of all possible
worlds.

In the field of mechanics, Leibniz discovered the special importance of a property possessed by a moving body which he called *vis viva*, or "living force," and which he defined as the product of the mass of a body and the *square* of its speed:

$$vis\ viva = mv^2$$

The modern term for the concept of *vis viva* is *kinetic energy*, the energy a body has by virtue of being in motion. Today, however, kinetic energy is defined as *one-half* the product of m and v^2. This definition was chosen because of the relation of the energy of motion to another kind of energy, *potential energy*.

In our discussion of the lever we showed that any object having weight and being at a certain height in the earth's gravitational field can, through a lever, be made to do work on another object, raising this second object to a greater height and itself descending in the process. The first object can be said to have "a potential to do work" determined by its mass, the strength of the earth's gravity (g), and its height. As it raises another object, doing work on it, and itself descending, the object loses some of its potential to do work while the second object gains, in being raised, an equal potential to do work. The "potential to do work" (we will use the word *potential* for short, and for the time being) of a body can be given a numerical value provided we choose some arbitrary reference level from which to measure its height; customarily mean sea level is used. The potential is thus *weight* $\times$ *height* above sea level. Obviously a body at sea level has a potential of zero, because its height above sea level is zero. This does not mean it can't do any work: we could be using this body to lift some other body in Death Valley or near the Dead Sea, well below sea level but still in the earth's gravitational field. The potential is now negative, according to our definition, but the body can still be made to do further work by decreasing its already negative potential still further. It is like having a bank account with extensive overdraft privileges (Figure 2.4). As an alternative to sea level, we could use the bottom of Death Valley as our reference level, but we rarely do so.

The Speed of Fall

Now an object at a certain height above the earth's surface can lose its potential without actually doing any work. It's easy: just drop it. No weight is lifted as it falls. But as it falls its speed of fall increases, and the further it falls the greater its speed becomes. Can the speed it acquires be predicted from the loss in its potential (its decrease in weight × height)? Would we expect that the speed v will increase in proportion to the height it has fallen, or is the relation more complicated?

This is the question Leibniz answered, with the introduction of his concept of *vis viva*. In brief, he was able to prove that as the quantity weight × height decreases, it is not the speed that increases in proportion to the distance fallen but the *vis viva*. Specifically, the decrease in potential equals one-half the increase in *vis viva*. To restate this in symbols: if v is the speed of a body of weight w after falling a distance h,

$$w \times h = \tfrac{1}{2}mv^2$$

The equation tells us that a falling body must fall four times as far to double its speed.

In modern terminology, what we have clumsily called "potential to do work" is now the *potential energy*, and the equation states simply that the body gains in kinetic energy what it loses in potential energy. Leibniz did not use the term *potential energy*, of course; the name he

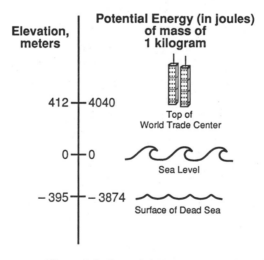

Figure 2.4 Potential Energy

For convenience, gravitational potential energy may be referred to sea level. If we choose sea level as the reference point, a body on top of a tall building above sea level has a positive potential energy, a body below sea level a negative potential energy.

used for the lost potential was *vis mortuum,* or "dead force." The object at rest had none of the energy of motion, but it did have the potentiality for *producing* motion. The relation between potential and kinetic energy as a body falls will be seen to be a special case of the first law, an application to a frictionless world.

Galileo, too, had observed that it is the square of the speed of fall, not the speed itself, that varies with the distance fallen. He did not make this discovery by studying the speeds of fall of dropped bodies, which cannot be measured accurately without modern instruments, such as high-speed cameras and stroboscopic lighting. Instead, he had the ingenious idea of sliding objects down gently sloping planes so that they accelerated much more slowly than do objects in free fall. He made the assumption that the speed of an object that has descended vertically 1 centimeter is the same whether it had fallen freely or slid down a very gently sloping plane (so gently sloping that the object may have traveled a meter on the plane in order to descend 1 centimeter vertically). The object on the plane takes much longer than the object in free fall to travel the same vertical distance, making observation and measurement easier (Figure 2.5).

The simple pendulum described in the beginning of this chapter is something like the objects on Galileo's inclined plane. When the supporting thread is kept taut and the body is moved away from its position of rest, it moves upward as well as to the side. The vertical distance the body is raised is, of course, much smaller than the distance it travels along the arc of motion, from one end of its swing to the other. It

Figure 2.5 Objects Falling Down Inclined Planes

Galileo was able to observe the speeds acquired by falling bodies by allowing them to "fall" down inclined planes. The speed of the body once it has reached the level portion of its path depends on the height through which it has fallen, but not the steepness of the path. Hence, once they reach level ground, both balls shown here will be traveling at the same speed, since they fell from the same height. In this experiment we assume a frictionless world; in the real world friction can be minimized, but not eliminated.

Flywheels

Potential energy—gravitational, chemical, or nuclear—can be stored almost indefinitely before it is used. In contrast, kinetic energy doesn't last long in that form—friction steadily diminishes it. In many applications, however, it is useful to store energy as motion for short periods of time—it can be put to use quickly when we need it. The engine of an automobile does not turn the wheels directly. Instead, it increases the speed of rotation of a heavy flywheel, and when we want the car to move, we connect the flywheel to the wheels via the transmission.

What is the kinetic energy of a rotating wheel? The formula $\frac{1}{2}mv^2$ still applies, but different parts of a wheel travel at different speeds (as anyone who has stared at a phonograph record on a turntable knows). If we imagine the wheel divided up into tiny pieces, each of the same small mass m, each piece has a kinetic energy $\frac{1}{2}mv^2$. Taking into account the fact that the speed v is different for each such piece, we must add all the kinetic energies of the pieces up to get the total for the wheel. This is easy to do with the aid of calculus. To cut a long and complicated story short, the kinetic energy of a flywheel depends on its mass, dimensions, and the square of its rate of rotation (expressed, say, in rpm).

follows that when we displace it to one side with our hand, we are doing a quantity of work $w \times h$ on it, where h is the change in vertical position we have produced, which increases its potential energy. When we let the object go, the object "falls"—it really swings in an arc, but the driving force for the motion is just the small upward displacement we gave it. As it descends, the potential energy we gave it by displacing it decreases and in turn the kinetic energy increases until the lowest position of the object (the supporting thread now vertical) is reached. The object, however, does not stop dead. It is in motion, and it continues in motion, swinging past the bottom position; now it is gaining potential energy at the expense of kinetic energy. When it reaches (nearly) the same height above the earth as it originally fell from, it has no more kinetic energy—the energy has been transformed from kinetic to potential—and it begins to fall back. From here on the process

In the flywheel shown here, a weight hangs from a rope wound around the shaft on which the flywheel turns. It illustrates the interconvertibility of potential and kinetic energy, just as a pendulum does. If the flywheel were at rest, it would be set into motion (thereby gaining kinetic energy) as the weight descends. If the flywheel were spinning rapidly, it could raise the weight, giving the weight potential energy at the expense of its own kinetic energy.

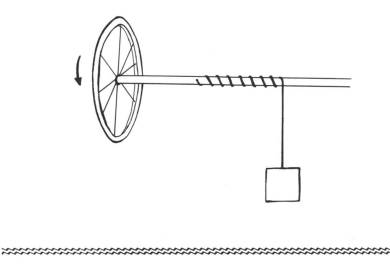

repeats itself, until friction, which acts during each swing to decrease the motion of the pendulum, finally stops it.

We have expressed the result for falling objects generally, and the pendulum specifically, in the form:

Loss of potential energy *(PE)* = Gain of kinetic energy *(KE)*

We can simplify this equation by introducing the idea that the whole system can be thought of as having at each moment both potential and kinetic energy, which we can add together to give a *total energy E:*

$$E = PE + KE$$
$$E = (w \times h) + (\tfrac{1}{2} mv^2)$$

Our statement now is that the total energy does not change as the system moves. Figure 2.6 illustrates this for the pendulum. Obviously, a

Pendulum

Total Energy

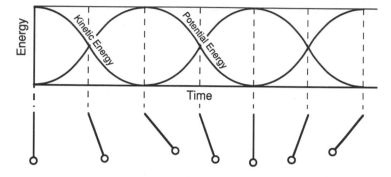

Figure 2.6 Energy and the Pendulum

The kinetic and potential energies of an ideal, frictionless pendulum vary with time as the pendulum swings, but total energy remains constant. The kinetic and potential energies are completely "out of phase," each reaching a maximum when the other reaches a minimum, but they always sum to the same total energy.

gain in kinetic energy must be compensated for by a loss in potential, or vice versa. The total energy E given to the pendulum depends on how far we moved the suspended object in the beginning: we can give it either a large or small amount of energy. But once we leave it alone and let it swing, the energy should remain constant.

But of course it doesn't. The amplitude gradually diminishes because of friction, the object comes to rest at the lowest point of the swing, and the energy it was given has "disappeared." This disappearance of mechanical energy was not particularly regarded as a problem or paradox of mechanics. Friction as a force was simply different from other forces in its consequences. However, a small minority of scientists, noting that friction produced a heating effect, suggested that the mechanical energy was still present, in an altered form. Proving this turned out to be more difficult than suggesting it.

3
Heat from Work: The First Law

~~~~~~

The study of the phenomena of heat began, as mechanics did, with the ancient Greeks. By the end of the eighteenth century, after the invention of the steam engine, a unified and comprehensive theory of heat—called the caloric theory—had won the support of most scientists. Heat was then considered a material substance ("caloric") whose presence was what made a hot body hot. The theory, which was developed independently of the laws of mechanics, provided coherent answers to such questions as why hot bodies expand, why water boils when heated, how much work expanding steam can do, and others. With its aid, precise experimental procedures for measuring quantities of heat, similar to those we use today, had been established. Both the name and concept of the Calorie, a unit quantity of heat, are legacies of the caloric theory.

Support for the theory, however, was not unanimous. That friction produces heat, an ancient observation, led some philosophers and scientists to conclude that heat was not material but a form of motion. Plato made the suggestion in the dialogue *Timaeus:* "heat and fire . . . are themselves begotten by impact and friction: but this is motion. Are not these the origin of fire?" Francis Bacon, aware of the production of heat by friction and of the violent motion of boiling liquids, concluded that "the very essence of heat . . . is motion, and nothing else." Galileo and Newton held this view also, as did the philosopher John Locke: "The axle-trees of carts and coaches are often made hot, and sometimes to a degree, that it sets them on fire, by the rubbing of the [hub] of the wheels on them . . . [H]eat is a very brisk agitation of the insensible parts of the object, which produces in us that sensation from whence we denominate the object hot." Leibniz explained the disappearance of *vis viva* in frictional processes in these words: "This loss occurs in appearance only. The forces [the *vis viva*] are not destroyed, but dissipated among the small parts."

In 1738 Daniel Bernoulli, a mathematician and physicist at the University of Basel, published an influential treatise, *Hydrodynamica,* on the flow of fluids. In the chapter "Concerning Properties and Motions of Elastic Fluids [Gases], but especially of Air," he proposed the hypothesis that a gas is a collection of rapidly moving particles and that its pressure arises from the impacts of the moving particles on the walls of the container. Bernoulli showed that the pressure of a gas should be proportional to the number of particles in the container and to their average *vis viva;* as the gas was warmed, he suggested, the average *vis viva,* and thus the pressure, would increase. But Bernoulli's "moving particle" theory did not explain anything the widely accepted caloric theory did not already explain, and in spite of the fact that *Hydrodynamica* was an influential work, remaining in print and in use throughout the eighteenth century, his views had no discernible impact.

The conflict between the two theories, the caloric and the kinetic, is of interest as a case study in scientific discovery, and we will describe it briefly in this chapter. It was definitively resolved in the period 1840–1850, primarily by the work of the English amateur scientist James Joule. Joule used the procedures for measuring quantities of heat developed during the ascendancy of the caloric theory to show that there is a quantitative equivalence between mechanical energy (either kinetic or potential) and heat. It had just been discovered that an electric current could be generated by mechanical motion, such as from falling water. Joule showed that an electric current in a wire produces heat, and that the quantity of heat produced is precisely predictable from the quantity of mechanical energy needed to generate the current. In other experiments he showed that when mechanical energy is caused to disappear by friction, the quantity of heat produced by the friction bears the same relation to the mechanical energy as did the heat produced by the electric current in his electrical experiments.

Joule's work led to two conclusions: First, heat is not a substance; the phenomena of heat can be accounted for in terms of mechanical energy: the energy of the invisible particles—atoms and molecules—of matter. Second, energy was recognized as an entity found in many different and interchangeable forms—mechanical, chemical, electric, magnetic, elastic, and thermal (heat), among others—but in all its transformations the total amount of it never changes: it is *conserved*. This is the first law of thermodynamics.

The Development of the Caloric Theory

Empedocles of Agrigentum in Sicily postulated in the fifth century B.C.E. that there are four elements—earth, air, fire, and water—and two moving forces—love and strife. This doesn't sound much like science, but Empedocles did establish one important fact by a scientific experiment: air is, at least, a substance, whether or not it really is an element. The experiment involved a closed vessel with holes at the top and bottom called a clepsydra (clepsydras, or "water clocks," measured the passage of time by the slow dripping of water out of the small hole in the bottom); Empedocles placed his finger over the hole in the top and immersed the vessel in water. Hardly any water came in through the hole in the bottom until he removed his finger. Conclusion: air is material. In the view of the historian George Sarton, this experiment earns him an honorable and permanent place in the history of science. According to Empedocles' theory, all substances were assumed to be made up of mixtures of the four elements, in varying proportions, and the more "fire" there is in a body the hotter it will be.

In the seventeenth and early eighteenth centuries, the theory of the four elements had been abandoned, but as alchemy was developing into chemistry the notion that there is a "principle" of fire or flammability trapped in certain kinds of matter was proposed. This "principle" was given the name *phlogiston* by the chemist G. E. Stahl around 1700. The process of combustion—coal or wood burning and giving forth flames and heat and leaving an ashy residue—was regarded as a liberation of phlogiston. The part played by air in the combustion process was not clearly recognized.

A systematic explanation of chemical reactions in terms of phlogiston developed by Stahl dominated chemistry for a century, until it was overthrown in the period 1780–1790 by Antoine Lavoisier. Lavoisier showed that combustion was not a release of phlogiston but a combination of the combusting substance with the oxygen of the air, air being shown by him to be a mixture of the elements oxygen and nitrogen. Lavoisier gave the first list of elements we recognize as elements today: in addition to oxygen and nitrogen, hydrogen, carbon, sulfur, phosphorus, and a number of the metals—gold, silver, copper, iron, and zinc.

Also in this list were two "substances" of quite a different sort—light, and the substance of heat, which he called *calorique*. To decide what was an element and what was not he employed a hypothesis, "mass cannot be created or destroyed" (it is conserved), along with a scientific instrument capable of verifying it: the chemical balance, even then highly accurate.

In Lavoisier's view caloric, like the other elementary substances, was composed of indestructible atoms. He was not the first to propose that both heat and light are substances. Earlier scientists considered light and heat to be "subtle fluids," substances without a particulate structure; they were called "subtle" because they could move easily through the interstices of even hard matter, as light passes through glass and heat through any substance whatever. The theory that light consisted of material particles had the prestige of Newton behind it at that time. Electricity, which moves easily through metals, was also regarded as a subtle fluid.

One important property of caloric was definitively established by Lavoisier and his colleague the mathematical physicist Laplace: it had no detectable mass. They showed this by burning such substances as carbon or phosphorus in sealed containers. These chemical reactions give out large quantities of heat, which escaped from the containers although the chemical products of the reaction could not. The masses of the sealed containers did not change during the course of the combustion in spite of the loss of "caloric" to the surroundings.

In spite of its lack of mass, once caloric was viewed as composed of indestructible atoms, it followed that caloric, like each of the other elements, should also be conserved. Just as Lavoisier could show with his accurate chemical balance that elements like carbon or oxygen could be carried through a whole series of chemical changes without any change in the *quantities* of the elements, the same must be true for the quantity of caloric. The fact that caloric had no mass meant only that some other means of measuring quantities of it had to be found, and, as we will see, Lavoisier was quick to find one.

What the Caloric Theory Explained

By the end of the eighteenth century the scientific community had largely accepted the caloric theory and was using it to explain and predict phenomena involving heat. Atoms of caloric were assumed to repel each other but to be attracted to the atoms of the ordinary kinds of matter. The atoms of ordinary matter attract each other, as shown by the fact that gases, whose atoms are dispersed, condense as they cool to form first liquid and then solid substances, whose atoms are packed closely together. It was believed that when cold solids were warmed, additional atoms of caloric entered the material, attached themselves to the ordinary atoms, and reduced their attractive forces. Result: expansion on heating, and eventual melting to a liquid. On the absorption of sufficient caloric, all attraction between the particles of matter is overcome by the self-repulsion of caloric; thus liquids boil to become gases, in which the caloric atoms attached to the molecules repel each other so strongly that unless they are confined by a container, gases tend to expand indefinitely. The higher the temperature of a gas, the greater the quantity of caloric around each atom, leading to a greater mutual repulsion and thus a greater pressure on the walls of a container. In fact gases do exert a greater pressure the warmer they are. It all fit together. The kinetic theory, in contrast, did not seem to offer any explanation of expansion on heating or the changes from solid to liquid to gas.

The central concept of the caloric theory was that heat, being an elemental substance, was *conserved:* the total amount of it had to remain unchanged. It could not be created or destroyed; what one body gained, some other body must have lost. This was one prediction that seemed amply confirmed by most experiments designed to test it.

Temperature, Heat, and the Equilibrium of Heat

We are all familiar with thermometers as a means of giving a quantitative measure to the sensations of "hotness" and "coldness," a measure we call *temperature.* We are aware also that the numerical magnitudes given by thermometers are somewhat arbitrary—we can use either the Fahrenheit or Celsius (centigrade) scales with equal ease: the melting point of ice is 0°C or 32°F, and the boiling point of water is 100°C or 212°F, both at normal atmospheric pressure. The thermometer was invented by Galileo, who used the expansion of air as the measure of temperature. His instrument consisted of a glass globe with a long tube attached, the open end of which was placed under the surface of water in a vessel. When the globe was cooled, water rose in the tube; on heating the globe it fell. A physician friend of Galileo's used this device to detect fever in his patients. Soon it was realized that the expansion of a liquid makes for a more reliable instrument, easier to construct. Thermometers manufactured in Florence, Italy, in the seventeenth century look much like the thermometers of today.

It is a commonplace observation that bodies hotter than their surroundings tend to cool down, and bodies that are colder tend to warm up. Let us imagine a more precisely controlled situation: a styrofoam picnic container provides quite good insulation, enough to keep a lunch cool for at least a few hours even on a hot day. If we place a warm object (a lukewarm can of beer) and a cold object (a block of ice) together inside the container, we expect a fairly uniform temperature to be reached by all objects in the container within a relatively short time. If the container were a perfect insulator instead of a pretty good one, we would expect no further change in temperature to occur. This is not a surprising state of affairs to us today, but it is not a self-evident one. What does a "thermometer" actually read? Why should different bodies placed in contact with each other eventually end up giving the same reading on that instrument?

When the tendency toward this equalization of temperature, after which no further change could occur, was first observed in the late eighteenth century, it was rightly recognized as telling us something important—it was not clear what, exactly—about the nature of heat and the meaning of temperature. This final state of uniform temperature was termed by the Scottish physician and chemist Joseph Black the *equilibrium of heat*. Why should there be an equilibrium state in which the reading of a thermometer should be the same in all parts of the system?

Analogs of Temperature

Part of the problem in providing an explanation is that the terms *temperature* and *heat* are not sharply distinguished in ordinary speech. The necessary distinction was first made with the aid of the caloric theory; though the caloric theory is dead, the distinction remains an important one. It can be clarified by analogy with some other familiar phenomena. One analog to temperature is summed up in the saying, "Water seeks its own level." The large coffee makers we see in restaurants are made of metal and are thus opaque; we would not be able to tell whether an urn is full or empty except that a vertical glass tube is attached to the side, with short horizontal tubes at the top and bottom opening to the inside. We know that the level of the coffee inside should be the same as the level in the glass tube, and if the level inside is changed (by drawing off a quantity of coffee), then the level in the tube adjusts very rapidly to the new level inside. The principle that the levels of liquid must match when an equilibrium is reached is analogous to the principle that temperatures will match at an equilibrium state. The vertical glass tube is an analog to a thermometer.

A similar principle applies to gas pressure. If two tanks of a gas such as oxygen or carbon dioxide are connected, gas will flow from the tank in which the pressure is higher until the pressures equalize, after which no further flow will occur. The air around us is at a pressure of about 15 pounds per square inch. In an automobile tire, the pressure is higher—typically 25–35 pounds per square inch greater—than that of the surrounding air. We all know what happens to the air in the tire when the tire is punctured. A pressure gauge is another analog to a thermometer.

Temperature differences thus *resemble* water-level differences or gas-pressure differences. They tell us in which direction flow will occur, and when flow will stop, but not the quantities that flow. Water seeks its own level equally in a coffee urn and an ocean: the principle of equal levels doesn't tell us whether reaching equilibrium will require the transfer of a million gallons or a cupful.

What is it that flows when there are temperature differences? For those who believed the caloric theory, there was an obvious answer: the flow of "caloric" leads to a temperature equalization and the end of flow. We know how to measure gallons of water; how are quantities of "caloric" measured?

Measuring Quantities of Heat

Black's principle of the equilibrium of heat tells us that the contents of our picnic container will eventually come to a single equilibrium temperature, but what will that temperature be? Our common sense and common experience leads us to expect it will depend on the quantities of the different substances—one ice cube or a trayful, a can of beer or a six-pack—placed in the container and their initial temperatures, but just how?

Let us try to guess the outcome of some simple experiments. Suppose we place 1 kilogram of water at a temperature of 80°C and 1 kilogram of water at 20°C side by side in an insulated container. The obvious guess for the final temperature is 50°C, and measurement confirms our guess. Suppose now we have 2 kilograms of 80°C and only 1 kilogram at 20°C? We no longer expect 50°C but a higher temperature, and a naive but plausible guess would be a temperature two-thirds of the way from 20°C to 80°C, or 60°C. Again this is what we find. There are no dramatic surprises so far, but we have in effect introduced a new concept into the discussion: we are assuming that 2 kilograms of water at 80°C has more of "something" than 1 kilogram at 80°C, even though both are at the same temperature. If we thought of heat as a substance, as did Lavoisier and most of his contemporaries, we would say that 2 kilograms of water at 80°C will have twice as much of it as 1 kilogram at 80°C, and that 1 kilogram of water at 80°C will have more of it than the same kilogram at 20°C. But how much more? How do we measure the quantities of a substance whose nature is mysterious, a substance which cannot be handled separately from ordinary kinds of matter like mercury or water?

Scientists, being pragmatic about such matters, are always willing to measure what they do not yet understand. The procedure of defining a unit quantity of anything always begins with an arbitrary choice of a unit of measurement: this lump of metal shall be 1 kilogram; the distance between these two marks 1 meter. The first quantitative measure used for heat was neither the BTU nor the Calorie, but rather the heat needed to melt one pound of ice. Why this represents a definite quantity of heat is an interesting story.

Melting Ice and Latent Heat

How or why do ice cubes cool a drink? The obvious answer is that ice is cold: if you put it in lukewarm water (or lukewarm gin) the principle of temperature equilibrium ensures an immediate and cooler temperature for the mixture. But there is more to it than that. Ice doesn't just cool a drink, it keeps it cold, even in a warm room or when clutched in warm fingers.

For simplicity let us think of pure water in a glass in a warm room—the alcohol in gin or the sugar in tea complicates the story a little. Put in ice and a thermometer. If a small amount of ice is used, it melts quickly and cools the water some, but the water will warm up again fairly quickly. If enough ice is added, the temperature of the mixture falls to 0°C and stays there. As long as there is some ice present and the mixture is stirred, the temperature does not depend on the proportion of ice to water. Now the glass with the mixture of ice and water is in a warm room, or in contact with a warm hand. Heat is flowing from the warm surroundings into the cold mixture: we can easily verify that the air around the glass, or the flesh touching it, is cooled by the cold mixture. Yet the temperature of the ice-water mixture doesn't rise. This sounds paradoxical: heat flows in but the temperature stays the same.

Is anything happening inside to indicate an inward flow of heat? Well yes, the ice is melting. Once it all melts, the temperature is no longer stuck at 0°C; it begins to rise and will reach that of the warm room in a brief period of time. What is happening to the heat flowing in from the surroundings to the cold mixture, heat which seems to disappear into the ice and water without raising the temperature of either? Is heat being destroyed in this experiment?

Joseph Black resolved the paradox by the introduction of a concept he called *latent heat,* as distinct from "sensible" heat. He proposed that when matter undergoes those changes of form we call melting or boiling—the first a change from solid to liquid, the second from liquid to gas—a definite quantity of heat is absorbed by each pound of ice that melts or water that boils, solely to produce the change of form. No rise in temperature (which we could detect either by touch—thus the term *sensible heat*—or by the thermometer) occurs during the process. It is only when all the ice is melted that the water can rise in temperature, which occurs on further absorption of heat (Figure 3.1).

The caloric theory provided a satisfying explanation of the phenome-

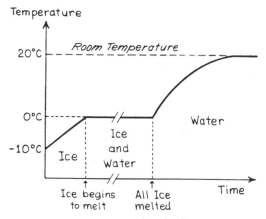

Figure 3.1 Ice Melts Slowly

In an environment warmer than 0° Celsius (32° Fahrenheit), ice originally at a temperature below its melting point will warm up rapidly until the melting point is reached and melting begins. Then the temperature will not change (provided the ice–water mixture is kept in equilibrium by rapid stirring) until all the ice is melted, in spite of the warmer environment. Because of the large quantity of heat that must be absorbed to melt ice, the melting process will take much longer than it took to bring the cold ice to the melting point or than it will take to warm the water, once all the ice is melted, to the temperature of the environment.

Joseph Black on Melting Snow

One of the important practical consequences of latent heat is that snow and ice don't just spontaneously melt when the air temperature rises above the freezing point. It takes much longer, because the quantity of heat required is so large. Here is Joseph Black's description of the phenomenon:

If we attend to the manner in which ice and snow melt, when exposed to the air of a warm room, or when a thaw succeeds to frost, we can easily perceive, that however cold they might be at the first, they are soon heated up to their melting point, or begin soon at their surface to be changed into water. And if the common opinion had been well founded, if the complete change of them into water required only the further addition of a very small quantity of heat, the mass, though of a considerable size, ought all to be melted within a very few minutes or seconds more, the heat continually to be communicated from the air around. Were this really the case, the consequences of it would be dreadful in many cases; for, even as things are at present, the melting of great quantities of snow and ice occasions violent torrents, and great inundations in the cold countries, or in the rivers that come from them.

But, were the ice and snow to melt as suddenly, as they must necessarily do were the former opinion of the action of heat in melting them well founded, the torrents and inundations would be incomparably more irresistible and dreadful. They would tear up and sweep away every thing, and that so suddenly, that mankind should have great difficulty to escape from their ravages. This sudden liquefaction does not actually happen; the masses of ice or snow melt with very slow progress and require a long time, especially if they be of a large size such as are the collections of ice, and wreaths of snow, formed in some places during the winter. These, after they begin to melt, often require many weeks of warm weather, before they are totally dissolved into water . . . In the same manner does snow continue on many mountains during the whole summer, in a sufficient time for its complete liquefaction.

non of latent heat: a molecule in a liquid has attached to it a considerably larger number of caloric atoms than would be attached to the same molecule in a solid, and a considerably larger number would be attached to a molecule in the gas state. The solid does not melt until this excess number is acquired, nor do the liquid molecules evaporate. The excess atmosphere of caloric around the liquid molecules explains the usual increase of volume on melting, and also why liquids flow easily. The still greater excess around the gas molecules accounts for the mutual repulsion that gives gases their tendency to expand and their increasing pressure when warmed.

Lavoisier, the chemist, used this concept of the latent heat of melting ice to measure the heat given out when various substances were burned in oxygen. As a quantitative standard, he and the physicist Laplace defined a unit quantity of caloric as that amount that would melt one (French) pound of ice. (The French pound—established at the time of Charlemagne—was slightly heavier than the English pound. The metric system came in with the French Revolution: Lavoisier's treatise was published in 1789, reporting work done in the preceding decade.)

For the experiments Laplace designed a *calorimeter* (Figure 3.2). With this instrument they measured the quantities of ice melted by the heat given out by the combustion of a number of substances. Lavoisier's treatise gives results on three elemental substances, phosphorus, carbon, and hydrogen, in pounds of ice melted per pound of the element burned. We have converted their figures to modern energy units (kilojoules per mole of substance; the mole is the chemist's unit for a definite quantity of matter) and compare them with values obtained with modern highly accurate instruments (Table 3.1). The accuracy of Lavoisier's results leaves something to be desired, but any modern chemist, recognizing that these were the first measurements of heats of chemical reactions ever performed, will look on them with respect.

Table 3.1
Lavoisier's Results Compared with Modern Measurements
of Heats of Combustion

Substance	Pounds of ice melted per pound of substance burned	Heat of combustion (kilojoules per mole[a] of product)		
		Lavoisier and Laplace	Modern	% Error
Phosphorus	100	2,077	1,530.5	+35%
Carbon	96.5	388	393.5	−1%
Hydrogen	295.8	198	285.8	−31%

a. The mole is a unit of quantity of matter used in chemical calculations. It corresponds to a definite number of molecules of the substance, rather than a definite mass. (See Chapter 4).

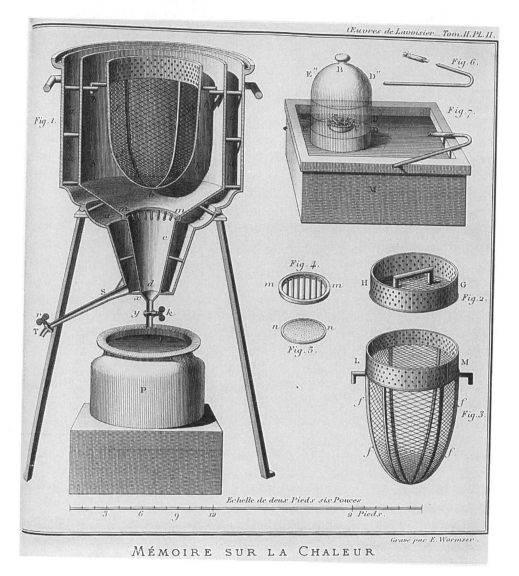

Figure 3.2 Laplace's Calorimeter
This is a reproduction of the original illustration of the ice calorimeter in
Lavoisier's treatise. The inner reaction compartment (*f*), the surrounding ice
compartment (*b*) with its spigot (*T*) to drain off the water formed by melting
(which was then weighed), and the outermost compartment (*a*) for the ice
shield are plainly visible.

The Calorie as a Unit of Heat: or, Is Hot Water Fattening?

The unit quantity of heat proposed by Lavoisier was not widely adopted, however. Instead, the conventional unit of heat was based on the quantity needed to raise the temperature of water. The unit used most widely by scientists (until recently) is the *calorie*, defined by the French after the Revolution and the adoption of the metric system. The calorie is that quantity of heat needed to raise the temperature of 1 gram of water 1 degree Celsius. (To be precise, we should specify that the temperature is to be raised from 14.5°C to 15.5°C, as the quantity of heat required per degree rise depends slightly on the initial temperature of the water.) According to the principle of conservation of caloric, it is also the quantity given out by a gram of water when it cools 1 degree Celsius. This is for many purposes an awkwardly small quantity of heat, and it is often more convenient to use a unit 1,000 times larger, originally called the *Calorie* (capital C). Scientists today call it the *kilocalorie,* a term less likely to lead to confusion. When we talk about the Calorie content of foods it is this larger unit we are using. A cracker "containing 60 Calories" will give out 60 kilocalories of heat when burned in oxygen, with all the carbohydrates and fat in the cracker converted to carbon dioxide and water. The heat given out by the combustion is enough to raise the temperature of 60 kilograms of water 1°C or, equally well, raise the temperature of 1 kilogram of water 60°C, or 5 kilograms 12°C. (Most readers will also have heard of the British thermal unit, or BTU, the amount of heat needed to raise one *British* pound of water 1° *Fahrenheit.* One calorie is equal to 3.97 BTUs.)

The same principle applies to melting ice as to burning crackers. When 1 gram of ice melts, it is found to absorb 80 (small-*c*) calories. Thus 1 gram of ice can cool 4 grams of water initially at 20°C to 0°C because it absorbs heat (80 calories of it) from the water, until both ice and water reach the same temperature. In this example, the 80 calories melt the gram of ice, so to keep the water (of which there is now 5 grams) at 0°C, one would have to add more ice.

If, instead, we drink the ice water, the body supplies enough heat to warm it to body temperature. When we drink hot water (or tea or coffee), the heat from the drink provides the body with heat as the liquid cools down to body temperature. How many calories do we gain just from the heat of the drink? Enough to compromise our dieting? A cup holds about 200 grams of water. If the temperature of the hot drink is 95°C and the body is at approximately 37°C, the water is cooled 60 degrees, so each gram of water supplies 60 calories and the whole cup 1,200 calories. These are small-*c* calories, so our gain from the temperature of the water is 1.2 Calories (kilocalories), much less than we get from one teaspoon of sugar in the cup (18 Calories) or one tablespoon of light cream (30 Calories).

Specific Heat: A New Property of Matter

The choice of ice or water as substances for defining the unit of heat is arbitrary; one must expect that other substances will require different quantities of heat to raise their temperatures and that they will have different latent heats of melting and boiling. We have implied the existence of a new property of matter here: the quantity of heat needed to raise the temperature of a unit mass of a particular substance by a given number of degrees. Joseph Black, who first defined this quantity, named it the "Capacity for Heat," but the modern term is *specific heat capacity*. If we use the units of calories for heat, grams for mass, and degrees Celsius, the specific heat capacity of a substance is defined as the number of calories needed to raise the temperature of 1 gram of that substance 1° Celsius.

The specific heat of (liquid) water is necessarily 1 calorie per gram per degree C, because the calorie was defined that way (strictly speaking, this is so only at 15°C, but it is close to 1 calorie per gram at other temperatures). We give some representative values of latent and specific heats for various substances at various temperatures in Tables 3.2–3.4. It is worth noting that latent heats of boiling—the heat that must be supplied to evaporate one gram of a liquid, or, by the principle of conservation of caloric, the heat given out when a gas condenses to a liquid—are much greater than those of melting. For example, 540 calories per gram is required to convert water to steam at the normal boiling point, compared with 80 calories to melt 1 gram of ice. We can suffer burns from water at or near the boiling point of 100°C, but steam at that temperature is much more dangerous. On condensing on one's skin, 1 gram of steam gives out 540 calories, almost 10 times the amount given out by the same quantity of boiling hot water cooled (when it touches the skin) to body temperature.

Another interesting point noted in Table 3.2 is the extraordinarily low value for the specific heat of "heavy" metals: for example, that of mercury is 0.03 calories per gram per degree C. When water at 80°C is

Table 3.2
Specific Heat of Various Substances at 20°C

Substance	Specific heat (cal/g/°C)
Solid metals	
Silver	0.056
Copper	0.092
Aluminum	0.136
Other solids	
Rock	0.2
Glass	0.2
Wood	0.3–0.4
Ice (at 0°C)	0.2
Liquids	
Water	1.0
Mercury	0.034
Benzene	0.41
Glycerine	0.54
Gases	
Air	0.25
Argon	0.12
Carbon dioxide	0.20
Hydrogen	0.20

equilibrated with an equal mass of mercury at 20°C, the final temperature of both is about 78°C: the quantity of heat needed to warm mercury is only one-thirtieth of that given out by the same mass of water on being cooled the same number of degrees.

The caloric theory gave no explanation for the extraordinary differences in specific heats and latent heats of different substances, but for that matter neither did the kinetic theory of heat, which in Black's time was a much vaguer idea, with very little predictive value. Black reasoned that the atoms of mercury are heavier than those of water (which is correct), so if heat *is* motion, why should it be easier rather than harder to get the heavy mercury atoms moving? He felt the observation made the motion theory less plausible. The caloric theory, although it could not explain the very small specific heat of mercury, at least did not

Table 3.3

Specific Heat of Copper at Various Temperatures

Temperature (°C)	Specific heat (cal/g/°C)
−223	0.023
−173	0.061
−123	0.077
−73	0.086
−23	0.090
27	0.092
227	0.097

Table 3.4

Latent Heat of Various Substances

Substance	Melting point (°C)	Latent heat of melting (cal/g)	Substance	Boiling point (°C)	Latent heat of boiling (cal/g)
Ethyl alcohol	−114	25.	Helium	−269	6.
Water	0	80.	Nitrogen	−196	48.
Benzene	5.4	30.	Ethyl alcohol	78.3	204.
Sulfur	119	13.	Benzene	80.2	94.
Lead	327	5.9	Water	100	540.
Sodium chloride	804	124.	Mercury	357	65.
Copper	1083	42.			

predict exactly the opposite of what is observed. Black should not have overlooked the fact that if mercury atoms are heavier, there must be fewer of them in one gram.

The Controversy Begins

The measurements used to establish the concept of a quantity of heat, and to determine specific and latent heats, all made use of that central principle of the caloric theory; caloric, being an elemental substance, cannot be created or destroyed. In measuring the specific heat of a metal, aluminum for example, a quantity of aluminum of a known mass and heated to an elevated temperature would be dropped into a known mass of water at 20°C. The temperature of the aluminum would fall and that of the water rise until they are equal. The number of calories gained by the water in being warmed would be calculated from its mass and its rise in temperature. By the principle of *conservation of caloric,* this is exactly equal to the number of calories lost by the aluminum. The specific heat of aluminum would then be calculated from its mass and its fall in temperature.

In all processes involving transfer of heat from one body to another, strict bookkeeping was both assumed and repeatedly confirmed: what one body loses, another must gain, down to the last fraction of a calorie.

One might have thought the apparent weightlessness of caloric would tell against a substance theory, but caloric, like light and electricity, was assumed to be a substance of a different kind from ordinary matter, and no one was yet prepared to rule out the possibility that the force of gravity may not act on it.

There were too many phenomena that the caloric theory could plausibly explain that the theory of heat-as-motion, then only a vague idea, could not. The fact that heating effects are associated with friction, an observation that suggested a kinetic theory, was not forgotten, but little attention was paid to it. It is not surprising that at the end of the eighteenth century most scientists believed the caloric theory. But not all.

One who did not was Count Rumford of Bavaria. Rumford was born Benjamin Thompson in Woburn, Massachusetts, in 1753. In the period just prior to the American Revolution, as tension between the colonists and the British was building up, Rumford concealed his Loyalist sympathies in order to provide information for the British army. He invented a secret ink for these communications which, in the opinion of his biographer, Sanford Brown, was of a quality not surpassed until the First World War. Fearing exposure, Rumford fled to the protection of the British army in 1775 and left for England shortly thereafter. His secret ink was not his only scientific contribution to the art of war; he developed methods for measuring the force of gunpowder by the motion imparted to a heavy pendulum by a rifle bullet (this method is still used in elementary physics courses). In 1784 he became an aide-de-camp to the Elector of Bavaria and eventually received the honorific title Count of the Holy Roman Empire. As part of his work on the organization and equipment of the Bavarian army, Rumford continued his scientific and technological investigations.

As a young man, he had read a treatise on chemistry by the Dutch physician and chemist Hermann Boerhaave, who believed that heat was a vibratory motion of matter. A large number of Rumford's experiments were designed to support this hypothesis and destroy the caloric theory. The most famous of these experiments began when he noted the great quantity of heat given out as cannons were being bored. That friction produces heat was, as pointed out earlier, an old observation, and it led some observers to connect heat with motion. Those who believed the caloric theory were aware of the phenomenon, but they had other explanations for it. Some of these explanations were anticipated by Rumford and tested by his experiments. As long as there was friction, he found, heat continued to be given off with no diminution in its rate of production. He concluded that if heat can be produced in unlimited quantities from friction it cannot be a substance, as *it is not conserved.*

Rumford put it as follows:

> By mediating on the results of all these experiments, we are naturally brought to the great question which is so often the subject of speculation among philosophers; namely—
>
> What is Heat? Is there any such thing as an igneous fluid? Is there anything that can with propriety be called caloric?
>
> We have seen that a very considerable quantity of Heat may be excited in the friction of two metallic surfaces, and given off in a constant stream or flux in all directions without interruption or intermission, and without any signs of diminution or exhaustion . . .
>
> It is hardly necessary to add, that anything that any insulated body, or system of bodies, can continue to furnish without limitation, cannot possibly be a material substance; and it appears to me to be extremely difficult, if not quite impossible, to form any distinct idea of anything capable of being excited and communicated in the manner the Heat was excited and communicated in these experiments, except it be MOTION.

Why Rumford Failed

Rumford's work was not ignored by his contemporaries, and he did have some influential converts, but he failed to convince the majority of his scientific contemporaries. There were good reasons for his failure.

First and foremost, he did not demonstrate any quantitative relation between the "motion" that gave rise to the frictional heating and the quantity of heat produced by it. Theories carry conviction to the extent that they can predict the results of an experiment, more so if the prediction is quantitative. Rumford's results were striking, but qualitative: he showed that friction produces heat in seemingly unlimited amounts, but he could not say how much heat, in calories or BTUs, would be produced from "motion," nor could he quantify the motion producing the heat. Caloric scientists, accustomed to measuring quantities of heat with reasonable accuracy, were not impressed.

Second, the principle of the conservation of caloric worked very well in predicting temperature quantitatively when heat was allowed to flow from one body to another. This principle followed quite naturally if heat were a material and elementary substance, but not so naturally if heat were a form of motion. Of course Rumford's whole point was that heat was *not* conserved, but could be created in inexhaustible amounts from motion, but he did not offer any explanation of why it was conserved in some kinds of experiments and not in others. Those who believed the caloric theory were able to defend the conservation principle by suggesting that the drill used in boring cannon could alter the properties of the layers of the bored metal in such a way as to produce a certain quantity of caloric when new metal surface was exposed. This hypothesis permitted them to deny Rumford's claim that there is no limit to the quantity of heat produced, but it does overlook some experiments Rumford did to forestall this objection.

Third, Rumford had no plausible explanation of the fact that heat can be radiated through a vacuum (as illustrated by the sun warming the earth). For those who believed in the caloric theory it was simple: hot bodies emit atoms of caloric that can travel through empty space and fall on cold bodies, warming them up. But if heat, as Rumford believed, is a vibration of a material body, how can it be transmitted through space empty of matter? He was forced to invent a fanciful hypothesis: vibrating matter can produce "undulations in the ether" and these undulations can set other matter into vibration. The idea that even space empty of all kinds of matter was filled with a tenuous substance (ether) had been proposed by Isaac Newton, to explain how the force of gravity could reach through the vacuum between the heavenly bodies. It was a speculative idea, and it was not taken seriously by scientists in Rumford's time. (See Chapter 10 for a detailed discussion of the properties of "heat" radiation.)

Shortly after Rumford's death in 1814, it was shown, first, that heat radiation is a form of light and, second, that light, rather than being a material substance (a "subtle fluid" or Newton's "corpuscles"), was a wave motion. As, in the view of scientists at that time, some sort of medium was required to carry a wave motion, the ether was reborn, and Rumford's fanciful hypothesis became a prophetic insight. Further, if radiant heat was a wave motion in the ether, and not the free flight of caloric particles through empty space, it was the caloric theory that began to sound fanciful. How can a *substance* be converted into a *wave motion* when a hot body radiates its heat, and how can the wave motion produce the substance when it falls on cold bodies and warms them up? Some ideas describing heat as a wave motion in matter were put forward at this time, but they were not formulated with sufficient clarity to allow clear-cut experimental tests of their validity. Many scientists continued to believe that heat was a substance anyway, in spite of these difficulties.

The Unification of Physics

Until the nineteenth century no clear relation was recognized among the various physical phenomena under study: mechanics (the science of motion and forces), heat, light, electricity, magnetism, and chemical reactions. This situation began to change as a result of new scientific discoveries toward the end of the eighteenth century. Partly in response to those discoveries, a philosophical doctrine called *Naturphilosophie* was developed in Germany at this time which stressed among other things the belief that a single unifying principle in nature underlies seemingly disparate phenomena. One of its proponents, Friedrich Schelling, wrote in 1799 that "magnetic, electric, chemical, and finally organic phenomena would be interwoven into one great association . . . [which] extends over the whole of nature." In addition, as we have mentioned earlier, it was discovered in the early 1800s that light was a wave motion, not a substance. At about the same time it was shown that heat radiation is a form of light, traveling with the same speed and capable of being reflected or refracted by matter, just as light is. But the strongest evidence of a unifying physical principle came from studies of electricity and magnetism.

In 1800 Alessandro Volta reported his discovery of the "voltaic pile"—what we call today the electric battery—to the Royal Society of Great Britain. It consisted of alternating plates of zinc and copper, interleaved with sheets of cardboard or leather moistened with brine (a solution of sodium chloride in water). When a metal wire was connected to the top zinc plate and the bottom copper plate, an electric current flowed in the wire. Soon after the discovery, this electric current was found by Humphry Davy and his assistant Michael Faraday to produce chemical changes: first the decomposition of water into hydrogen and oxygen, and then the production of such elements as sodium and chlorine from their compounds. The source of the electric current from the pile was not at first correctly identified, but shortly it became clear that it was produced by a chemical reaction between the brine and metallic zinc, and that it stopped when the zinc was used up. In brief, a chemical reaction can produce an electric current, and an electric current can produce a chemical reaction.

In 1820 the Danish scientist H. C. Oersted, influenced by the doctrines of *Naturphilosophie*, instituted a search for a connection between electricity and magnetism. He found, after some preliminary stumbling, that the needle of a compass appropriately positioned near a metal wire would change its direction when an electric current flowed in the wire. This led A. M. Ampère to infer that if a wire carrying an electric current can affect a magnetic needle, it must itself in some sense be a magnet: it followed that two wires carrying electric currents should repel or attract each other, as two magnets do. Ampère, who realized that a stronger magnetic effect can be produced by winding the electric wire into the shape of a helix (a coil), thus invented the electromagnet.

The realization that an electric current produces a magnetic field led to a search for the opposite effect: can a magnet (or the magnetic field of an electric current) induce an electric current in a loop of wire in its vicinity? Experiments failed to show any such current until Faraday in 1831 discovered almost by accident that a current is produced not by a magnetic field but by a *change* in the magnetic field. The simplest way to visualize this is by imagining a wire, in the form of a coil, connected to a sensitive current-measuring device. A magnet held stationary near or inside the coil produces no current, but when the magnet is moved in or out of the coil, there is a transitory current while it is in motion (see Figure 3.3). A changing magnetic field can be produced not only by physically moving a magnet but also by starting or stopping the current flowing in a stationary electromagnet, which can be done by opening or closing a switch or by using a current that changes in direction in the coil of the electromagnet periodically (an alternating current). From Faraday's discovery we have the *electric generator,* which produces electric current from motion; the *electric motor,* which produces motion from electric current; and the *transformer,* in which an alternating current in one circuit produces another alternating current in a second circuit, usually of a different voltage from the original current.

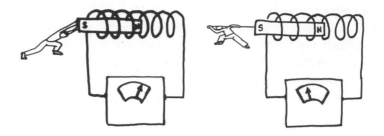

Figure 3.3   Work Required to Make a Current Flow
To induce an electric current by moving a magnet into *(left)* or out of *(right)* a
coil of wire, a force must be exerted and work done on the magnet. If the circuit
were open, so that no current could flow, moving the magnet would not require
a force.

Let us consider for a moment the current induced in a coil when a
permanent magnet is thrust inside the coil. The current in the coil
makes the coil temporarily a magnet also: how does this magnet act on
the permanent magnet that is inducing the current? The magnetic
force on a moving magnet exerted by a current induced in a coil by
that magnet was found always to be in such a direction as to *oppose* the
motion of the magnet; to keep the magnet moving required the exer-
tion of a force and therefore the performance of work by some external
agent. In brief, *generating electricity requires work to be done* (Figure 3.3).
To cite modern examples, a bicyclist can feel the extra force needed to
operate a light powered by a generator attached to the bicycle wheel
when the light is turned on, and hydroelectric plants use falling water
to do the work needed to generate the electric current.

Neither last nor least in this list of scientific advances, the invention
of the steam engine had already shown that heat could produce motion.
And motion, by the mechanism of friction, had been known to produce
heat, a point made forcefully by Count Rumford but not much in the
consciousness of the scientific community generally.

What Is Conserved?

Let us summarize: previously scientists had studied a number of phenomena of nature they had regarded as unconnected—mechanics, electricity, magnetism, light, chemical reactions, and heat. What became
apparent in the early nineteenth century was that these phenomena
were in a sense interchangeable: one of them could be used to produce

another. This led very naturally to the question: among all these transformations, is there something that doesn't change? If electricity can be used in a motor to produce motion, and the motion used to power a generator, so that the motion is converted back to electricity, what relations are there between the initial and final parameters of the electric current? Are the currents the same? Are the voltages the same? A chemical reaction can be used (in a voltaic pile) to produce an electric current, and the current in turn can produce a chemical reaction. For example, a pile in which zinc dissolves in brine to form a salt of zinc produces a current which can be used to electroplate dissolved zinc out in metallic form onto the surface of a metallic electrode. What relation is there between the quantity of zinc used up to provide the current, the amount of current produced, and the quantity of zinc electroplated? What are the economics of these interconversions?

A number of scientists drew the conclusion that there is some quantity of what they called *force* that must remain the same in spite of the changes in form: they were thus anticipating the law of the conservation of energy. Some were famous for other discoveries, such as Faraday and the chemist Justus von Liebig; others, such as William Grove and C. F. Mohr, are known primarily to historians of science. No one formulated the concept of a conservation of "force" in quantitative terms; it was not yet possible to compare a numerical measure of the "force" stored in the chemicals of a battery with the "force" of the electric current coming from the battery or with the "force" of the motion produced by an electric motor run from that electric current.

The Emergence of Energy

James Joule, born in 1818 and privately educated, began a serious study in 1837 of electromagnets and electromagnetic engines, initially with the intention of finding an inexpensive source of power for the family brewery. He was aware that work must be done to produce an electric current, and he also knew, from the laws of mechanics, that work must be performed to set a body into motion and that the kinetic energy gained by the body is exactly equal to the work done to get it moving. Having adopted the kinetic theory's view that what makes a hot body hot is not a substance but the kinetic energy of invisible molecules, Joule was looking for a relation between the work done to produce motion and the quantity of heat produced by the motion, either when the motion is directly slowed by friction or when it is used to generate an electric current that gives out heat as it passes through a wire. Studying the production of heat by electric currents, he established that the heating effect is proportional to the product of the resistance and the square of the current, the famous I^2R law.

The fact that a current in a wire causes heat to be given out does not on the face of it contradict the caloric theory: the electric current may merely be transporting caloric from one place in the circuit to another, for example from the battery to the wire. It occurred to Joule to design an arrangement in which the heat given out when the current flows could not have been transferred from another point in the circuit.

As Faraday had shown, a current could be made to flow in a closed loop of wire by moving the loop between the poles of a magnet, and a force must be exerted to keep the loop moving while the current is generated, thus doing work. Joule placed such a loop inside a closed container surrounded by water in which a sensitive thermometer was immersed (Figure 3.4). The container was then rotated between the poles of an electromagnet, causing an electric current to flow in the loop. The force required to rotate the container to generate the current

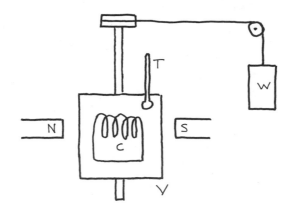

Figure 3.4 Joule's First Experiment
In this experiment an electric current was induced in a wire *C* immersed in water inside the calorimeter by rotating the calorimeter *V* between the poles *N* and *S* of a magnet. The force needed to induce the current is exerted by a descending weight *W.* The product of height descended and the weight gives the work. The heating produced inside the calorimeter by the current is determined from the temperature rise of the water measured by a thermometer *T.* The increase in heat (measured in BTUs) is compared with the work done to generate the current to obtain "the mechanical equivalent of heat."

was exerted by a pair of weights hanging by fine strings passing over two pulleys and turning a vertical axle attached to the rotating container. The amount of work was, as usual, the product of the weights and the distance they descended. Joule thus could calculate the work done, and with the thermometer he could measure the very small temperature rise of the water surrounding the loop of wire: knowing the quantity of water, he could calculate in turn the quantity of heat produced, which he measured in BTUs. This heat, by the nature of his apparatus, could not have been transferred from somewhere else. In a way it was similar to Rumford's experiment, except that Rumford produced heat directly from work, whereas Joule performed work to produce an electric current which then produced heat. The important difference was that Rumford failed to make any quantitative prediction of the amount of heat produced. Joule measured the work in units of "foot-pounds": he gave his result in terms of the number of foot-pounds of work that produced 1 BTU of heat. His first result was 823 foot-pounds per BTU, which he called the *mechanical equivalent of heat*.

Joule went on to other experiments more like Rumford's. When we think of friction, we think of solids being rubbed together, but if liquids are set into motion by stirring, they soon come to rest. There is thus friction within liquids: its strength is measured by a property called *viscosity*. Motor oil is more viscous than water, cold molasses more

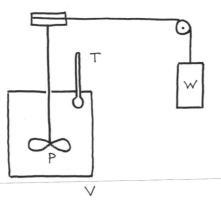

Figure 3.5 Joule's Second Experiment
This experiment is simpler to understand: the water in the calorimeter *V* is stirred by a paddle wheel *P* and the temperature rise measured on the thermometer *T*. The force to turn the paddle wheel is again supplied by a descending weight *W*.

The Modern Calorimeter

The discovery of the conservation of energy greatly simplified the determination of the specific heat of a body and thus its thermal energy. Instead of comparing its change of temperature with the change of temperature of a quantity of hot water the body is placed in contact with, we now measure the amount of energy that is supplied electrically. A modern calorimeter consists of an insulated container containing the body (S) being studied, a thermometer (T), and an electric heating coil (H). A wattmeter (W) measures the wattage (in joules per second) supplied to the heater, and an accurate clock (C) the time elapsed (in seconds) during which the temperature rise is being observed. The product of the time and the wattage is the total number of joules supplied to S producing the observed temperature rise, after correction for the heat necessary to warm the calorimeter itself.

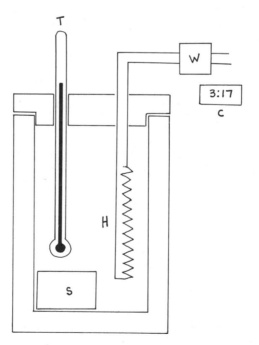

viscous than warm. In one of Joule's experiments, a paddle wheel, turned by a descending weight, stirred water in which he had immersed a sensitive thermometer. The temperature rose very slightly on stirring. There was thus a direct conversion of motion, through friction, into heat (Figure 3.5).

Still other experiments compared the work needed to compress a gas with the ensuing rise in temperature. The heating of gases by compression is not an effect of friction but a direct conversion of the kinetic energy of the moving piston into kinetic energy of the gas molecules. When gases pushing against a piston make it move, the kinetic energy of the molecules is transferred to the piston and the gas is cooled.

Joule's Conclusion

Joule's measured value of the mechanical equivalent of heat varied somewhat from one experiment to another, but after a long series of studies he settled on the figure of 772 foot-pounds of work done for each BTU of heat produced, within a percent of the modern value (778 foot-pounds per BTU). In an era of alternating-gradient magnetometers and electron-spin resonance spectrometers, Joule's experimental methods may not sound impressive, but with them he discovered one of the most fundamental generalizations of science.

Whether one measures energy in foot-pounds or kilowatt-hours is not crucial. What *is* crucial is that heat can be measured in the same units as work: for each BTU of heating produced, 778 foot-pounds of work must have been performed. Heat and work are two examples of the same thing. A useful analogy is the variety of currencies: it does not matter whether a bank expresses its assets in U.S. dollars or Swiss francs—either will do to check if the books balance. Unlike currency, though, the conversion rate between foot-pounds and BTUs doesn't fluctuate daily. It has the same value regardless of the substance it is measured on, the kind of measurement used, or the particular forms of energy involved in the conversion. It is what is called a *constant of nature,* and it is the first of several we will encounter.

Because heat and work are found to be examples of the same thing, a term is needed to cover both: the term is, of course, *energy.* Joule's work showed that the concept of "heat in a body," which made sense when heat was thought of as a substance, is misleading; a body can have its temperature raised either by bringing it into contact with a hotter body or by doing work on it: the results for the body are identical. What

Energy Conversions

Once we understand that the many forms of energy are interconvert-ible, we learn to recognize the astounding number of energy conver-sions taking place all around us. Examples of familiar processes (and some unfamiliar ones) in which one form of energy is converted to another are listed below. Some processes are technological, others bio-logical or physical. We omitted nuclear energy, which does not readily convert to other forms except to thermal energy and radiation, and a few others. The list could be expanded almost indefinitely.

From chemical to:	*Example:*
Electromagnetic	Discharging a battery
Mechanical	Animal muscles (see Chapter 12)
Radiative	Firefly
Thermal	Burning coal

From elastic to:	*Example:*
Electromagnetic	Piezoelectricity (as in a quartz crystal)
Mechanical	Wind-up toy or spring-driven watch
Thermal	Snapped rubber band

From electromagnetic to:	*Example:*
Chemical	Charging a battery; electroplating
Elastic	Loudspeaker

used to be called the "heat in a body" is now called its *thermal energy*. (The term *heat* was not abolished from the vocabulary of science, however. It is now reserved only for the thermal energy transferred when two bodies at different temperatures are in contact.)

Earlier we mentioned the intuitive feeling of some nineteenth-cen-tury scientists, aware of how one phenomenon of nature can be used to produce another, that during the transformations *something* must remain the same. We have an answer now: what remains the same is the *total energy*. The performance of work is often used as the yardstick for energy, and energy may be defined in terms of it: energy is the *capacity to do work*.

Mechanical	Electric motor; electric fan
Radiative	Electric light
Thermal	Electric heater; electric frying pan

From radiative to:	*Example:*
Chemical	Photosynthesis in plants
Electromagnetic	Solar voltaic cell
Mechanical	Sun's radiation repelling the tails of comets
Thermal	Sun warming the earth

From mechanical to:	*Example:*
Elastic	Winding a watch
Electromagnetic	Electric generator
Radiative	Triboluminescence (light produced by friction)
Thermal	Friction

From thermal to:	*Example:*
Chemical	Fixation of nitrogen by the Haber process (see Chapter 11)
Elastic	Bimetallic strip used in thermostats
Electromagnetic	Thermocouple
Mechanical	Heat engine
Radiative	Glowing coal

Joule saw that the capacity to do work is the *something* that remains the same. The kinetic energy of a moving body, the potential energy of the same body in a gravitational field, the chemical energy of fuels, the electrical energy of a current, the thermal energy we determine with the aid of a thermometer, all can be converted from one form to another, but the total capacity to do work does not change.

The caloric theory's principle of the conservation of caloric and the conservation of total mechanical energy in a frictionless world were thus united into a single principle, that of the conservation of energy, the first law of thermodynamics. In the one hundred and fifty years since it was discovered it has been applied successfully to forms of energy unknown to Joule. It holds when neutrons bombard atomic nuclei, when red blood cells absorb oxygen, when refrigerators cool beer, when volcanoes erupt, and when stars collapse.

Appendix 1: The Nuts and Bolts of Energy

Any macroscopic body—a 50 kilogram sphere of rubber, a steel tank filled with oxygen gas, or a refrigerator—can have energy simultaneously in several different forms. If it is in motion, it has kinetic energy. If it is 100 meters above sea level, it has (gravitational) potential energy. If it is placed in a magnetic field (and anything on the surface of the earth is in a magnetic field), it has magnetic energy. Solid bodies are usually elastic—if we twist or stretch them they exert an opposing force tending to restore their original shape—so bodies held in a stretched or twisted state have elastic energy. What we call chemical fuels store chemical energy until we burn them. Nuclear energy is stored in atomic nuclei and released when those nuclei undergo fission or fusion.

What was once called "heat" and we call from now on *thermal energy* arises from the invisible motions of the molecules of which matter is composed.

Since energy is conserved, a body cannot gain it unless some other body near enough to interact with the first body loses it. Transfers of energy can take place in a limited number of ways. Heat can flow from hot bodies to cold ones, increasing the thermal energy of the latter at the expense of the former. Work can be done by one body on another: for example, a flywheel (body 1) can be put into motion by a weight (body 2), which loses gravitational potential energy as the flywheel gains kinetic energy. Work done on a body can increase its thermal energy also, as Joule showed.

We can express the first law as follows:

Net change in energy of a body =
Net inflow of heat to the body
+ Net quantity of work done on the body

It is customary to symbolize the energy transferred by heat flows as Q, work done by W, energy by E, and "change in" by the Greek letter capital delta (Δ), so the above equation can be written

$$\Delta E = Q + W$$

The symbols Q and W must be interpreted algebraically: for example, when Q is *negative*, heat has flowed *out* of the body to some colder body, and when W is negative, some other body has done work *on* the body to which the equation is being applied.

This relation tells us how to measure any *changes* of energy a body undergoes, but it doesn't tell us how much energy there was in the body in the first place. From the point of view of nineteenth-century science, that quantity couldn't be known. Only Einstein's theory of relativity, with its equation $E = mc^2$, provided an answer. Until then scientists were limited to saying how much more or less energy a body has than before something was done to it.

In Chapter 2 we said that measuring the potential energy of a body requires an arbitrary choice of reference level—for the earth's gravity we chose sea level—and then the potential energy in some other position is determined by measuring the work done to move the body from the reference level to its present level. For thermal energy we must also make an arbitrary choice of a reference state. This state is specified by a particular temperature, a particular pressure, and, if relevant to the situation, the electric, magnetic, and gravitational fields acting on the body. A common choice is 25°C (or 67°F) one atmosphere (1 atm) pressure, and the absence of electric or other fields—the condition of a typical laboratory. There is nothing sacred about this choice, and others are often used.

Once we have chosen the reference state, we measure the sum total of energy inputs and outputs needed to bring the body from this reference state to any other state—say, 100°C and 1,000 atm pressure. The total energy input gives us what we call "the energy of this body at 100°C and 1,000 atm pressure," though of course it is really the energy *difference* from the reference state.

Let us examine the details of an actual energy calculation. Let us heat 1 kg oxygen initially at 25°C and 1 atm until its temperature increases to 150°C (we will keep pressure at 1 atm, in balance with the earth's atmosphere outside the containing vessel). The oxygen gas can be heated with an electric heater of known wattage. The wattage multiplied by the time, measured in seconds, needed to raise the temperature the desired amount gives the total joules of energy supplied by the heater, which for 1 kg oxygen under these conditions is found to be 1,270 joules. However, there is a complication: as the gas is heated, it expands, pushing back the earth's atmosphere and therefore doing work on it. The work done is 295 joules and must be taken into account. The net energy change in the gas (ΔE) is the amount of energy input (the 1,270 joules of heat) minus the amount of energy output (the 295 joules of work the gas did on the atmosphere), or $1,270 - 295 = 975$ joules.

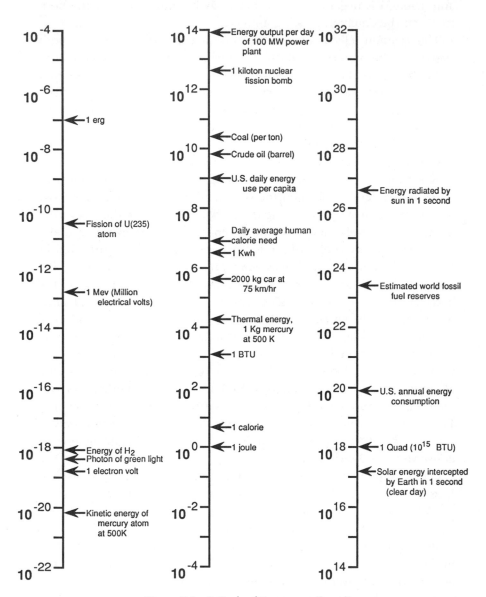

Figure 3.6 A Scale of Important Energies
Various quantities of energy of human significance are indicated on the scale
in joules. A logarithmic scale is used because we need to cover an enormous
range of energies. (One joule corresponds to 10^0 on the scale.)

Another way to express this is to use the equation given above: $\Delta E = Q + W$. Q, of course, is the heat added to the gas, 1,270 joules. Since the gas expands and does work *on* the atmosphere, W is expressed as a negative number, -295 joules. Therefore,

$$\Delta E = Q + W$$
$$= (1{,}270 \text{ joules}) + (-295 \text{ joules}) = 975 \text{ joules}.$$

We now tabulate the thermal energy of the gas at 150°C and 1 atm as 975 joules, keeping in mind that what we really mean to say is that the gas now has 975 joules more than it had in the reference state.

For bookkeeping purposes, all forms of energy used in a calculation must be measured in the same units. Which unit to use in any problem is a matter of personal choice, made with due regard to the intended audience. Our own preference is for the scientist's Standard International Unit, the joule, a unit never used by the man it was named after, but we have also used here calories, Calories, BTUs, kilowatt-hours, and foot-pounds.

In Figure 3.6 we represent graphically the magnitudes of various energy units relative to the joule. Because the range of magnitudes is enormous (an electron-volt is about 10^{-19} joules, a Quad 10^{18} joules), we use a logarithmic scale. Watts and horsepower do not appear, for they are not units of energy but of *power*, the rate at which energy is supplied: 1 kilowatt is 1,000 joules per second. We have noted on the scale various quantities of energy of practical importance, such as the U.S. annual energy consumption, the energy of a nuclear bomb, and the energy needed to tear a single molecule of water apart into its constituent atoms.

Appendix 2: Outsiders, Insiders, and the Reception of Scientific Contributions

Joule reported the results of his successively more accurate measurements of the mechanical equivalent of heat both in scientific journals and at scientific meetings. Neither his published reports nor his talks aroused any interest in England, though they began to attract attention in the rest of Europe. In May 1847 he gave a public lecture in Manchester, with the title "On Matter, Living Force, and Heat," and published the text in the *Manchester Courier*. In June 1847 he gave a talk to the British Association meeting in Oxford: this time there was a response. He had been asked by the chairman of the meeting to keep his talk short, because his previous communications had not excited much interest. This, Joule later said, he "endeavored to do, and, discussion not being invited, the communication would have passed without comment if a young man had not risen in the Section, and by his intelligent observations created a lively interest in the new theory. The young man was William Thomson, who had two years previously passed the University of Cambridge with the highest honor, and is now probably the foremost scientific authority of the age."

William Thomson (later Lord Kelvin), then twenty-two years old, was at the beginning of a long and distinguished career as a physicist. What was admirable about his initial reaction to Joule's "new theory" was not that he immediately believed it—for reasons to be discussed in a later chapter he was at this time a devoted supporter of the caloric theory—but that he understood its implications and recognized its challenge to the theory he preferred.

In the same year the German physicist and physician Hermann von Helmholtz, drawing on published results of Joule and of others, presented a definitive and completely general statement of the law of conservation of energy to the Physikalische Gesellschaft of Berlin: his essay was entitled "Über die Erhaltung [Conservation] der Kraft" (*Kraft* means "force"; the term *energy* was generally adopted shortly afterward). Joule's struggle for recognition was over.

Joule's story was a happy one, recognition and acceptance coming only after a short delay and all the sweeter for it. Other scientists had, as we have mentioned, previously put forward ideas similar to Joule's but less clearly and quantitatively stated. Some recognized the significance of the various transformations possible between phenomena that were usually considered distinct. Others were guided by the principles of *Naturphilosophie,* then influential in German-speaking countries. One proponent of this philosophy was Julius Robert Mayer, a German physician whose work on heat and energy was equivalent to Joule's but whose struggle for recognition was less successful.

Mayer, while serving as a doctor on a Dutch merchant vessel on a voyage to the East Indies, noted that venous blood was a brighter red in the tropics than in Europe. He ascribed this difference to a lesser demand by the body for oxygen in warm climates, since the body needs to burn less fuel to maintain normal temperatures, and concluded that the performance of work by a human being and the production of heat to maintain body temperature, both made possible by the food consumed, were in some sense equivalent. In 1841 he extended his ideas beyond physiology to the rest of nature and proposed as an axiom the indestructibility of "force." A respected scientific journal in Germany, however, declined to publish his work. There were good reasons: Mayer knew very little physics, and he presented his thoughts in an obscure and confused manner, making them sound more like metaphysics than physics. Aware of his limitations, he started studying physics and mathematics and soon realized that kinetic energy was the best measure of motion in the frictional conversion of motion to heat. He succeeded in publishing his revised ideas in 1842 in Liebig's journal, *Annalen der Chemie*. The physics is better, but the metaphysics was still there:

> [Energies] are causes: accordingly, we may in relation to them make full application of the principle—*causa aequat effectum*. If the cause *c* has the effect *e*, then *e* = *c* . . . In a chain of causes and effects, a term or a part of a term can never, as plainly appears from the nature of an equation, become equal to nothing. This first property of all causes we call their indestructibility . . . If after the production of [effect] *e*, [cause] *c* still remained in whole or in part, there must be still further effects [*f,g,* . . .] corresponding to the remaining cause. Accordingly, since *c* becomes *e*, and *e* becomes *f,* etc., we must regard these various magnitudes as different forms under which one and the same entity makes its appearance. This capability of assuming various forms is the second essential property of all causes. Taking both properties together, we may say causes are quantitatively indestructible and qualitatively convertible entities. [Energies] are therefore indestructible, convertible entities.

The physical ideas are stated strongly enough and, most strikingly, Mayer gives a figure for the mechanical equivalent of heat less accurate but not far from the value painstakingly obtained later by Joule. He performed no experiments—today we would call him a theoretical rather than an experimental scientist. How did he do it?

Briefly, Mayer had the physical insight to recognize that a well-known observation about gases could be explained simply by his concept of the equivalence of heat and work. It was known that when gases expand against an external pressure—the pressure of the earth's atmosphere, for instance—they tend to cool, and heat must be supplied to keep their temperature from falling. Mayer realized that work is being done by the expanding gas against the external pressure, and that the amount of heat input needed to keep the gas from cooling must be in proportion to the work done. He compared data on the heat intake and work output of an expanding gas, information that was already available in the scientific literature: no laboratory was needed.

In 1845 Mayer published a more comprehensive treatment of his ideas and applied them to the physiological problems he had started with years earlier. By this time Joule was beginning to publish the results of his more definitive and precise studies of the mechanical equivalent of heat. Mayer's work was ignored—he had to publish his 1845 paper at his own expense—and he saw Joule receiving the credit for ideas he had proposed first. In 1849, Mayer suffered a mental breakdown, attempted suicide, and underwent periods of confinement in a mental hospital. For the next ten years he did no more scientific work. But restitution came eventually. In 1852 Helmholtz, who had been unaware of Mayer's work when he wrote his 1847 memoir, came across his papers. Another German physicist, Rudolf Clausius, of whom we will hear more later, also learned of them and began to correspond with Mayer. For a time the question of who should have the credit for the discovery took on a nationalist tinge: England versus Germany. William Thomson and a colleague P. G. Tait supported Joule with more patriotic ardor than is considered seemly today among scientists, criticizing Mayer for never having performed an experiment (which was true but not really relevant). It took the intervention of John Tyndall, a more fair-minded man than some of the other English scientists, to get Mayer the recognition in England that he deserved. In the 1860s his papers were translated into English, and in 1871 Mayer received the Copley Medal of the Royal Society. He had already been awarded the Prix Poncelet of the Paris Academy of Sciences in 1870. He died in 1878.

So Mayer achieved due recognition at the end of his life, but reading his story one is reminded of the legal dictum: "Justice delayed is justice denied."

There were obviously many differences between Joule and Mayer—differences in educational background, experimental skill, and scientific success. But they had at least one thing in common, with each other and with other important figures in the early history of the energy concept. Both were outsiders.

With the discovery of the law of conservation of energy, the study of heat became a branch of mechanics. That science was almost two hundred years old: Newton's originally simple statements had been given elaborate mathematical formulation that permitted their application to an extraordinary range of problems, and they were repeatedly confirmed in the laboratory. The law of conservation of energy and the recognition that heat was a particular manifestation of energy are easily seen in retrospect to be logical extensions of mechanics. Yet these discoveries were made not by the outstanding scientists of this discipline but by outsiders.

There is a popular myth about great scientific discoveries that they are made by people outside the scientific establishment, who are initially rewarded with contempt and scorn when not actually ignored by the establishment. As the saying has it, "They thought Einstein was crazy." Well, they *didn't* think Einstein was crazy. Although only a civil servant—Technical Expert (Third Class)—in the Swiss Patent Office, he had no difficulty getting his three revolutionary papers of 1905 accepted by the editor of a prestigious physics journal. The papers did not create an overnight sensation, but their worth was recognized in a short time. In 1908 Einstein was awarded an honorary doctorate by the University of Geneva and appointed Associate Professor of Theoretical Physics at Zurich. Later he received appointments to the Chair of Physics at Prague (1911) and to the Kaiser Wilhelm Institute in Berlin in 1913.

Other major scientific discoveries either were made within the establishment or received prompt recognition by it: Maxwell's electromagnetic theory, quantum mechanics, radioactivity and the nuclear atom, evolution by natural selection.

There are two striking features about the discovery of energy conservation and the "mechanical" theory of heat. First is the extraordinary number of individuals who proposed it in a primitive or incomplete form in the twenty years preceding its acceptance after Joule's work: Sadi Carnot, Colding, Holtzmann, Ampère, Grove, Mayer, among others. If ever it can be said that the time was ripe and the intellectual climate ready for a new theory it is here. Second is the outsider status, and lack of formal training in the highly mathematical field of mechanics, of most of these men. In a classic essay on the discovery, Thomas Kuhn points out that the major figures in this story fell into three groups. One consisted of those actively concerned with the various transformations among such phenomena as electricity, magnetism, motion, heat, and light. A second group, mostly German or living in German-speaking countries, were influenced by the ideas of *Naturphilosophie*. The third was composed of engineers working with heat engines, for whom the concept of work—force times distance—was of vital importance. Academic physicists trained in the science of mechanics, like Kelvin and Clausius, at most gave the idea of energy conservation, once it was brought to their attention, recognition and formulation in a more general and mathematical form.

Reading Kuhn's essay, one understands readily why this particular discovery was made, by these particular groups of individuals, at this particular time. But the story raises a haunting question: could it happen again, and if so, how? No one having the least familiarity with research at the frontiers of science in particle physics, say, or molecular biology would expect individuals without rigorous academic training to contribute anything at all to these fields, much less to make discoveries that have an impact comparable to the impact of energy conservation on nineteenth-century physics.

The popular myth that major scientific discoveries are often made by outsiders serves a psychological need. Science, with its mathematics, its expensive and elaborate instruments, and its incomprehensible jargon, is intimidating. It is comforting to believe that an ordinary person using ordinary common sense can sometimes beat the scientists at their own game. While neither Joule nor Mayer were "ordinary" persons, and their insights could scarcely be described as the application of "common sense," they fit the description of the outsider closely enough to keep the myth alive as a possibility.

4
The Microscopic View of Energy

~~~~~~

The distinction between macroscopic and microscopic approaches in physical science is like the distinction between those who can't see the trees for the forest and those who can't see the forest for the trees.

The macroscopic properties of matter are those we can easily observe and measure on portions of it large enough to be visible; we need not make any hypotheses about the invisible atoms that constitute matter. Examples of such properties are mass, force, density, temperature, pressure, color, hardness, electrical resistance, and so on. In addition to measuring the macroscopic properties of matter, we can observe in the laboratory that there are relationships among them, relationships which become "laws" if they are widely applicable. Newton's laws, describing the motions of visible samples of matter in terms of their masses and the forces between them, are macroscopic laws; so is Ohm's law, which describes (sometimes inaccurately) the relation between the electrical current in a wire and the applied voltage; and so is the ideal gas law, a particularly simple relation among the temperature, the pressure, and the density of gases (provided they are not too cold and not too compressed). For many purposes we can use these macroscopic relations without ever invoking the atomic theory—sometimes we want to focus on the forest and leave the trees out of the picture.

At other times, though, we *do* need to see the trees. Matter is made up of atoms and the relatively stable combinations of atoms we call *molecules,* and the properties of matter are ultimately determined by the properties of the individual atoms, the forces they exert on each other, and the way they move about. The behavior of matter must ultimately be understood in terms of atoms and molecules. This is the *microscopic* approach to the physics of matter (though we do not mean to imply that ordinary microscopes are used—we cannot "see" molecules with them). It is called the *kinetic-molecular theory* of matter and is a more advanced version of the kinetic theory of heat discussed in Chapter 3.

We have already discussed an experiment that illustrates the macroscopic and microscopic approaches—Joule's studies of the mechanical equivalent of heat. Joule was guided by a microscopic view: he believed that there are atoms that move, and that moving atoms have kinetic energy; he drew pictures of molecules in motion in his notebooks. Such ideas led him to the conclusion that the rise in temperature produced by friction was a manifestation of an increased kinetic energy of the molecules. Joule's experimental tests of the new theory were macroscopic, however, not microscopic. The determination of quantities of thermal energy produced by friction or electrical heating required weighing balances for the measurement of masses of the water or other substances used and thermometers for the measurement of increases in temperature. Measuring the work done either to turn the paddle wheel that stirred the water or to generate an electric current required a determination of a weight and of distance—a balance, again, and a meter stick. All these instruments are those of macroscopic physics: to use them we do not need to know or believe in a molecular theory.

Of course it is true that Joule's conclusion, that there is a fixed relation between the work done and the amount of heat produced by it, is confirmatory evidence for a kinetic-molecular picture. It occurred to some scientists, once the significance of Joule's discovery was recognized, to ask if other corroboration could be found. If molecules can move, would they not move according to Newton's laws? If so, should it not be possible to use these laws to deduce other consequences testable in the laboratory?

Nineteenth-century scientists encountered several problems in their attempts. Aside from mathematical difficulties, the very existence of molecules was speculative. They could not be seen, and the forces they exert on one another were not known, nor were their masses. How can Newton's laws be applied when we are that ignorant? How can we calculate any properties of matter accurately enough to compare with a laboratory measurement?

In spite of these problems, some scientists were brave enough, or foolhardy enough, to try. Their first step was to make as simple a hypothesis as they could of how forces between molecules act and see how well macroscopic properties could be explained by it. If the hypothesis worked, fine. If not, they would try a slightly more complex but still simple one. They did not expect exact agreement, and they did not find it. There were some striking successes of this theoretical approach, but there were also some striking failures, which could only be explained with the twentieth-century discovery that molecules obey the laws of quantum mechanics, that they obey Newton's laws only as an approximation, one that works better in some situations than in others.

We have three goals in this chapter. One is to give a simple picture of how molecules move and how they can gain or lose energy as they do. Second, using this picture, we want to help the reader visualize what happens on a molecular level when, in the macroscopic world, friction converts the energy of moving bodies into thermal energy, manifested by a rise in temperature of the bodies and their surroundings. Our final goal is to give examples of quantitative predictions of macroscopic properties of matter made by the nineteenth-century kinetic-molecular theory and describe how they fared in laboratory tests. Scientific theories, if they are to earn our confidence, must do more than provide plausible but qualitative pictures of phenomena; they must predict with reasonable accuracy what we will find when we measure things quantitatively.

A Model of Molecular Motion

We will base our molecular model on a simple analogy to molecules: hard, elastic balls in motion. Such a model can account quantitatively for the properties of hot gases and not much else, but it provides a qualitative picture of molecular motion applicable to real molecules. The game of pool, for those who may not know it, is played with a number of hard, elastic balls on a flat table covered with smooth green felt, with walls around the sides of the table to keep the balls from falling off. A stick is used to set in motion one of the balls, the cue ball, which collides with the other balls and transmits its motion to them. The object of the game is to "sink" the balls into pockets along the sides of the table. Our version of pool will be idealized. First, we eliminate the pockets, so that the balls remain on the table. Second, we assume no frictional losses, so balls set in motion remain in motion indefinitely. The result is a fair first model of molecules confined in a container and obeying Newton's laws of motion. This model allows only for motion in two dimensions (on the flat table), but if we were to escape the earth's gravitational field we could as easily construct a three-dimensional model from a felt-lined box. Molecules feel the force of gravity just as pool balls do, but their speeds are so much greater than the usual speeds of pool balls that for our present purposes we can neglect gravity. Finally, we will simplify the kinds of motion the balls can undergo, at least at the start, so we will disregard the fact that pool balls not only travel in different directions on the table (a motion called *translation*) but spin about their centers as well *(rotation)*. We will discuss this latter kind of motion later on.

It is necessary to introduce some definite amount of kinetic energy into the system at the start, but after that we leave it alone, so that the total kinetic energy no longer changes. There are different ways we can visualize doing this, but the simplest is to imagine giving all the energy to one ball initially and allowing it to transmit it to the others. This is a little artificial—it is not so easy to do with real molecules (though not at all impossible)—but it will serve.

Now let us begin our game. We place only one ball on the table and set it into motion by a thrust of the cue stick. It moves in a straight line until it bounces off a wall and then continues in another direction (see Figure 4.1). Because we have hypothesized the absence of friction, it must retain the same kinetic energy, $\frac{1}{2}mv^2$, for all time, and thus the

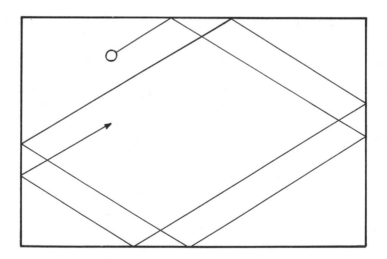

Figure 4.1 Path of One Pool Ball
A pool ball travels in a straight-line path until it is reflected when it hits the wall of the table. Note that the angle between the path of the incoming ball and the wall is the same as that between the path of the outgoing ball and the wall.

same speed v, since the mass m of the ball doesn't change. All that changes is the direction of motion, at each collision with a wall.

Next we use two balls, but we give only one an impetus, so that it has a speed v_1. The other ball, initially at rest, has a speed $v_2 = 0$. At first the path of the first ball is the same as in the first example. But at some point it collides with ball number 2. At once ball 2 is set in motion with a speed v_2 no longer equal to zero, and the speed v_1 of the first ball is now less than it was before the collision. On the basis of the limited information given we cannot say *exactly* what the new speeds are; they will depend on such details of the collision as whether it is head-on or a glancing one. But one thing we can be sure of: the *sum* of the kinetic energies of the two balls after the first collision is the same as the kinetic energy of the first ball prior to the collision (Figure 4.2). That is,

$$\tfrac{1}{2}m(v_1)^2 \text{ before collision} = \tfrac{1}{2}m(v_1)^2 + \tfrac{1}{2}m(v_2)^2 \text{ after the collision}$$

With each subsequent collision there will be changes of speeds and directions of both balls. The precise changes are determined by factors we have not bothered to specify: the angles of approach, the prior speeds, and the nature of the impact, but again whatever the new speeds are, the total kinetic energy is unchanged.

Order from Disorder

The next example moves closer to a real game of pool, played with 15 balls, which at the start of the game are placed in a triangular array on one side of the table, and a cue ball, placed on the opposite side of the table. The cue ball is then impelled toward the array and after collision the other balls scatter. In a real game, the next play begins when the balls stop rolling, but on our frictionless table they never stop. In our

(A)   (B)

Figure 4.2   Conservation of Energy in a Collision
Before the collision only one ball is moving, and it has all the kinetic energy. After the collision both are moving, and the sum of the two kinetic energies is equal to the kinetic energy of the first ball before the collision.

game, the initial impact of the cue ball with the array, which transfers some of the kinetic energy to the other balls, is followed by further collisions—balls with walls, balls with each other—which continue indefinitely. Each ball follows a complicated course, traveling in short, straight-line segments, with abrupt changes in speed and direction at each encounter with one of the other balls or the walls (Figure 4.3). It is a fairly disorderly situation.

If we increase the number of balls to correspond to the number of molecules in a liter of gas and change our table to a box, things became even more disorderly, but out of this disorder some order comes.

Let us use the term *molecules* rather than *billiard balls* from now on. Initially, in our model, one molecule travels at a high speed and has all the kinetic energy; the others in this large collection are stationary— they have no kinetic energy. With the initial collision of the first molecule with the pack, a redistribution of energy begins, and after a sufficient number of additional collisions, the collection gets as disorderly as it can. Each individual molecule pursues its own erratic course. Its speed (and thus its kinetic energy) and its direction of travel change abruptly at each collision with another molecule, and it wanders in the course of time all over the container, in a zigzag path somewhat like the billiard ball's (Figure 4.3) but in three dimensions. No two molecules

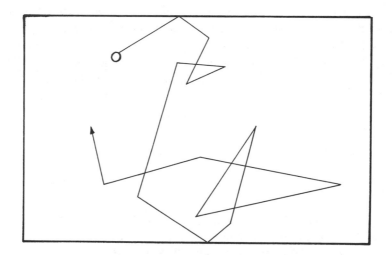

Figure 4.3 Path of One Ball among Many
Only one ball's path is shown; it changes direction after collisions with other balls (not shown) and with the sides of the table.

have the same detailed history: the speed, direction, and position of each follows a course unlike any other.

But in spite of the disorder, the whole collection has certain uniform properties, which depend on the average behavior of all the molecules and which settle down to constant values after an initial period and then no longer change. The spatial distribution of the molecules is an example. Initially, the molecules were bunched up in one region, but as time goes on they move about, spreading themselves uniformly over the space available, so that *on the average* half are on the left side of the container and half on the right.

Another example is the total force due to molecular impacts on the walls of the container. Initially, when only one molecule was moving, there were no impacts and no force on the walls. After the first impact, some of the molecules were driven against the walls, and after many collisions have occurred, all walls are bombarded at about the same rate and with about the same average force.

Equilibrium

Thus in time a stable state of affairs is reached—it is called an *equilibrium state*—after which certain properties (the macroscopic properties which depend on all the molecules at once) no longer change.

It might be expected that in this uniform state of equilibrium each molecule would tend to have about the same speed and thus the same kinetic energy. This isn't what happens. Instead, at equilibrium the individual molecular speeds differ considerably. There are always some moving much faster than the average, and some moving much slower. Each molecule undergoes sudden changes of speed at each collision, so the *identity* of the molecules that are traveling at, say, twice the average speed is constantly changing, but the *number* traveling at that speed remains about the same once equilibrium is reached. The molecules are said to have a *distribution* of speeds.

One may reasonably wonder why a nonuniform distribution in speeds is characteristic of this disorderly equilibrium state: why should not all molecules tend to come to the same speed? After all, when a fast molecule collides with a slow one, won't there be a tendency toward equalization of the speeds? Here our intuition misleads us. While more often than not a collision between fast and slow molecules leads to speeds that are more nearly alike than before the collision, it is not inevitable. The actual transfer of kinetic energy that occurs in a collision depends not only on the speeds but also on the kind of impact. In

Figure 4.4 we illustrate a collision between a fast and a slow molecule in which the fast molecule ends up with all the kinetic energy and the slow molecule with none. Once we recognize that collisions leading to greater differences in speeds rather than lesser can also occur, we should find it plausible that the molecules will not all have the same speed at equilibrium.

We can give an even stronger argument, but one a little more abstract. Suppose that a fast and a slow molecule collide and that the two molecules travel at nearly the same speed afterwards. Suppose further we make a motion picture of this collision and then play the film backwards. It would show two molecules traveling at nearly the same speeds colliding, after which one is moving much faster than the other. If we object that playing the film backwards does not describe a possible occurrence in the real world, the physicist answers that it really does: the film run backwards shows events that are as consistent with Newton's laws of motion as what we see when the film runs forward.

In a later section we will describe how the detailed form of the distribution of the molecular speeds was calculated theoretically, and how well it agreed with experiment.

Kinds of Motion

In describing the motion of molecules so far, we have considered only one kind of motion, that in which a molecule travels in straight lines between collisions at a speed v and thus has a kinetic energy $\frac{1}{2}mv^2$ until the next collision. This kind of motion is called *translation*, and its

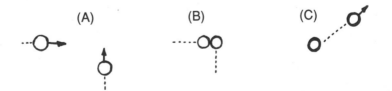

Figure 4.4 Collision Leading to Unequal Energies
At first the two balls are traveling at equal speeds *(A)* and at right angles, then one ball strikes the other squarely at the middle *(B)*. After the collision, one ball is standing still, and the other is now traveling at a 45-degree angle to its previous path, with all the kinetic energy *(C)*. This collision is equivalent to the collision of Figure 4.2 played backward. In a collection of many balls moving at equilibrium, both kinds of collision occur equally often.

kinetic energy *translational kinetic energy.* We already noted that pool balls also spin, a motion also called *rotation.* Molecules, too, may spin rapidly about their centers of gravity while they travel; it is only the center of gravity that travels in a straight line. Let us consider the oxygen molecule, which consists of two identical oxygen atoms. If an O_2 molecule rotates as it travels, the path traced out by one of the oxygen atoms will follow a course like that shown in Figure 4.5A. The kinetic energy

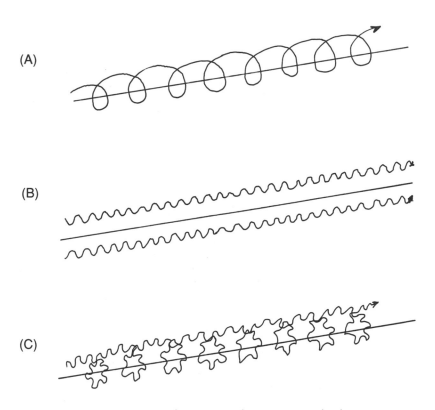

(A)

(B)

(C)

Figure 4.5 The Motions of an Oxygen Molecule
A two-atom molecule, like O_2, can simultaneously engage in translational, rotational, and vibrational motion. In *(A)* we show the path of one of the two atoms for a molecule that is not vibrating; each atom travels in a path consisting of motion in a straight line combined with a circular motion about the center of the molecule (midway between the two oxygen atoms). In *(B)* we imagine the paths of the two atoms in a molecule that is vibrating and translating but not rotating. Each atom performs an in-and-out motion combined with the straight line motion of translation. In *(C)* we show the path of *one* of the oxygen atoms when all three motions are taking place.

States of Matter: Gases, Liquids, and Solids

The model we have described here corresponds to matter in the gase-ous state at temperatures high enough for the attractive forces between the molecules to be disregarded and with so much empty space in the system that the distances molecules travel between collisions are much larger than their own diameters. That there *are* attractive forces is shown by the fact that all gases on sufficient cooling will condense to liquids

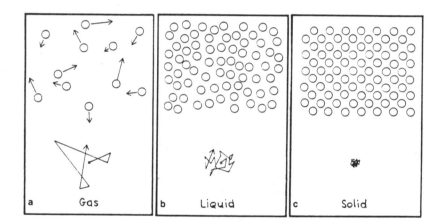

of this rotational motion must be added to the translational kinetic energy to get the total energy of the oxygen molecule.

A more complicated additional kind of motion occurs if the distance between the two oxygen atoms can change. Now a new kind of energy enters the discussion: if there are forces that hold the two atoms in place in the molecule—and there must be, otherwise the molecule would fly apart as it rotates—then changing the distance between the two atoms requires work to be done and thus increases the potential energy of the molecule. One may visualize an elastic spring holding the atoms to-gether; the atoms alternately stretch and compress the "spring" as the molecule moves in its path (Figure 4.5B). This oscillatory motion is called a *vibration;* it has something in common with the oscillatory motion of a pendulum, which we considered in Chapter 2: there is an and that on further cooling, liquids freeze to form solids. The difference between these three states of matter—gas, liquid, and solid—is the arrangement of molecules and the way they move.

In the gas, the molecules are arranged almost randomly in space and travel long distances (relative to their sizes) in straight-line paths. In the liquid, the molecules are much closer together; the density of a liquid is about a thousand times greater than that of a gas. Although the molecules are not packed in an orderly way, they are not placed at random, but more like the individuals in a large crowd who do their best to stay a certain distance away from each other. The paths of molecules in a liquid are not made up of straight-line segments, as the molecules are so close to their neighbors that in a sense they are always involved in collisions. As time goes on, a liquid molecule will wander all around the container, but it will take longer to do so than a gas molecule. In a solid the arrangement of molecules is orderly and periodic in space, like a tiled floor. Each molecule is confined by its neighbors and can only move back and forth around a fixed position (at least in an idealized solid; in real solids, with occasional imperfections, molecules do wander, but only slowly).

alternation of the energy in the molecule from kinetic to potential and back again.

Anyone who has had the dismaying experience of witnessing the collision of two autos on a highway knows that autos after a collision often spin around as they recoil from the collision. What was purely translational energy before is now a combination of both rotation and translation. In molecular collisions the colliding molecules transfer not only translational energy but rotational and vibrational as well, so that in the disorderly state of equilibrium all three kinds of energy are present. When the energy of a quantity of oxygen gas is increased (for example, by allowing heat to flow into it from a hotter body), the incoming energy is distributed among the translational, rotational, and vibrational motions of the molecules. To raise the temperature of oxygen gas, more energy must be supplied than if only translational motion were possible, making the specific heat of oxygen higher than it would otherwise be. Atoms of helium and argon do not combine to form molecules; they have only translational energy under ordinary conditions and no rotational or vibrational energy, for reasons too complicated to explain.

Friction Viewed Microscopically

In order to understand what effect the force of friction has on a body, we must distinguish between two kinds of translational motion that molecules in matter undergo. The disorderly motion we have been discussing above has no preferred direction. On the average, as many molecules are moving to the left as to the right, as many up as down. But molecules usually come to our attention assembled together in material bodies such as chunks of iron, streams of water, or jets of air. These macroscopic bodies may themselves be in motion, which means that in addition to the disorderly motion associated with their thermal energy, there may be a *net* motion of all the molecules, in, say, a northwesterly direction at 15 miles per hour. It is no longer true that as many molecules are moving in one direction as are moving in the opposite direction (see Figure 4.6). The moving body can be thought of as having two kinds of energy: *first* the kinetic energy of its overall motion (at a speed V) equal to $\frac{1}{2}MV^2$, where M is the total mass and V, in this case, is 15 mph; and *second* the thermal energy of the disorderly motion of its molecules, the sum of their kinetic and potential energies.

We will use the term *orderly motion* for that component of the molecular motion that manifests itself as a macroscopic motion of the body. Keep in mind, however, that the orderly motion of a macroscopic body can easily be converted to *disorderly* molecular motion. This is what friction does, although the image usually conjured up by the word is specifically the friction between two macroscopic bodies. Let us describe a simpler example of this conversion first.

Let us go back to the pool table, on which we place the 15 balls in the rack that is used to set the balls in a triangular array. Next we move the rack with the 15 balls rapidly and, just before the rack would hit one of the walls, we deftly remove the rack and allow the balls to continue in a straight line. The balls have thus been given an orderly component of motion: they can be regarded as a "body," triangular in shape, moving in a definite direction. In this idealized experiment they possess no disorderly motion yet.

What happens when the "body" hits the wall? It could be, if we had taken sufficient care to align the leading edge of the rack exactly parallel to the wall toward which the rack and balls were moving, that the collection of balls may be reflected from the wall and still keep its triangular arrangement. But this is unlikely. It is much more likely that some irregularities of motion were introduced and that the coherence of the motion of the collection of balls will be lost (see Figure 4.7). Initially the individual balls may tend to travel after contact with the wall in the direction opposite to their initial approach, but they will also tend to spread out on the table. After individual balls have made several collisions with the walls and with each other, the motion will become the disorderly kind we have been describing and an equilibrium state will be approached.

What was once the orderly motion of a macroscopic body (the collection of balls in the rack) has become the disorderly motion of molecules (the individual balls on the table) moving in no preferred direction.

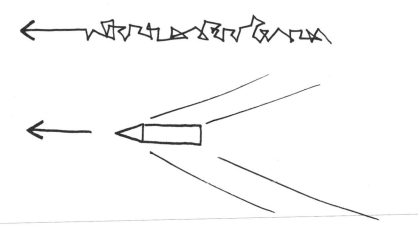

Figure 4.6 Path of an Atom in a Speeding Bullet
Although this drawing is schematic, it approximates reality in that the speed of an individual atom or molecule at ordinary temperatures is comparable to that of a bullet shot from a rifle. The atom has a disorderly motion superimposed on the net motion that it shares with all the other atoms in the bullet and that constitutes the macroscopic motion of the bullet.

Elasticity and Friction

This example represents an extreme of behavior in that the disappearance of orderly motion takes place more rapidly than common experience leads us to expect. Rubber balls do bounce up and down quite a few times before they come to rest; a swinging pendulum swings even longer, going through more cycles than the rubber ball, before its orderly motion disappears.

The attractive forces between molecules, which we omitted from our pool-ball analogy for the sake of simplification, tend to keep the orderly motion orderly, so that often only a small amount is degraded at a time. The precise nature of such forces and how they operate to resist the degradation would get us into some complex questions; for now it is easier to switch viewpoints from the microscopic to the macroscopic.

Some substances, such as rubber or steel, are *elastic,* and some, such as lead, putty, and oil, are not. Elastic bodies store orderly motion: when their shape is deformed because of an outside force, most of the kinetic

(A)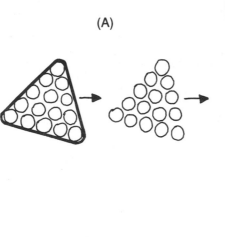

(B)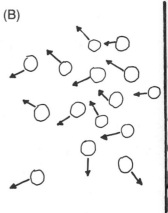

Figure 4.7 Orderly Motion to Disorderly
The ordered motion of a set of pool balls in triangular array becomes disorderly motion after the array strikes the wall. The energy of this disorderly motion is called thermal energy.

energy is temporarily converted to potential energy (for example, a wound-up spring or a stretched rubber band); and when that force is removed, most of it is converted back to kinetic energy again. That is the way the ball bounces. Inelastic substances convert kinetic energy rather quickly to disorderly motion manifested macroscopically by a rise in temperature. For example, if you drop a ball of putty to the ground, it will not bounce; it will remain, somewhat flatter in shape, on the ground, and the kinetic energy it acquired by falling is converted on the impact to thermal energy, manifested by a slight rise in temperature.

Friction between Solids

What goes on on the molecular level when two solids are rubbed together? Let us simplify the picture by imagining that the molecules of the two solids (call them A and B) are initially at rest; in other words, they are without any disorderly motion at all. Now we set the two bodies in motion relative to one another, one moving to the left with a speed V, the other to the right with the same speed. It follows that each molecule in body A is also moving to the left with speed V, each molecule in B to the right with the same speed. Each molecule in either body has a kinetic energy $\frac{1}{2}mV^2$, each body a total kinetic energy $\frac{1}{2}MV^2$, where M, the mass of the body in question, is the sum of the masses of its individual molecules. As two molecules on the surfaces of the two different bodies pass close to each other when the bodies are rubbed together, they exert forces on each other that change the motion of each: in brief they collide. Each molecule is deflected from its initial direction of travel, away from the surface of the body, and is no longer moving in the same direction as its immediate neighbors. Let us assume, unrealistically, that their speeds are unaffected: all that happens is a change of direction of the speeds of the two molecules in question, setting each on a collision course with other molecules of the body it is part of. The kinetic energies of the two molecules are the same as before, but their motion has been deflected in toward the second layer of molecules in each body. Once they collide with molecules in the second layer, they lose some of their original kinetic energy and the molecules that have acquired it pass it on to others. The consequence is the diversion of a small amount of kinetic energy into disorderly motion. This process, repeated with all the molecules in the surface layers and repeated by each surface molecule many times, leads to a slowing down and ultimate cessation of the relative motion of the two bodies, an apparent disappearance of the kinetic energy of their orderly motion, and a heating of the two bodies detectable with a thermometer.

In macroscopic as opposed to microscopic language, motion has been converted by friction into heat.

Do Molecules Encounter Friction?

We introduced in Chapter 2 the idea of an idealized, frictionless world in which motion continues forever without loss. We spoke about it as though it were a useful fiction, but now we see that it is exactly the world of molecular motion.

As molecules move, they do encounter other molecules, and one might expect that such encounters slow them down. The expectation is wrong. In any collision, a molecule has as much chance of increasing its speed as decreasing it. No energy is lost in a collision: what one molecule loses the other must gain.

Making and Testing a Kinetic Theory

Once the principle of the conservation of energy was accepted, and molecular motion recognized as a form of mechanical energy, a number of scientists (Rudolf Clausius, James Clerk Maxwell, and Ludwig Boltzmann were the leaders) began to explore the question: how well can a molecular theory account for the properties of matter?

They began with a simple and reasonable hypothesis: planets, billiard balls, and dust particles all moved according to a single set of laws, Newton's laws of motion. Why not molecules too? Very little was known about the forces acting between molecules that determine their motion, but plausible guesses could be made about them. The real difficulty came from the large number of molecules of which any real sample of matter is composed. Physicists had already run into mathematical difficulties in trying to describe the motion of the sun and seven or eight planets. How to deal with one million molecules at once? How about 10^{20} molecules?

A solution was found: combine Newton's laws with the theory of probability. It was an odd coupling: the two approaches appeared completely incompatible, and there has been friction between them ever since, but the marriage has lasted.

Probability, developed initially to help gamblers gamble intelligently, assumes a degree of ignorance: the next card upturned, the next roll of the dice cannot be predicted in advance. Newton's laws, on the other hand, leave nothing to chance. The motion of 10^{20} billiard balls can be predicted exactly by solving 10^{20} equations. Of course we don't (yet) have computers large enough for the task. If use of the theory of probability requires ignorance we can surely claim to be ignorant, but some have argued that our problem is not the right kind of ignorance, it is more like laziness. We will return to this question in a later chapter. Here we will describe some of the early results of the combination of classical mechanics and probability, and how well they agree with what can be measured in the laboratory.

In the nineteenth century the kinetic-molecular theory made a number of dramatically successful predictions of certain properties of gases, the only kind of matter the mathematical techniques of that time could handle. The molecules of gases under low pressure are, on the average, far apart. Collisions occur, but their duration is much shorter than the time a molecule spends traveling in straight lines between encounters. Attractive forces, those that cause molecules to stick together and form liquids and solids, have much less effect on the properties of gases when the temperature is high and the molecular speeds are great, though what constitutes "high" temperature depends on which molecule we are studying: helium under ordinary atmospheric pressure condenses to a liquid at −269°C, mercury at 357°C. A pretty good model of a high-temperature gas is one in which we think of the molecules as having properties like pool balls; no attraction, repulsion only on contact. Quantitative predictions of such properties as temperature, pressure, and total energy based on this model were readily made early on and showed an encouraging agreement with the properties of hot, dilute gases.

As we will see, the kinetic-molecular theory gave almost all the right answers, but on some questions it failed badly, and its failures had important implications for nineteenth-century science.

"Ideal" Gas Pressure and Absolute Zero

The pressure of a gas, the force it exerts on a unit area of the wall of its container, can be measured in a variety of units. The most familiar, in English-speaking countries, is pounds per square inch (psi). The "pound" here is the gravitational *force* the earth exerts on a mass of one pound at sea-level gravity; it is a weight and therefore a force, not a mass. The "square inch" is the unit area for this system of units. Average atmospheric pressure at sea level is 14.7 psi. The scientific unit, the one we would use if we measure energy in joules and areas in square meters, is the *pascal* (Pa). Sea-level atmospheric pressure is fortuitously very close to 100,000 Pa.

The pressure of a gas in a container of fixed volume decreases when its temperature is lowered. In gases initially at high-enough temperatures and low pressures, the decrease is slow and steady, and one might therefore expect the pressure to reach zero at a low-enough temperature. Most remarkably, however, this temperature is the same for all such gases, −273°C, regardless of their identity. Helium, oxygen, carbon dioxide, and steam all act the same, though the temperature range in which steam shows this behavior is much higher than for the other gases. Now in fact gas pressures do *not* become zero on extreme cooling, because on cooling forces between molecules become important, and other things intervene: an even more rapid fall of pressure, at first, and finally condensation to a liquid. Figure 4.8 shows schematically how the

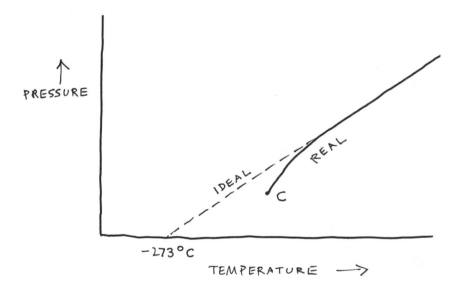

Figure 4.8 Real vs. Ideal Gas Behavior
This representation of the pressure-temperature relationship for a real gas is greatly exaggerated for emphasis. In an accurate drawing for pressures around one atmosphere, the deviation from ideal behavior as the gas is cooled to the point *C* (where it begins to condense to a liquid) would not be noticeable.

pressure of a representative gas depends on temperature. In particular, we see that the vanishing of pressure at $-273°C$ is only a hypothetical possibility, not a real event. It can be conjectured that if a gas of nonattracting molecules really existed, its pressure would vanish in this fashion, but no such gas is known. This hypothetical gas is therefore called an "ideal" gas; as noted, real gases behave like the ideal gas only when their temperatures are "sufficiently" high and their pressures "sufficiently" low.

The behavior of real gases under such conditions was well known before the development of the kinetic-molecular theory began. Its explanation in terms of that theory is deceptively simple: $-273°C$ is a temperature at which all molecular motion ceases. If there is no motion, there are no impacts on the walls, and if there are no impacts, there is no pressure. This leaves unanswered the question of just what the relation is between the kinetic energy of translation of a gas molecule, a microscopic property, and the temperature read by a thermometer immersed in the gas, a macroscopic property.

Before we go further, we need a mathematical description of the relation between gas pressure P and Celsius temperature t for gases behaving "ideally," one that shows the pressure going to zero at a temperature $t = -273°C$. An equation that gives this result is:

$$P = A(t + 273) \qquad \text{(valid at high temperatures only)}$$

where A is a numerical factor whose value depends on the identity of the gas, the mass of it in the container, and the volume of the container, but not on the temperature. The pressure of a hot, dilute gas changes with temperature in accord with this equation, but cooling eventually destroys the agreement even before condensation occurs. (See Figure 4.8).

The expression $(t + 273)$ is equal to zero when $t = -273°C$, but is not zero when $t = 0°C$. That makes sense; gas pressures do not fall to zero at temperatures of zero Celsius, the temperature of a moderately cold winter day in northern latitudes. But it does remind us of the arbitrariness of the calibration points of the Celsius scale: the reference temperature $0°C$, at which solid ice melts to form liquid water, was chosen for convenience; it was not based on any theory of molecular motion. This arbitrariness that suggests a more fundamental scale could be established by calling the temperature at which molecular motion is expected to cease "zero." It is, after all, hard to imagine anything colder than that.

For convenience and consistency with the Celsius scale, we can keep the degree the same size, so that there will still be 100 degrees between the melting point of ice and the boiling point of water. Table 4.1 shows the relation between these two scales and the Fahrenheit scale. Put simply, the new scale, called the *Kelvin scale,* is obtained by adding 273 to the Celsius temperature. The degrees in the new scale are called *kelvins* (abbreviated K). Ice thus melts at 273 K, water boils at 373 K.

We see immediately that the equation we have written for gas pressure simplifies to

$$P = AT$$

where T is the Kelvin temperature. The equation, which states that the pressure is proportional to the absolute (Kelvin) temperature, can be tested and confirmed for gases at high temperatures and low pressures: it is a general property of such gases.

The Probabilities of Various States

We said earlier that in the equilibrium state that a collection of colliding molecules will eventually reach, the molecules will be traveling not at the same speed but with many different speeds. One of the first major steps in the development of the molecular-kinetic theory was the calculation of how many molecules are traveling at each possible speed. In other words, what is the distribution of speeds among the molecules? It occurred to James Clerk Maxwell that the problem could be simplified by the use of the theory of probability. This meant ignoring the fact that Newton's laws make numerically precise predictions and pre-

Table 4.1
A Comparison of Temperature Scales

	Celsius	Kelvin	Fahrenheit
Absolute zero	−273	0	−459
Melting ice	0	273	32
Typical ambient	20	293	68
Boiling water	100	373	212
Boiling mercury	357	630	675
Melting gold	1,063	1,336	1,945

Note: All melting and boiling points are given at atmospheric pressure.

tending instead that the motion of molecules is like a game of chance. After all, the motion of the little steel ball on a roulette wheel obeys Newton's laws, but we never have enough information—about the speed of the wheel, the initial speed and spin of the ball, and just where on the fast-turning wheel the operator drops it—to apply those laws and predict the outcome. Out of ignorance we assume any outcome is as likely as any other and let the laws of probability take over. They do not tell us what will happen on the next spin of the wheel, only what the average result of a large number of plays will be.

We are usually even more ignorant about the details of real molecular motion. We don't know, at any moment, which particular molecules in a liter of gas are moving at any given speed, nor in what direction; further, for many purposes, particularly for the calculation of macroscopic properties, we do not need to know the detailed history of each of the 10^{20} or so molecules in the liter, but only the *average* properties of the collection. These are precisely the properties that the laws of probability can tell us something about.

As we said, to apply probability theory to roulette we had to make one plausible but unprovable assumption: any of the 38 possible outcomes is as likely as any other. We do the same with the tosses of a coin when we assume that heads and tails are equally likely on each toss. What shall we assume for molecules? The law of conservation of energy tells us that if the system we are studying is isolated, so that no heat flows into or out of it and no forces are exerted between it and other systems (that is, so no work is done), the total energy remains constant. The molecules can be distributed in the container (our isolated system) in an enormous number of ways and the given total energy can be divided up, or *distributed,* among them. The simplest assumption to make is that any distribution is neither more nor less likely than any other. It has not yet been shown that Newton's laws can justify this assumption, but on the other hand no better suggestions have been made. It is sufficient to note here that the use of the theory of probability together with the assumption that *all arrangements with the same total energy are equally likely* gives the same result as that obtained experimentally more than half a century after Maxwell's calculation.

Maxwell's Speed Distribution

Maxwell first applied the assumption of equal probability of all states having the same total energy to the distribution of the translational kinetic energy of gas molecules, and later he and Boltzmann extended

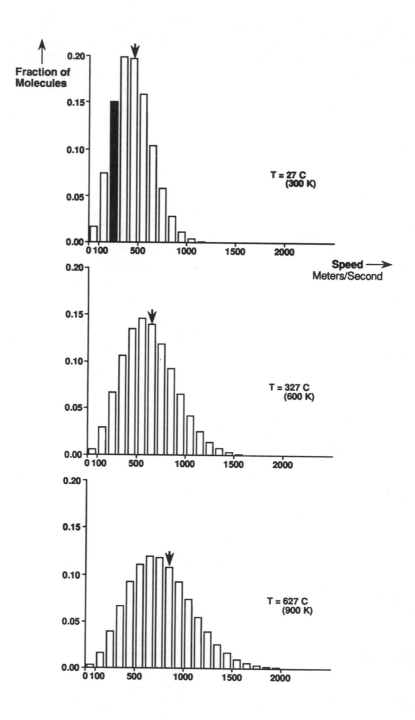

Figure 4.9   Maxwell-Boltzmann Distribution of Molecular Speeds
The results for oxygen molecules at 300 K, 600 K, and 900 K are shown, with
speeds in meters per second (m/s) indicated on the horizontal axis. The heights
of the rectangles give the fraction of molecules traveling within a given speed
range: for example, the shaded rectangle at top has a height of about 0.15 units,
representing the fraction of molecules (15 percent) traveling with speeds between 200 and 300 meters per second at 300 K. The arrows point to the average
speed in the collection at each temperature. Maxwell's calculation shows that
when the absolute temperature doubles from 300 K to 600 K, (a) the average
speed of the molecules increases about 40 percent, from 483 m/s to 684 m/s;
(b) the average translational energy doubles; and (c) the fraction of molecules
having speeds greater than 1,000 m/s increases 200-fold. (This dramatic increase in the number of fast-moving molecules as temperature is raised is related
to the rates of chemical reactions, as discussed in Chapter 11.)

it to other kinds of energy and to states of matter other than the ideal
gas. Because the translational kinetic energy of a molecule, $\frac{1}{2}mv^2$, determines the speed v (or vice versa), if we know how many molecules
have a given translational energy we know how many are traveling with
the corresponding speed. Figure 4.9 shows the speed distribution calculated by Maxwell.

Direct experimental confirmation of Maxwell's result did not come
until the twentieth century, but Maxwell and others were able to use
the theoretically calculated distribution to predict various averages of
the collection of molecules. These average values permitted the calculation of macroscopic properties of gases that could be compared with
experimental values. Considerations of space, relevance, and ease of
exposition limit our discussion to just a few properties, those concerned
with the relation between the total translational energy of the gas, its
temperature, and the number of molecules.

The Average Energy and the Equipartition Theorem

One of Maxwell's most significant results was that when two different
gaseous substances are compared at the same temperature, the *average
translational energy* of each kind of molecule is the same. This average
translational energy, which we symbolize by ε, depends *only* on the
temperature, not the mass or number of atoms in the molecule:

$$\varepsilon = KT$$

This equation applies only when the collection of molecules is at equilibrium and therefore has the equilibrium speed distribution. The

Experimental Corroboration of Maxwell's Speed Distribution

The distribution of molecular speeds was determined experimentally only in the twentieth century, when vacuum pumps able to produce very "high" vacuums were developed. A small number of molecules of a substance *S* is allowed to escape from an enclosure into a high vacuum by opening a shutter briefly. The few molecules that escape from the enclosure travel outward in various directions toward a wall with a small hole in it. Very few molecules get through the hole; those that do form a train with the faster ones in the lead. The train, or beam, is then passed through a *speed selector.* (This device operates much as the traffic lights do on one-way avenues in New York City, which allow cars to travel long distances without being stopped by a red light, provided they are driven at 25 miles per hour.) A shaft bearing two circular disks, each with a pie-shaped wedge cut out of it, is rotated at a high speed in the path of the beam. Only a part of the beam gets through the wedge-shaped opening in the first disk, and in this truncated beam the mole-

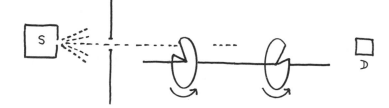

quantity *K* is a universal constant, the same for all kinds of molecules (and in this context is not the symbol for kelvins), and *T* is the absolute (kelvin) temperature.

The constant *K* is another example of a constant of nature. Like the factor Joule found for converting heat to work, its value is the same for all substances and all kinds of measurements. Another such constant is the speed of light in a vacuum, symbolized by *c,* to be discussed in Chapter 15. Constants of nature differ from constants of mathematics, like π, in that their numerical values depend on the units we choose to cules are still traveling with the same distribution of speeds. Only those traveling at a certain speed, however, will get through the opening in the second disk; this speed is set by adjusting the relative position of the two openings, the speed of rotation of the shaft, and the distance of separation of the two disks. The number of molecules that get through the second opening are counted by a detector *D* placed behind the second disk. Once the number traveling at a given speed is determined, the speed of rotation of the shaft is changed and another count is made, and so on until a complete profile of the different possible speeds is determined.

An alternative "experiment" is really a calculation. Newton's laws of motion can be solved on powerful computers not only for 16 balls on a flat table but for several thousand molecules in a three-dimensional box. No use of the theory of probability is necessary. The molecules are assumed to have some definite amount of kinetic energy to start with, and the computer solves the equations of motion that describe how the molecules would move about and collide until no further changes in the average properties occur. The computer is programmed to tally the numbers of molecules that would be traveling at various speeds once this equilibrium state is reached.

While several thousand molecules is a smaller sample of matter than we might like to study, there is no obvious reason why the speed distribution of this relatively small number should differ from that for 10^{20} molecules, and in fact the results of the experiments on real molecules and the computer simulations agree within computational error and both agree with Maxwell's theoretical calculation.

express them in. Aside from this difference, however, they have the usual mathematical properties that the word *constant* implies. In particular, it is arithmetically obvious that if we multiply a constant by 2, or divide it by 3, the result is simply another constant. For reasons that will soon become apparent, we find it convenient to follow what has become the conventional practice and replace K by another constant k, just two-thirds of K:

$$k = \tfrac{2}{3} K \quad \text{or} \quad K = \tfrac{3}{2} k$$

Hence Maxwell's result becomes:

$$\varepsilon = \tfrac{3}{2}kT$$

Unfortunately, the numerical value of k could not be determined unless the number N of molecules in the collection could be determined first; as we noted, this work lay some fifty years in the future.

It might seem as though an impasse had been reached. The goal was to compare theory and experiment for the average translational energy of molecules, but it was inaccessible by either theory or experiment, because N and therefore k were not known. Fortunately, theory and experiment for at least some properties of gases could be compared without knowing either. Before we describe how these comparisons were made, it will be useful to go into more detail concerning Maxwell's result.

Maxwell's calculation of the average translational energy is one application of a more general consequence of the kinetic-molecular theory, the *equipartition theorem*. This theorem describes the relations between the average molecular energy and the absolute temperature, and it does so for each of the kinds of molecular motion we have mentioned. (We will give the results for rotational and vibrational energies later.) For our discussion of the theorem, let us attach subscripts to the symbol to denote the specific type of molecule. ε_{H_2O} is thus the average energy of the water molecules in some collection we are studying. The equipartition theorem can be thought of as implying two separate statements:

1. Molecules of *different* substances, when compared at the same temperature, have the *same* average translational energy. Now different kinds of molecules will usually have different masses. Hydrogen gas is composed of molecules with two hydrogen atoms joined together—the chemical symbol is H_2—and it is the lightest molecule commonly encountered. A water molecule (H_2O) is nine times as heavy as a hydrogen molecule, an oxygen molecule (O_2) sixteen times as heavy, and a uranium hexafluoride molecule (UF_6) more than 150 times as heavy. Yet when collections of these different molecules are at the same temperature, their average translational energies are, according to the equipartition theorem, the same:

$$\varepsilon_{H_2} = \varepsilon_{H_2O} = \varepsilon_{O_2} = \varepsilon_{UF_6}$$

This implies that heavier molecules must have lower average speeds than light ones.

2. The average translational energy—the same for all kinds of molecules at any given temperature—is directly *proportional* to the absolute temperature *T*. This means, for example, that it doubles when the temperature doubles (see the appendix at the end of the book for a review of some mathematical concepts, such as proportion).

The following equation summarizes *both* statements:

$$\varepsilon_{H_2} = \varepsilon_{H_2O} = \varepsilon_{O_2} = \varepsilon_{UF_6} = \tfrac{3}{2}kT$$
$$= \text{average molecular translational energy}$$

where the constant *k* is the same for all kinds of molecules.

Although, as we mentioned, the average translational energy could not be calculated in Maxwell's day (because *N* and therefore *k* were not known), this equation certainly was not useless. If we know the total energy *E* (a macroscopic quantity) and the number of molecules *N* (a microscopic quantity, which we may or may not know), then the average energy of a molecule ε (a microscopic quantity) is obtained by dividing *E* by *N*:

$$\varepsilon = \frac{E}{N} \quad \text{or} \quad E = \varepsilon N$$

Though neither *N* nor ε was then known, their product was a known quantity, the total (macroscopic) thermal energy *E*. As a result, even with *k* unknown, some of the consequences of the equation could be tested experimentally in the nineteenth century. Furthermore, even without a solution for ε, the equation had important implications for the current understanding of temperature.

A New View of Temperature

In the eighteenth century Joseph Black had discovered a concept he called the "equilibrium of heat": when two bodies at different temperatures are brought into contact, the temperatures tend to equalize; once they have equalized no further changes take place. The first law tells us that what leads to the equalization of temperature is a transfer of energy—a flow of heat—from the hot body to the cold one. The contribution of molecular theory was to explain how this is done: energy is transferred by collisions between molecules, and temperatures change because the thermal energy of each body is the sum of the kinetic and potential energies of its molecules.

Experimental Comparison of Molecular Masses and Speeds

If two different kinds of molecules, with different masses, have the same average translational kinetic energies, the heavier of the two must, on the average, be moving more slowly. If $\frac{1}{2}mv^2$ is the same for both, and m_A is greater than m_B, then v_A must be less than v_B. Let us consider a specific example. The mass of an oxygen molecule, O_2, is about sixteen times that of a hydrogen molecule, H_2. The equality of average translational energies requires that

$$\tfrac{1}{2}m_{H_2}(v_{H_2})^2 = \tfrac{1}{2}m_{O_2}(v_{O_2})^2$$

where v_{H_2} and v_{O_2} are average speeds. Let us rewrite the above equation:

$$\frac{m_{O_2}}{m_{H_2}} = \frac{(v_{H_2})^2}{(v_{O_2})^2}$$

If m_{O_2} is 16 times m_{H_2} we know immediately that

$$\frac{m_{O_2}}{m_{H_2}} = \frac{(v_{H_2})^2}{(v_{O_2})^2} = 16 \quad \text{or} \quad \frac{v_{H_2}}{v_{O_2}} = 4$$

In other words, the velocity of the H_2 molecule is 4 times that of the O_2 molecule.

How to test this? If a container of gas has a tiny orifice opening to an evacuated space, the gas molecules will leak out slowly. The phe-

Let us imagine two separate containers, each with a different gas, and each at equilibrium. The temperatures of the two containers need not be equal. For simplicity let us consider hydrogen and (gaseous) water: H_2O molecules are nine times more massive than H_2 molecules. In each gas initially the molecules are traveling at a range of speeds given by Maxwell's speed distribution. If we bring the containers into contact and remove the adjacent walls so the gases mix, collisions take place between hydrogen molecules and water molecules. Because of the speed distributions, the relative speeds of the colliding molecules vary

nomenon is called *effusion*. The rate of leakage—measured, say, in cubic centimeters of gas per hour—will depend on the size of the orifice, the number of molecules per cubic centimeter in the container, and, finally, on their average speed. If we compare the rates of effusion of two different substances from the same orifice, and if the number of each kind of molecule in the container are the same, the rates will be in proportion to the average speeds of each kind. In other words, hydrogen should leak out 4 times as fast as oxygen, other things being equal.

Experiments that permitted a test of the kinetic theory's prediction had been performed by a British scientist, Thomas Graham, in the 1840s, before the theory had been developed. His results, examined several decades after he had obtained them, were in complete agreement with the theory.

The dependence of effusion rate on molecular mass was applied to powerful effect in the Second World War, when it was used to separate the lighter isotope of uranium (relative mass 235) from the more abundant isotope of relative mass 238, to make the first nuclear bomb. A gaseous compound of uranium and fluorine (UF_6) was prepared from natural uranium, a mixture of the two isotopes, in which the two kinds of UF_6 molecules differed in mass by only about 0.5 percent. Repeated passage through an effusion apparatus produced sufficient separation of the lighter molecule to permit accumulation of the amount needed to fuel a chain reaction.

over a wide range; so also do the angles at which they meet. In some of the collisions a hydrogen molecule will gain energy at the expense of the water molecule, in others it will lose it. To decide whether heat flows from hydrogen to gaseous water or vice versa, we must ask what happens to the energy *on the average* in an enormous number of collisions.

Maxwell, whose model of molecular motion was similar to our example of rotationless billiard balls, was able to prove, by analyzing all possible collisions, that when the *average* translational kinetic energy of the lighter molecules is the same as the *average* translational kinetic energy of the heavier molecules, there is no *net* transfer of energy, or in other words, no flow of heat either way. Since temperature in the macroscopic world determines the direction of the flow of heat, he concluded that the temperature of a gas is a measure of the average translational kinetic energy of its molecules. That is what the equation $\varepsilon_{H_2} = \varepsilon_{H_2O} = \frac{3}{2}kT$ implies: if temperatures are equal, so are average translational kinetic energies, and if the temperature of one substance is higher than another, its average molecular translational energy is higher in proportion.

This result seems to imply a simple molecular meaning for the concept of "temperature." Unfortunately, it is limited in its applicability to gases under conditions when the equipartition theorem works. As we will see, the theorem often fails, and we must find a more universal explanation of the molecular meaning of temperature.

Ideal Gas Behavior

The kinetic-molecular theory attributes the pressure of an ideal gas to the combined effects of all the molecular impacts on the container walls. We can make a crude estimate of the force exerted on 1 square meter of wall by these impacts, and thus the pressure. Let us assume a cubical container, 1 meter on each side and thus with a volume of 1 cubic meter, holding N molecules, each with mass m. Let us further simplify by pretending that all molecules move with the same speed v. Three factors must be taken into account to calculate the pressure.

1. Newton's laws show that the average force on the wall due to one impact depends on the mass m and the speed v toward the wall; specifically, it is proportional to the product mv.
2. Any one molecule makes repeated impacts on the wall, and makes them more often the faster it is moving, so the frequency of impacts is proportional to the speed v.
3. The total force must be proportional to the number of molecules in the container.

The pressure is proportional ($\propto$) to each of these quantities and is therefore proportional to their product. That is, P is proportional to the mean force of one impact × the frequency of impacts × the number of molecules, or

$$P \propto mv \times v \times N$$

Hence P is proportional to Nmv^2.

Since the average translational kinetic energy $\varepsilon = \frac{1}{2}mv^2$, we can replace mv^2 by 2ε, and, since $N\varepsilon = E$ (as defined above),

$$P \propto 2N\varepsilon = 2E$$

where E is the total translational energy of all the molecules. We derived this estimate for a volume of 1 cubic meter. If we have the same number N of molecules in a larger volume V, the frequency of impacts is reduced in proportion, and our estimated pressure is thus:

$$P \propto \frac{2E}{V}$$

Note that the (unknown) number of molecules N has disappeared from the formula. Only the macroscopic property E, the total translational energy, and the macroscopic volume V determine the pressure.

A less crude treatment must take into account, among other things, the fact that the molecules are moving in all directions with equal probability, so that most of them do not strike the wall perpendicularly but at smaller angles, thus reducing the force of impact below the above estimate, and further that they are not all traveling at the same speed. The exact result of a full mathematical treatment is

$$P = \frac{2E}{3V}$$

This result was obtained by Daniel Bernoulli in the early eighteenth century, but it was ignored for one hundred years.

Experimental Tests

Gas pressure and gas energy. The relation $P = 2E/3V$, though obtained from a molecular theory, is a relationship between only macroscopic quantities: the pressure of a gas, the volume of the confining container, and the total translational energy. The only difficult step is the determination of the energy. The total energy of a gas can be inferred straightforwardly from specific heat measurements, as described in Chapter 3, but with the following qualification: most molecules can rotate and vibrate also, movements that contribute to the total energy of the gas, and it was not, at that time, a simple matter to decide what share of the total energy is translational kinetic energy.

The problem was solved with the aid of an assumption: gases in which the molecules are composed of single atoms rather than two or more, as in O_2 or H_2O, were believed then to be incapable of rotation and vibration. We know today, for reasons too technical to give here, that this is not strictly correct, but it is a good enough approximation. The only monatomic gas known in the nineteenth century was mercury vapor (helium, neon, and argon gases were discovered later). The relation between the translational kinetic energy of mercury, as determined from its specific heat, and the pressure was found to match the predicted value.

Gas pressure and temperature. Consider the following derivation.

$$P = \frac{2E}{3V}$$

$$\text{(since } E = N\varepsilon) \quad P = \frac{2N\varepsilon}{3V}$$

$$\text{(and since } \varepsilon = \tfrac{3}{2}kT) \quad P = \frac{2N(\tfrac{3}{2}kT)}{3V} = Nk\frac{T}{V}$$

This equation contains those two quantities whose values could not be determined in the nineteenth century, the number of molecules N and the constant k that tells us the relation between the average translational energy of the molecules in a gas and the absolute temperature. This does not, however, rule out the possibility of an experimental test of the equation.

If we keep both the volume V and the number of molecules N fixed by confining the gas in a closed container, the equation predicts correctly the observed proportionality between the pressure and the absolute temperature, and the pressure is predicted to fall to zero when $T = 0$. We can bring this out more clearly if we rearrange the equation:

$$P = Nk\frac{T}{V} = \left(N\frac{k}{V}\right)T$$

This is just the relation we expressed earlier by the equation

$$P = AT$$

Note that without direct knowledge of either N or k we cannot tell whether the numerical value of the constant A is correctly given by the theory, but even without that knowledge the proportionality of pressure to temperature is predicted.

Avogadro's forgotten hypothesis. Another conclusion of Maxwell's theory is that when equal volumes of different gases are compared at the same pressure and temperature, the (then-unknown) numbers of molecules must be the same in each. We can show why by rewriting the equation

$$P = Nk\frac{T}{V}$$

in the equivalent form

$$P\frac{V}{T} = Nk$$

Now consider two kinds of molecules, A and B. For a gas of each kind, we have

$$\frac{P_A V_A}{T_A} = N_A k \qquad \frac{P_B V_B}{T_B} = N_B k$$

Now suppose we compare equal volumes of the gases, so that $V_A = V_B$, and at the same pressure and temperature, so that $P_A = P_B$ and $T_A = T_B$. Then it follows that

$$\frac{P_A V_A}{T_A} = \frac{P_B V_B}{T_B}$$

Thus we can conclude that

$$N_A k = N_B k$$

and hence that

$$N_A = N_B$$

showing that under the conditions of the comparison, the numbers of molecules of the two kinds must be the same.

This is an interesting result, but was there any experimental evidence for it at a time when molecules could not be counted? The answer was provided by an Italian chemist, Stanislao Cannizzaro, who remembered an overlooked hypothesis offered a half-century earlier by a compatriot of his.

In the late eighteenth century a French chemist, Joseph Louis Gay-Lussac, had discovered some surprising regularities concerning volumes when two gaseous substances react. For example, when steam is formed from hydrogen and oxygen, exactly two liters of hydrogen are needed for each liter of oxygen, and they form exactly two liters of steam. The *weights* of hydrogen and oxygen when they react to form steam bear no such simple relationship.

In 1811, Amedeo Avogadro, an Italian physicist, had conjectured, on the basis of these regularities, that the number of molecules in equal volumes of different gases (when the pressures and temperatures of the two gases were equal) must be the same (Figure 4.10). Avogadro did not have a kinetic-molecular theory in mind. He assumed that the volume occupied by a gas molecule was determined by the quantity of "caloric" attached to it, and that this quantity would be the same for all molecules regardless of their chemical nature. His hypothesis had two things against it when it was first proposed. First, it implied some things about the structure of molecules of elementary substances that contradicted the chemical theories of the time (but not those of today). In particular, he suggested that certain elements, such as hydrogen and

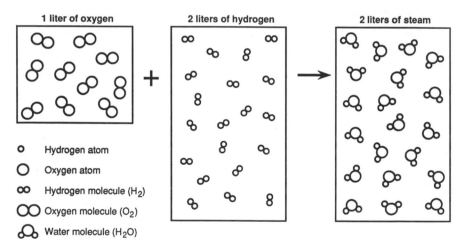

Figure 4.10 Avogadro's Hypothesis

The hypothesis that equal volumes of all gases contain the same number of molecules (when compared at the same pressures and temperatures) was suggested to Avogadro by experimental observations of the volume changes when gases react. For example: in the reaction in which gaseous water (steam) is formed from hydrogen and oxygen, for each liter of oxygen, exactly 2 liters of hydrogen are needed. If there were more hydrogen, some would be left over. If there were less hydrogen, not all the oxygen would be used up. Exactly 2 liters of steam are formed for each liter of oxygen and for each 2 liters of hydrogen consumed. In general, the volumes of the reacting and resulting gases are in the proportion of small whole numbers, such as 1 to 1, 2 to 1, 3 to 2, and so on.

oxygen, existed as diatomic molecules: H_2 and O_2. Second, it was an isolated hypothesis, unsupported by any experimental or theoretical evidence other than the regularities of volume changes in chemical reactions that had inspired Avogadro to suggest it in the first place. So although Avogadro's hypothesis did explain Gay-Lussac's surprising regularities, and also suggested new structures for the molecules of gaseous elements, it had been almost forgotten by the time Maxwell published his results.

Counting Molecules

Cannizzaro not only revived Avogadro's hypothesis but also pointed out that it provided a kind of counting procedure for molecules. Scientists did not then know the number of molecules in a cubic centimeter of either hydrogen or oxygen, but Cannizzaro realized that, if Avogadro was right, the number had to be the same in both. For example, in 1 cubic centimeter of hydrogen gas at 1 atmosphere pressure and 0° Celsius there are a certain number of molecules. This number was then unknown: let us call it L. Then it would follow that in 1 cubic centimeter of oxygen gas under the same conditions the number of molecules must also be L.

The determination of the relative masses of molecules follows in a straightforward way. It was then no problem to weigh accurately 1 cubic centimeter of gas, and thus determine its density, defined as the mass of 1 cubic centimeter (g/cc). Using the symbol L for the (unknown) number of molecules in a cubic centimeter, we can express the total mass of hydrogen in a cubic centimeter of hydrogen as L times the mass m_{H_2} of a single molecule of hydrogen, and the mass of oxygen in a cubic centimeter of oxygen is L times the mass m_{O_2}. The ratio of the gas densities is therefore:

$$\frac{\text{Density of oxygen}}{\text{Density of hydrogen}} = \frac{\text{Mass of 1 cc of oxygen}}{\text{Mass of 1 cc of hydrogen}} = \frac{L \times m_{O_2}}{L \times m_{H_2}} = \frac{m_{O_2}}{m_{H_2}}$$

This equation gives the *ratio* of molecular masses with high accuracy, although we do not know either individual molecular mass at all.

With this ratio, the masses of all molecules relative to an arbitrary reference molecule can be obtained. The original choice for the reference molecule was the hydrogen atom, the lightest element, whose mass was set at 1.000. The mass of the modern choice, calculated as a fraction of the mass of an isotope of the element carbon, differs from that of the early choice by less than 1 percent.

Cannizzaro used Avogadro's hypothesis to determine relative atomic and molecular masses, a necessary step toward Dmitri Mendeleev's proposal of the periodic table of the elements about a decade later. Mendeleev showed that if the known elements were arranged in the order of their atomic masses, starting with hydrogen, a remarkable periodicity in properties manifested itself. Mendeleev's periodic table led directly to the discovery of previously unknown elements and enabled chemical knowledge to be organized into a system that is still in use.

It was also the first major break in the wall that up to that time separated chemistry from physics. The development of quantum mechanics in this century tore down what still remained of that wall, and there is no longer any real boundary between the two sciences, other than the administrative ones set up for the convenience of universities and other scientific institutions.

Triumph and Failure

The demonstration of the interconnections among pressure, temperature, translational energy, and numbers of molecules in a gas was one of the great triumphs of the kinetic-molecular theory. But the theory had its failures too. Maxwell's method of predicting the average translational kinetic energy of a molecule had been extended by him to the average energies of other kinds of motion also. The results, in all cases predicting an energy proportional to the absolute temperature, are known collectively as the *equipartition theorem*, and they are a direct consequence of the application of the theory of probability to Newton's mechanics. The theoretical prediction for the average rotational and vibrational energies are a little more complicated to state than for translational energy, as they depend on the number of atoms in a molecule and how the atoms are arranged in space (the "structure" of the molecule). But for diatomic molecules, like O_2 or HCl, the predictions are particularly simple: both rotational and vibrational energies should be kT. On examination of the data on specific heat, it was immediately apparent that the observed energies were always less than the predicted ones.

A careful examination of data on a variety of molecules with different numbers of atoms and different structures showed that the average rotational energies were usually correctly given by the kinetic theory, but the vibrational contribution was almost always much less than predicted. In some molecules it appeared to be totally absent.

One might have been tempted to conclude that some molecules don't vibrate, but there were direct methods, using infrared light absorption, to test whether the molecules were actually capable of vibrating; and by this criterion, the molecules really were. Where was the missing energy then?

The discrepancies were first observed for molecules in the gaseous state, but the study of solid substances made matters more tantalizing.

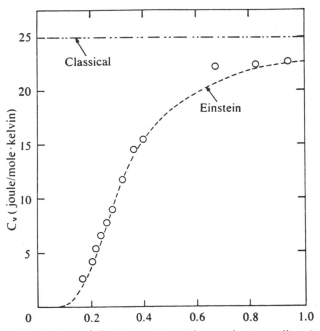

Figure 4.11 Failure of the Equipartition Theorem for Crystalline Solids
The equipartition theorem predicts that the thermal energy of a crystalline solid should be proportional to the absolute temperature, and that therefore the specific heat of the solid (sybolized here by *Cv*) should be the same at all temperatures. This graph compares the measured specific heat of diamond, shown as the open circles, with the specific heat predicted by the equipartition theorem, the horizontal line labeled "Classical." The dashed line labeled "Einstein" is a theoretical prediction to be discussed in Chapter 14. The horizontal scale is proportional to the absolute (Kelvin) temperature; 1.0 corresponds to 1,320 kelvins, and 0 to absolute zero.

The atoms in solids should also vibrate about their resting places in the crystal and the equipartition theorem predicted an energy of $3kT$ per atom. Indeed most solids, when studied around ambient laboratory temperatures (20–25°C), have specific heats close to but usually a little less than predicted, which was reassuring. The specific heat should have remained unchanged as the solid is cooled, however, and instead the specific heats of solids were found to decrease rapidly to low values on cooling. They seemed to approach zero as absolute zero was approached, an observation totally at odds with the equipartition theorem (Figure 4.11).

This was one of several serious discrepancies between the early kinetic-molecular theory and experimental reality that, toward the end of the nineteenth century, led a number of scientists to turn away from the hypothesis that matter is made up of atoms and molecules capable of moving and of having kinetic and potential energy. To them, postulating the existence of unobservable small particles of matter with mysterious and often contradictory properties was more speculation than science.

The law of the conservation of energy remained unchallenged: its power and range of applicability became the standard for what science should be. The quantities with which it dealt—heat inputs and outputs, work done on or by the system under study—were macroscopic quantities easily measured in the laboratory. It did not need to be justified by a fanciful atomic theory. Best of all, it never failed.

5
Engines and Refrigerators: The Second Law

~~~~~~

A heat engine is a device for converting thermal energy into mechanical energy. Most commonly the source of the thermal energy is the combustion of a fuel, though nuclear, solar, and geothermal energy have been used. Heat engines require at least two temperatures for operation; a steam engine, for example, needs a hot boiler and a cold condenser. A considerable portion of the thermal energy supplied by the fuel is discharged to the condenser and eventually wasted on heating the surroundings. This was true at the beginning of the Industrial Revolution, and it is true today, for all forms of heat engine, including the internal combustion engine of the automobile. If an engine could be designed to operate without this waste of thermal energy, the consequences would be tremendous: we would no longer need fuel to operate our engines, and there would never be an "energy shortage"—more precisely, a shortage of inexpensive fossil fuels.

In the nineteenth century Sadi Carnot combined the old idea that perpetual-motion machines are impossible (an early and incomplete version of the first law) with the caloric theory (heat is a substance) to prove that in a heat engine a discharge of heat to the cold surroundings was inevitable. From this starting point, he and others deduced some remarkable conclusions about the efficiency of heat engines and about the properties of matter, conclusions that were repeatedly confirmed in the laboratory.

By the time the first law had been discovered and the caloric theory discarded, scientists were faced with a paradox: Carnot's conclusions about the properties of matter were almost certainly true, but the caloric theory he derived them from was certainly false.

The paradox was resolved by the conjecture that there must be a hitherto unknown law of nature, now known as the second law of thermodynamics, that states as an axiom that a *cyclically operating* heat engine cannot convert thermal energy to mechanical energy unless the device uses at least two temperatures and wastes some of the thermal energy at the lower temperature. The meaning of the phrase *cyclically operating*, and the reason for the restriction, will be given later. From this new starting point Carnot's conclusions about the efficiencies of heat engines and the properties of matter still follow. There really was no paradox: it is logically possible for a correct conclusion to follow from false premises.

In this chapter we will analyze the operation of heat engines using the first and second laws and derive Carnot's conclusions. The analysis, although subtle, relies more on verbal reasoning than on mathematics, and it is worth going through for the sake of its scientific importance.

Previously we showed how the first law implied the existence of a property of matter called *thermal energy*, which could be determined in the laboratory. The second law can be shown to imply the existence of an additional property of matter, called *entropy*, which can also be determined in the laboratory. The first law states that the energy of the universe is conserved: the total amount of it remains the same for all time. The second law states that the entropy of the Universe is *not* conserved but that the total amount of it can never decrease.

The second law provides a criterion of *possibility:* only processes that cause a *net increase* in entropy are possible, though not everything that is possible need happen. It follows that if there is an upper limit to the entropy that the universe can have, then when that limit is reached nothing further can happen. The universe, having exhausted its capacity for change, would have run down like a clock.

Heat Engines, Possible and Impossible

In Chapter 3 we explained how various forms of mechanical or electrical energy could be converted to thermal energy through friction or through the electrical resistance of a conducting wire. Now we want to look in some detail at the *reverse*—the conversion of thermal energy to mechanical energy, a conversion that takes place in the engines of automobiles and in power plants that generate electricity from fossil or nuclear fuels. What, if any, are the limitations on the conversion of thermal energy to other forms?

Before we address that question, let us examine how some specific examples of heat engines actually work. The first successful cyclical steam engine was put into operation by Thomas Newcomen in England in 1712, to pump water out of coal mines. It included a cylinder with a moving piston, similar to the cylinder and piston of a modern automobile engine. The back-and-forth motion of the piston operated a lever, the "great beam," which in turn operated the pump. In one cycle of operation, the cylinder was first filled with steam from the boiler, pushing the piston outward. Then cold water was sprayed into the cylinder, condensing the steam and creating a partial vacuum. The pressure of the atmosphere then forced the piston downward to its starting point—this was the power stroke. The freshly condensed, still-warm water was returned to the boiler to be reused. The repeated return of all components of the engine, including the working substance, to the same state in which they began is what is meant by *cyclical operation*.

Although Newcomen's engine made grossly inefficient use of the energy of the fuel, it had one crucial feature shared by all other heat engines invented since: it did not operate at a single temperature. Instead, it received thermal energy at a high temperature, provided by the boiler, and gave a portion of it out at a low temperature, usually the temperature of the ambient environment or close to it, so that a portion of the thermal energy supplied by the boiler was wasted on warming the environment.

The cylinder of the Newcomen engine undergoes large changes in temperature during a cycle of operation: it is heated to 100°C by the incoming steam, then cooled down when the cold water is injected to condense the steam. James Watt recognized that the repeated cooling and reheating of the necessarily thick metal walls of the cylinder was a source of inefficiency, and in 1769 he patented an improved engine, in which the steam was condensed in a different chamber from the cylinder in which the piston moved. In his engine the cylinder could be hot all the time, and the condensing chamber always cold (see Figure 5.1). This innovation led to a dramatic improvement in the work the engine could do per bushel of coal burned for fuel, and Watt's career, and much else, was launched. One feature that Watt's engine shared with Newcomen's was the return of the water produced by condensation of steam to the boiler: the water is thus *recycled*. (In some steam locomotives, low-temperature steam was discarded in each cycle, which is why such locomotives had to stop from time to time to take on more water.)

The Internal Combustion Engine

We can compare and contrast Watt's engine with the internal combustion engine of the automobile. In the cylinders of the automobile engine the power comes from the explosive combustion of a gasoline-air mixture, which drives the piston forcefully outward to increase the rotational speed, and thus the kinetic energy, of a heavy flywheel. Valves operated by the flywheel then open to allow the gaseous products of the combustion to be driven out of the cylinder on the return stroke of the piston and discharged through the exhaust system; then the valves close and the cylinder is ready to receive new fuel and air.

The internal combustion engine differs from the steam engine in a number of significant ways. One is that the combustion products of the

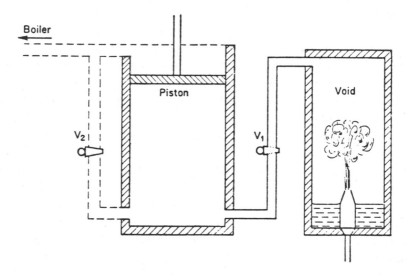

Figure 5.1   Watt's Steam Engine
In the first step of operation of the Watt engine, steam from the boiler enters the cylinder, allowing the piston to move outward. In the second step the valve to the boiler (V_2) is closed and that to the condenser (V_1) is opened. A jet of cold water sprayed into the condenser condenses the steam, thus creating a partial vacuum in the condenser and in the cylinder. The pressure of the atmosphere forcefully drives the piston down: this is the power stroke.

fuel are also the working substance of the engine, while in the steam engine the fuel (coal, usually) only supplies heat to the working substance (water), the combustion products being discharged directly to the environment. A second difference is that in the automobile the working substance is not recycled, while in Watt's engine the water is converted to steam, condensed, and reused over and over again. This cyclical operation has a useful side effect: it simplifies the bookkeeping on energy inputs and outputs of the engine. Since the working substance is repeatedly brought back to its starting state, all its properties, including its thermal energy, must repeatedly return to the same values.

The two engines do share one important feature: in both, only a part of the thermal energy supplied by the combustion of fuel is converted to the mechanical energy of motion, the rest being discharged to the cool environment. Now this waste in the operation of a heat engine sounds like a drawback. Is there no way to design a heat engine that would convert thermal to mechanical energy without waste?

A Solution to the Energy Shortage

Suppose it *were* possible to extract the thermal energy of matter and convert it to useful work without a discharge of some of that thermal energy to a second, and lower, temperature. We would never again have to think of "energy shortages." There is an enormous amount of energy around us, in the form of atomic and molecular motion. Each atom, moving with a speed v, has a kinetic energy of $\frac{1}{2}mv^2$. The mass m of any one atom is of course very small, but there are enormous numbers of them in any substantial amount of matter, and their velocities are large, about the velocity of sound. For example, at a temperature of 20°C—about room temperature—air has about 122,000,000 joules of energy for each metric ton (1,000 kilograms) in the translational motion of its molecules and almost as much additional energy in their rotational and vibrational motion. Now if we could extract this energy from the air, we would necessarily cool it. The first law requires this. While it might be unreasonable to expect to extract all the energy of motion and cool the air to absolute zero (which would solidify it), we would do very well if we could extract as much energy as would cool it by only 10°C. On this much cooling the metric ton of air would supply about 10,000,000 joules, equivalent to nearly 3 kilowatt-hours.

So we need only to invent some engine that draws in air rapidly, extracts this much energy from each metric ton, expels the cooled air, and uses the extracted energy to power a vehicle. We could then drive from New York to San Francisco in this fashion without needing, or paying for, gasoline.

What happens to the energy we have extracted from the air to operate this device? It has first been converted to the motion of our vehicle and then, by the action of friction (in the moving parts of the engine, in the tires moving over the pavement, in air resistance, in the brakes of the vehicle when we need to stop), into a heating of the surrounding environment. Eventually, as the heated bodies cool down in the ambient air, all the energy returns to the air from which it was extracted and is thus ready to be reused for the return trip. So the energy supply is inexhaustible as well as free. All we have to do is invent the energy-extracting engine, and we do not have to worry about the energy shortage any longer.

By now we begin to feel a little uneasy. It is not obvious why we cannot make a heat engine that takes thermal energy from the surrounding environment and uses it all to do work. The first law certainly does not prohibit it: we are not creating energy from nothing, only recycling it—converting it from one form (thermal energy) to another (kinetic energy of a moving vehicle) and back again. But still we feel instinctively that it must be impossible. If it isn't impossible, why hasn't it been invented?

A New Law of Nature

This statement, "Such an engine is impossible," doesn't sound like a law of science: scientific laws should be experimentally testable, and how do we test this one? Certainly if anyone succeeded in building such an engine, we would know at once that the statement is false. But suppose a hundred years go by and no one succeeds, in spite of many attempts. Would we then have enough confidence in the statement to declare it a scientific law, or should we wait another hundred years?

Fortunately, testing this statement is not as difficult as it may sound. Logical analysis often permits us to draw an extraordinary number of inferences from a simple-sounding statement: if the inferences can be experimentally tested, if there are enough of them, if at least some of them are surprising, and if they are all confirmed in the laboratory, we become very confident in the truth of that statement. The reader who remembers how many theorems, corollaries, and homework problems follow from the simple axioms and definitions of plane geometry will have had vivid experience of the power of logical analysis. The first law itself can be stated in a few words: energy is conserved. Yet its consequences are far-reaching. Other major scientific theories can also be stated simply—sometimes in words, sometimes in a few mathematical equations—and yet generate a wide range of experimentally testable implications by logical and mathematical reasoning.

The statement that it is impossible for a cyclically operating device to convert thermal energy to mechanical energy without wasting some thermal energy on cold surroundings is only one example. It is called the *second law of thermodynamics*. We have as much confidence in it as in any scientific law yet discovered, not because no one has succeeded in building the engine since Kelvin first formulated the law in essentially these terms in 1848 ("It is impossible . . . to derive mechanical effect from any portion of matter by cooling it below the temperature of the coldest of the surrounding objects") but rather because an extraordinary number of logical consequences have been drawn from it that have been repeatedly confirmed in the laboratory. The reach of the second law is vast—it makes predictions we can test about the metabolism of a living cell, how the circulation of molten rock below the surface of the earth moves continents, and what happens when stars collapse—but its roots, as we discuss below, are embedded in a very particular problem of a very particular technology: that of the heat engine.

We will describe in this chapter why the scientists involved were led to make this statement as a law of nature; how they were able to draw conclusions from it for both the efficiencies of heat engines and the properties of matter; and what kinds of confirmatory evidence led to its acceptance as one of the greatest of scientific generalizations.

Wrong Theories and Right Answers: Carnot's Contribution

We will begin by going back in time to a period shortly before the discovery of the first law. The period is the 1820s, with the Industrial Revolution in full swing, powered both literally and figuratively by the steam engine. A question then (and now) of considerable economic importance was how to get the maximum output of work for a given quantity of fuel. The steam engine was the prevailing technology: there was constant competition to improve it, and experiments using substances other than water as the working substance were being made.

The French engineer Sadi Carnot addressed this question in a monograph published in 1824. Carnot was the son of a famous father, Lazare

Carnot, himself both a scientist and a prominent figure in the French Revolution. The elder Carnot's scientific work had a direct influence on his son's reasoning about heat engines. His goal was to discover the general principles by which machines in general operated, rather than to focus only on specific kinds such as water-powered pumps or wind-mills. He started from the definition of a machine as an intermediary body transmitting motion between two or more other bodies not otherwise connected, and assumed the principle of conservation of *vis viva* and its equivalency to work done. In effect, he used the laws of mechanics in the absence of friction to determine the maximum efficiency possible to any machine, and he recognized that friction reduces efficiency below that maximum. The impossibility of perpetual motion was his guiding principle.

That perpetual-motion machines are impossible is an old idea, as we have noted earlier. It is regarded today as an alternative way of stating the first law, although it was originally applied only to purely mechanical devices, as Lazare Carnot did. Sadi Carnot's achievement was to extend this idea, erroneously, to heat engines.

Sadi Carnot (named after a medieval Persian poet popular in French translation at that time) was born in 1796. He was educated by his father until he was sixteen, when he entered the Ecole Polytechnique. In 1814 he and his fellow students participated in the unsuccessful defense of Paris after Waterloo. In the 1820s, when not on military service as a captain, he studied physics and economics at the Sorbonne and at other schools, but he did not neglect the practical applications of his interests, visiting factories and studying the organization of industries and trades in France and elsewhere. In 1824 his one published work, *Reflections on the Motive Power of Fire, and on Machines Fitted to Develop that Power,* appeared. It received one favorable review but sold very few copies and had very little immediate influence on the practical aspects of heat engine design. In 1828, dissatisfied with the reactionary policies of the monarchical regime in France (he was his father's son in politics as well as science), he resigned from the army to devote full time to his studies in physics and economics. In 1832 he caught scarlet fever and suffered what was then called "brain fever" (meningitis, probably) as a sequel. Greatly weakened, he was convalescing in the country but contracted cholera when an epidemic broke out and died in one day. He was thirty-six years old when he died.

Carnot's monograph was written for a popular rather than a technically trained audience. It began:

Every one knows that heat can produce motion. That it possesses vast motive-power no one can doubt, in these days when the steam-engine is everywhere so well known.

To heat also are due the vast movements which take place on the earth. It causes the agitations of the atmosphere, the ascension of clouds, the fall of rain and of meteors, the currents of water which channel the surface of the globe, and of which man has thus far employed but a small portion. Even earthquakes and volcanic erup-tions are the result of heat [Carnot was guessing here, but he guessed right].

From this immense reservoir we may draw the moving force necessary for our purposes. Nature, in providing us with combusti-bles on all sides, has given us the power to produce, at all times and in all places, heat and the impelling power which is the result of it. To develop this power, to appropriate it to our uses, is the object of heat-engines.

The study of these engines is of the greatest interest, their impor-tance is enormous, their use is continually increasing, and they seem destined to produce a great revolution in the civilized world.

Heat Engines and Waterwheels

In his analysis Carnot used the common view of his time that heat was a substance just as water is a substance. He assumed that a heat engine was analogous to a waterwheel. As the "motive power," the work obtain-able, from falling water is the product of the weight of the water and the height through which it falls, so the motive power of heat should depend on the quantity of heat, which Carnot expressed in calories, and the "height" (the difference in temperature) through which it falls. Just as no work is obtainable from water at a single level—there must be a lower level to which it can fall—so work can be done only when heat is allowed to flow from a high to a low temperature. Of course, water can fall from a high level to a low level and do no work in the process; it will take place spontaneously in the absence of a waterwheel to make use of the potential for work. Carnot assumed that heat behaves the same: it flows spontaneously from a hot body to a cold one. It will do work if we interpose an appropriate engine. If we don't, it flows anyway.

Because Carnot thought of heat as a substance that could not be created or destroyed, he took for granted that all the heat supplied by the fuel was eventually given out to the surroundings, just as all the water falling on the wheel of a water mill flows out at the bottom.

Carnot carried his analysis further. He knew that the maximum work that can be done by falling water depends only on the product of the water's weight and the height of the fall, not on the design of the water wheel, and that this is true for all liquids, not just water. If alcohol or olive oil rained from the skies and formed rivers whose flow we wanted to use for power, the maximum work obtainable would still be the product of the weight of the liquid and the height through which it falls. In other words, the determination of maximum work is independent of the particular working substance used. Carnot argued by analogy that the maximum work a heat engine can do must also be expressed in terms independent of the working substance, that it must depend only on the quantity of heat used and the difference in the two temperatures of operation. As noted by Carnot, engines had been proposed and constructed using working substances other than water, with no dramatic improvement in efficiency:

> Wherever there exists a difference of temperature, wherever it has been possible for the equilibrium of the caloric to be reestablished [by this Carnot meant the tendency of "caloric" to flow from hot to cold bodies to produce a uniform temperature], it is possible to have also the production of impelling power. Steam is a means of realizing this power, but it is not the only one. All substances in nature can be employed for this purpose, all are susceptible of changes of volume, of successive contractions and dilatations, through the alternation of heat and cold. All are capable of overcoming in their changes of volume certain resistances, and of thus developing the impelling power. A solid body—a metallic bar for example—alternately increases and diminishes in length, and can move bodies fastened to its end. A liquid alternately heated and cooled increases and diminishes in volume, and can overcome obstacles of greater or less size, opposed to its dilatation. An aeriform fluid [a gas] is susceptible of considerable change of volume by variations of temperature. If it is enclosed in an expansible space, such as a cylinder provided with a piston, it will produce movements of great extent. Vapors of all substances capable of passing into a gaseous condition, as of alcohol, of mercury, of sulphur, etc., may fulfill the same office as vapor of water. The latter, alternately heated and cooled, would produce motive power in the shape of permanent gases, that is, without ever returning to

a liquid state. Most of these substances have been proposed, many even have been tried, although up to this time perhaps without remarkable success.

All Substances Give the Same Efficiency

The ability of a substance, whether air, water, or solid iron, to be used in a heat engine to produce work depends on its ability to change shape or volume as its temperature is changed and to exert forces on pistons or other restraining bodies. We know that different substances expand on heating by different amounts and exert different forces, and we would expect therefore that some would work better in heat engines than others. Carnot's reasoning implies that they do not: to the extent that it is possible to eliminate frictional losses and other sources of inefficient operation, all substances will do the same work for the same heat input and the same temperatures of operation. The importance of his conclusion is not so much that it tells us something we didn't know about heat engines, but rather that it tells us something we didn't know about substances. It tells us that those properties of *each* substance that determine its performance as the working substance in a heat engine must be interrelated in such a way as to give the same efficiency as *all* other substances.

Carnot accommodated his analysis to the accepted theory of the time, the caloric theory. He believed incorrectly that heat was an indestructible substance: the heat extracted from the high-temperature source (the boiler) must eventually all be delivered to the low-temperature condenser; the condenser is usually kept cold by contact with an outside source of cold water, such as a river, or else with the ambient air, so all the heat given out by the fuel eventually finds its way to this ultimate coolant, where it can no longer be used for doing work. Later, when the first law was discovered, it was realized that "heat" is *not* conserved, that in the operation of a heat engine some of the "heat" is converted to work.

After Carnot's death some notes he had taken on ideas he intended to pursue were found: their significance, like that of his published work, was realized much later. Aware of the challenge to the caloric theory in Rumford's work, he was becoming skeptical of the caloric theory and giving serious thought to the kinetic theory and was designing experiments similar to those done by Joule a decade later. Carnot's early death does not affect us like the premature deaths of Mozart, of Schubert, of

Keats, in which we have lost something we will never have and cannot even imagine. At most he might have discovered both the first and second laws ten or so years earlier than they were actually discovered, and instead of being merely one of a large number of scientific geniuses of the nineteenth century he would have been among its few towering figures, together with James Clerk Maxwell and Darwin. His story leaves us not with the sense of something irreplaceable lost forever, but rather with a sense of the potential of a remarkable human being unfulfilled.

The Second Law and Its Consequences

Joule's experiments showed that energy, not heat, is always conserved. It is essential to keep the difference between the two in mind, especially when even today we talk loosely about the "heat" in a body. It is always correct to use the term *thermal energy* in place of *heat,* but *heat* may properly be used for the thermal energy transferred from one body to another because of a difference in their temperatures. The thermal energy transferred between the working substances of heat engines and boilers or condensers is transferred in just this way, so we will often use the term *heat* for it. Heat engines can be simply described as devices that convert heat into work, though in more precise language they convert thermal energy to mechanical.

Once Joule showed that it is energy that is conserved, it was necessary to reexamine Carnot's conclusion, based on the now-discarded caloric theory, that the maximum work a heat engine could do does not depend on the working substance. It became apparent that the analogy on which Carnot's reasoning rested was a false one. A heat engine and a water wheel are not examples of the same kind of thing. The quantity of water is not changed by its passage through a water wheel: as many gallons leave as enter. In the operation of a heat engine, however, the first law tells us that the quantity of heat flowing out to the cold environment is *less* than that flowing in from the burning fuel. The difference is equal to the work done by the engine: energy must be conserved. If a heat engine could really operate without losing heat to a cold environment, it would merely mean that *all* the thermal energy supplied by the fuel is converted to work: energy is not being created from nothing. So it would not be a perpetual-motion machine.

That Carnot's analogy between a heat engine and a water wheel was inappropriate does not mean that his conclusion—that the maximum work a heat engine can do does not depend on the working substance—

must be false. In fact many of the logical implications of that conclusion, although surprising, had been confirmed by experiments. Kelvin, taking this convincing experimental support as equally convincing support for the caloric theory used by Carnot, was reluctant to believe Joule's results, though to his credit he recognized how revolutionary they would be if true.

It was the German physicist Rudolf Clausius who saw how to reconcile Carnot's reasoning with the first law. To do so he proposed a new law of nature, independent of the first law. His wording of it is logically equivalent to Kelvin's given earlier, which we can paraphrase as follows: *It is impossible for a heat engine operating cyclically, and gaining heat from an outside body at a single temperature, to do any net work.* As noted earlier, cyclical operation means that the engine, including the working substance, is repeatedly brought back to its starting state as the engine operates.

Why Cyclical Operation at a Single Temperature Does No Work

Let us imagine a system consisting of a flywheel, free to rotate, in close proximity to a large tank of water equipped with a sensitive thermometer, so that any change in the temperature, and hence in the thermal energy of the water, can be noted and measured. We would not be surprised to observe that if the flywheel is set in motion it will gradually slow down and the temperature of the water would rise in proportion, as the first law requires, but we would be surprised to observe the reverse of this sequence—the wheel gradually speeding up while the temperature falls—even though these events would not violate the first law. Even if we were unaware of Kelvin's statement, we would find it hard to believe that thermal energy, the random, disorganized motion of molecules, can spontaneously convert itself into the organized motion of the wheel.

Now let us imagine that some clever engineer places a device whose nature he refuses to reveal, enclosed within the traditional black box, next to the wheel and the tank, and asks if we would still be surprised to find the wheel set into motion at the expense of the thermal energy of the water. We would of course decline to answer until we know what is inside the box. Our engineer still refuses to tell us, but offers instead to guarantee that at the end of the experiment, its contents will be in exactly the same state as at the beginning. The positions of any pistons, the state of charge of any batteries, the temperature, pressure, and volume of any substances, and the energies of all components will be the same as before. Given this much information, can we now tell if the wheel will acquire kinetic energy at the expense of the thermal energy of the water?

It may seem rash to insist that under these conditions the wheel can't begin to move, but Carnot, Kelvin, and Clausius were just that rash. They could not conceive of a complete conversion of thermal energy into the mechanical energy of a macroscopic body unless *something* changed elsewhere to account for it, some transfer of energy from one body to another: some flow of heat to a lower temperature than that of the water in the tank, for example, or else a battery operating a motor to start the wheel moving and itself being discharged as a result. If nothing else in the universe had been found to have changed, the wheel could not have started moving.

Clearly, if the device in the black box has operated cyclically, therefore returning to its initial state, it could not be the site of that necessary rearrangement of energy elsewhere that would make the motion of the wheel possible. *Something* has to change, and with a cyclically operating engine the only other *something* in the universe of this experiment, nothing has.

Why Two Temperatures Enable Work to Be Done

Carnot, and later Clausius, used the properties of an ideal gas, rather than steam, to calculate the efficiency of a heat engine because its properties were better known then. We have described them in our discussion of the kinetic-molecular theory in Chapter 4. Unlike steam, an ideal gas doesn't condense, but its pressure changes when its temperature is changed. The way an ideal gas would actually operate in a heat engine shows simply why heat engines need more than one temperature to operate.

Since the engine, like all those to which Kelvin's statement applies, is operated in a cycle, the gas, after expanding and doing work, must be recompressed to its original unexpanded state. This requires work to be done on the gas, and therefore reduces the net work done in the cycle. If the gas is compressed at the same temperature it expanded, the work done on it in compression must at least equal, and in real operation exceed, the work the gas did on expansion.

There is a way to get a net performance of work by the gas in the cycle: cool the gas before compressing it, reducing its pressure and thus the work needed to compress it. The net work done by the gas in one cycle of operation would then be the *difference* between the work it does on expanding while warm and the work we must do on it to recompress it while it is cool. That difference is clearly larger the more the gas is cooled for the compressive step, and therefore larger the greater the temperature difference between the expansive and compressive steps.

Efficiency and the Ideal of Reversibility

Efficiency is another one of those words that are used in common speech and scientific discourse with different meanings. In the science of heat engines the meaning is spelled out very precisely in quantitative terms. First of all, the definition takes into account the fact that both thermal energy and the increase in potential or kinetic energy when mechanical work is used, say, to raise a weight or accelerate a flywheel are both forms of energy and can be measured in the same units. Further, the net quantity of work done by a heat engine cannot be greater than the quantity of heat the fuel has supplied; it will, in fact, invariably be less. The efficiency (we use the symbol *Eff.*) is defined as the ratio of the work done, W, to the heat Q_H supplied at the higher temperature of operation:

$$Eff. = \frac{W}{Q_H}$$

and must necessarily have a value between 0 and 1.

We have mentioned that the first law requires that the work done by a heat engine is the difference between the heat (Q_H) flowing in at the high temperature (T_H) and the heat (Q_L) discarded at the low temperature (T_L):

$$W = Q_H - Q_L$$

which permits the formula for efficiency to be written:

$$Eff. = \frac{W}{Q_H} = \frac{Q_H - Q_L}{Q_H}$$

We will need this formula later in the chapter.

Obviously engines can vary in efficiency, depending on flaws in design and ordinary wear and tear. They may have leaky valves or rusty cylinders; the mechanical connections to the valves may be poorly timed; the spark plugs may misfire; the boiler may lose heat because of poor insulation. All sorts of things will reduce efficiency, but even with the best design possible, the efficiency would be limited by the frictional losses that inevitably occur when there is motion and by the slowness with which heat flows. We can reduce these losses, though we cannot eliminate them, by operating the engine slowly, which requires keeping the various opposing forces as close to a state of balance as possible. This will ensure that the work done is close to the maximum possible. In practice we can never eliminate frictional losses completely, but we can often estimate their influence on the work and heat inputs and outputs and calculate the efficiency of an idealized, frictionless machine.

It is easier to visualize the idealized operation of a simpler device than a heat engine. Imagine the seesaw discussed in Chapter 2 with two riders perfectly balanced. To raise either person and increase his or her potential energy, we could add a small additional weight to the other side of the seesaw. If there were no friction, a grain of sand would be more than enough—provided we did not care how much time the raising took. In this scenario of no friction and unlimited time, the gain in potential energy of one person is just matched by the loss in potential energy of the other (if we neglect the loss in potential energy of the grain of sand). A process run in this idealized fashion is said to be *reversible.*

A heat engine can be *imagined* to be run reversibly, and if an engine were run in this manner, the work done for a given input of heat, and therefore the efficiency, would be the maximum possible. As a practical matter we do not run heat engines this way because we need to get things done in a reasonable time and we are prepared to pay the price in reduced efficiency. A reversible process, because of the near balance continuously maintained between the forces driving it, may have its direction reversed at any moment with negligible effort: just shift the grain of sand from one side to the other. When the process is run backward instead of forward, what were previously gains become losses and losses gains: the person who gained potential energy when the grain of sand was on one side loses the same amount when it is on the other.

Carnot pointed out that a heat engine run backward is a *refrigerator:* in return for the work input to operate it, it removes heat from a cold region and discards it to already warm surroundings. Inputs have become outputs. If an engine is run backward reversibly, each flow of heat changes direction, but the amounts of heat flowing are the same. Instead of work done *by* the engine, we must do the same quantity of work *on* the engine to run it backward reversibly. This is not true for real engines and real refrigerators: the engines do *less* work than the maximum possible, the refrigerators require *more* work to run them than the minimum possible.

The reversible process is one more example of the usefulness in science of an idealization that is never achieved in practice but that provides a standard to which real processes can be usefully compared. We have already encountered other examples: the idealized frictionless world imagined by Galileo, in which moving bodies never stop moving, and the ideal gas, which behaves like no real gas ever does. These idealized entities can often be approximated closely by real ones, and how the ideal would behave if it really existed inferred by an extrapolation from the real. The idealized process sets limits that the real process cannot overcome.

In the nineteenth century, it was difficult to calculate the reversible efficiency most substances would have if used as the working substance of a heat engine, because it requires detailed knowledge of a number of physical properties not usually available then. Ideal gases do follow certain simple regularities of behavior, however, so Clausius was able to calculate the reversible efficiency of a heat engine operating cyclically with an ideal gas as the working substance. His result is expressed by the equation:

$$Eff. = \frac{T_H - T_L}{T_H} \quad \text{(reversible, ideal gas)}$$

T_H and T_L are the higher and lower temperatures of operation. The equation shows that if the higher and lower temperatures are equal ($T_H = T_L$), the efficiency is zero: no work at all is done.

Now we come to the most subtle part of the argument. We have stated as our basic hypothesis that it is impossible for a heat engine operating cyclically, and exchanging heat with an outside body at a single temperature, to do any net work. We can see that an ideal gas engine satisfies this hypothesis, for it would do no net work when $T_H = T_L$. Any substance not an ideal gas will have properties different from an ideal gas: its pressure may change with temperature and volume in a different way, the heat it absorbs on expanding may be different. We will expect it therefore to have a different reversible efficiency also. All that our basic hypothesis seems to require is that its efficiency should also become zero when $T_H = T_L$. Thus it does not seem that a substance whose efficiency is one-and-a-half times the ideal gas efficiency or two-thirds the ideal gas efficiency should violate the hypothesis.

Carnot's argument, as modified by Clausius, proceeds as follows. Suppose there were a substance X with a reversible efficiency different from that of the ideal gas. If so, it should be possible to create a new heat engine, one whose net effect would be to convert heat gained at a single temperature into work while operating in a cycle, by combining a reversible heat engine using the ideal gas with one using substance X. This, however, would contradict our basic hypothesis that such an engine is impossible. We therefore conclude that if the basic hypothesis is correct, all working substances must give engines with the *same reversible efficiency*. Since we know the efficiency given by an ideal gas (because it happened to be easy to calculate), we know it for all other substances as well.

The Paradox of the Refrigerator

A heat engine and a water wheel are both devices for doing work, and both can be run in reverse. A water pump is in a sense a water wheel run in reverse. It doesn't do work *for* us; instead, work must be done *on* it to make it pump the water uphill. A heat engine run in reverse also requires an input of work, in return for which it removes heat from a cold reservoir and gives out heat to a warm reservoir; in other words it does the job of a refrigerator or an air conditioner. The generic term for all such devices is, unsurprisingly, *heat pump*. The heat output of a refrigerator can be detected by placing one's hand on the side or back of the unit; the heat output of an air conditioner outside the cooled room is perhaps more familiar.

Heat flows into a heat engine from what is often referred to as a *heat source*, which in turn is usually heated by burning fuel, and a lesser quantity of heat flows out to the surrounding cold environment, which is called a *sink* for heat. When we consider both engines and heat pumps operating together, the "source" and the "sink" can both give heat and receive it. As a more neutral term for source and sink, we will use *heat reservoir*.

There is an apparent paradox when we compare the reversible efficiency of an ideal gas heat engine with its performance when it is run in reverse as a heat pump. From a heat engine we want the *maximum* work output for a given heat input, the heat being supplied by the reservoir at the upper temperature of operation. The Clausius formula for the ideal gas engine tells us the efficiency, defined as W/Q_H, is greater the greater the difference of temperatures. To operate a heat pump we want to use the *minimum* work input to withdraw a given quantity of heat from the body (the inside of a refrigerator or an air-conditioned room) we are trying to keep cool, which is at the lower temperature of operation. To characterize how well the refrigerator does its job we use a "coefficient of performance" *(C.O.P.)*, which is defined as the *ratio* of the *heat* extracted from the colder reservoir, Q_L, to the *work* we must do to extract it, W. For an ideal gas heat pump operating reversibly it can be shown that:

$$C.O.P. = \frac{Q_L}{W} = \frac{T_L}{T_H - T_L} \qquad \text{(reversible, ideal gas)}$$

A comparison of the formula for the coefficient of performance of the heat pump with the formula for the efficiency of a heat engine shows that both depend on the difference between the two temperatures T_H and T_L but in opposite ways: the *greater* the difference in the operating temperatures, the *greater* the efficiency of the reversible heat engine and the *lesser* the coefficient of performance of the reversible heat pump. Put more simply, a heat engine performs better and a heat pump worse when the temperature difference between the heat reservoirs are greater. This sounds paradoxical, but on second thought it is not: it means we use more electrical energy to operate an air conditioner or refrigerator on a very hot day than on a merely warm one. The importance of this fact for the Carnot-Clausius argument is that if working substances could differ in their reversible efficiencies, the less efficient substances, although they make worse reversible heat engines, would make better reversible heat pumps.

An Impossible Engine

Figures 5.2 and 5.3 both show reversible, cyclically operating machines—a heat engine and a heat pump, respectively. In the engine, heat from the high-temperature reservoir expands the gas to drive the engine's piston; the outputs are the work done on the flywheel which the piston drives and the heat transferred to the low-temperature reservoir. Because the operation is cyclical there is an exact energy balance: the input is equal to the output ($Q_H = W_{ENG} + Q_L$). The heat

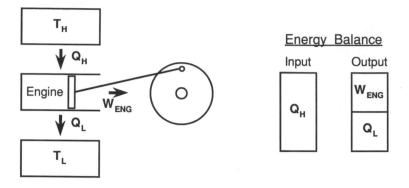

Figure 5.2 Energy Balance of a Reversible Heat Engine

In this highly schematic representation, a heat engine is shown as a cylinder with a moving piston that imparts motion to a flywheel on the *right;* it operates both cyclically and reversibly. The high-temperature reservoir (at a temperature T_H) is represented by a rectangle above the engine, the low-temperature reservoir (at a temperature T_L) by a rectangle below. The quantities of heat flowing between the reservoirs and the engine are symbolized by arrows labeled Q_H and Q_L, which show the direction of flow: *from* the reservoir at T_H *toward* the engine (an *input* to the engine), and *from* the engine *toward* the reservoir at T_L (an *output* from the engine). In addition, an arrow labeled W_{ENG} (for net work done *by* the engine) points from the engine toward the flywheel, symbolizing that the engine does work on the flywheel (an *output* from the engine).

The energy balance of the engine during one cycle of operation is shown by the two rectangles labeled *Input* and *Output.* We find it convenient to diagram *work* inputs and outputs differently from *heat* inputs and outputs. When a heat engine goes through a complete cycle, the working substance varies in temperature in different phases of the cycle, exchanges heat with external reservoirs, and is at times doing work on external bodies (during expansion of the working substance) and at others having work done on it (during compression). We will lump together work done *by* the engine in some parts of its cycle and work done *on* the engine in other parts of the cycle to obtain a single *net* work, which for an engine appears as an output. On the other hand, heat inputs and outputs will be kept separate according to the temperature of the working substance during the input or output. In the engine, Q_H represents a heat input at the higher temperature of operation, Q_L an output at the lower temperature. The diagram for the energy balance therefore shows a heat input at the higher temperature, a heat output at the lower temperature, and a net work output.

Since we only consider cyclical operation here, and the energy is restored to its original value once in each cycle, net energy inputs must equal net energy outputs. In other words, the net work and heat outputs must be combined to give a total energy output, which must equal the heat input.

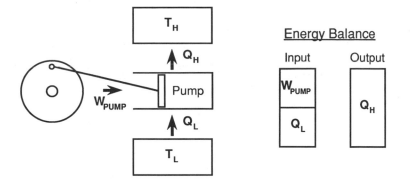

Figure 5.3 Energy Balance of a Reversible Heat Pump
In this corresponding diagram for a heat pump, the high-temperature and low-temperature reservoirs are represented as they were for the engine. The flywheel, which must do work to operate the pump, and which slows down as it does so, is shown to the *left* of the pump. The arrows showing heat flows Q_H and Q_L and work done on the pump W_{PUMP} point in directions opposite from those in the previous diagram: Q_H toward the reservoir T_H, Q_L and W_{PUMP} toward the pump.

The energy balance diagram shows a net work input W_{PUMP} and heat input Q_L, which combined give a total energy input that equals the heat output Q_H.

pump is the reverse of the heat engine. Work is done by the flywheel to drive the pump's piston, which pumps heat from the low-temperature reservoir; these make up the energy input. The pump then delivers an output of heat to the high-temperature reservoir. Again the energy input is exactly equal to the output ($W_{PUMP} + Q_L = Q_H$). Since a heat engine can do the work to set a flywheel into motion, and a moving flywheel can do the work to operate a heat pump, we could combine them into a composite engine. We will find that if the working substances of engine and pump differed in their reversible efficiencies, the second law would be violated.

We show a hypothetical "composite engine" in Figure 5.4. Our starting hypothesis is that there is a substance X with a lower reversible efficiency than the ideal gas. We imagine an ideal gas heat engine providing the work needed to operate a substance X heat pump. There is both a high-temperature reservoir at T_H, from which the engine extracts heat and to which the pump delivers it, and a low-temperature reservoir at T_L, to which the engine delivers heat and from which the

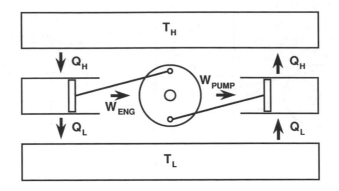

Figure 5.4   Energy Balance of a Composite Engine:
Can Heat Be Converted Completely to Work?

The composite engine combines a reversible heat engine and a reversible heat pump. The energy inputs and outputs are drawn so that we may apply them to two different cases: first, we consider the operation of the engine and the pump when both use the ideal gas as the working substance; and, second, we consider what would happen if the pump used the less efficient substance X. The operating conditions are chosen so that in both cases the heat extracted from the high-temperature reservoir is exactly equal to the heat delivered to it by the pump.

In the first case, all the work output of the engine is needed to operate the pump. The heat input from the low-temperature reservoir to the pump is exactly equal to the heat output of the engine to the T_L reservoir. The net effect of the operation of the first composite engine is *nothing at all:* (1) all heat given out by either reservoir is restored to it; (2) no net work is done on any other bodies. It is exactly as though 100 liters of water falling 20 meters was used to operate a reversible pump that pumped the same 100 liters of water back up the 20 meters it fell. (Obviously, only idealized, reversible operation could accomplish this net cancellation of all effects.)

In the second case we come to the heart of the argument. As before, the heat pump, this time using substance X, is operated so as to deliver, during one cycle of operation, the *same* quantity of heat Q_H to the high-temperature reservoir as the ideal gas engine received from it in that same cycle. This ensures that the

pump extracts it. We imagine operating the engine and pump in such a way that the pump delivers the *same* quantity of heat to the high-temperature reservoir in each cycle as the engine receives from it, hence the high-temperature reservoir is restored in each cycle of operation to its initial condition. *Because it is operated cyclically we can consider it part of the cyclically operating engine.* We can imagine it as being inside a black

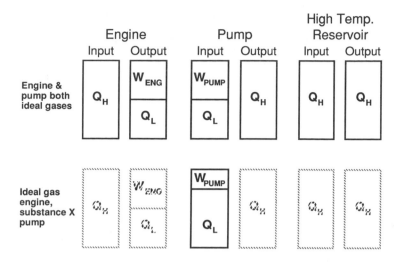

Figure 5.4 continued

high-temperature reservoir is operated cyclically. Because of the lower reversible efficiency of substance *X, less* work input is needed to deliver the *same* quantity of heat to the high-temperature reservoir than an ideal gas heat pump would need. (As we noted in the text, if there were a less efficient substance than an ideal gas, it would make a worse reversible engine but a better reversible heat pump.)

The energy balances for this composite engine (consisting of an ideal gas heat engine, a substance *X* heat pump, and the high-temperature reservoir, the whole operating in a cycle) are identical with those of the first case *except for the input to the heat pump.* The ideal gas engine now has a work output large enough to operate the substance *X* heat pump and *do other work besides.* Energy conservation is satisfied because a net heat equivalent to this extra work is extracted from the low-temperature reservoir. The composite engine has therefore performed work and extracted thermal energy from a reservoir at a *single* temperature, T_H. There is no lower-temperature reservoir to which heat is delivered at all. This arrangement clearly violates Kelvin's hypothesis.

box with the engine and the pump, the whole constituting the composite heat engine.

The detailed analysis shows that the composite engine does the impossible. Operating cyclically, it withdraws heat from a reservoir at a single temperature T_L and converts it all to work. No heat is discharged to a temperature lower than T_L.

Table 5.1
Energy Calculations for a Composite Engine

Heat pump operated with ideal gas	Input	Output
Engine		
Heat	100 (from T_H)	50 (to T_L)
Net work	0	50
Pump		
Heat	50 (from T_L)	100 (to T_H)
Net work	50	0
Composite		
Net heat	0	0
Net work	0	0

Heat pump operated with substance X	Input	Output
Engine		
Heat	100 (from T_H)	50 (to T_L)
Net work	0	50
Pump		
Heat	75 (from T_L)	100 (to T_H)
Net work	25	0
Composite		
Net heat	25 (from T_L)	0
Net work	0	25

Note: The composite engine consists of an ideal gas heat engine, a heat pump, and a heat reservoir at T_H. Energy calculations are shown first for the heat pump operated with an ideal gas as the working substance and then for the pump operated with substance X, whose reversible efficiency is half that of an ideal gas. Temperatures T_H = 600 K and T_L = 300 K are assumed. Heat and work are measured in joules.

In Table 5.1 we use a numerical example to demonstrate the same conclusion. To get simple numbers, we assume upper and lower temperatures at 600 K (327°C) and 300 K (27°C), giving an ideal gas reversible efficiency of 0.50. We further assume that the reversible efficiency of substance X is 0.25 for these temperatures. Finally, we assume that in one cycle of operation Q_H will always be 100 joules. The net effect is that 25 joules of thermal energy are extracted from the reservoir at 300 K and converted to work, with no other changes anywhere else.

Can substance X have a higher reversible efficiency? If it did, we would use it for our heat engine and the ideal gas for the heat pump. The logic is the same, and the result is equally impossible.

From this analysis we find that Carnot's conclusion was correct after all, even though his argument was flawed by his use of the caloric theory. All substances must have the *same* reversible efficiency in a heat engine operated cyclically, an efficiency that depends only on the temperatures at which heat exchanges occur.

The Universal Efficiency

Since we know this efficiency for the ideal gas, we know it for every other substance as well—namely:

$$Eff. = \frac{T_H - T_L}{T_H} \qquad \text{(reversible, any substance whatever)}$$

This is the crucial result of the argument, and a necessary step to exploring all the consequences of the second law.

In following the Carnot-Clausius argument that *if* any substance had a different reversible efficiency from an ideal gas *then* Kelvin's hypothesis would be violated, we have done a careful analysis of energy inputs and outputs for a complicated device. It is easy to get lost in the details and lose sight of the overall purpose of the argument. Let us give an example of something analogous and more familiar: the commodity prices reported for different markets. For example, we may read in the paper that the price of gold in London this morning is $465.50 per ounce, while in New York it is $467. It may occur to us that if we could buy a large quantity of gold in London and sell it in New York on the same day, we could get rich. While this ignores brokers' commissions and time delays during which price changes could wipe out potential profits, clever people do sometimes get rich on similar transactions.

In the work of Carnot and Clausius, the reversible efficiencies of different working substances in heat engines are somewhat like the price of gold, but there are no quick ways to get rich: the price is the same in all markets.

Improving Efficiency

Our analysis has focused on the efficiency of idealized reversible engines. But efficiency is a goal in the operation of real heat engines also, both for reasons of economy and for the protection of the environment.

The formula for the reversible efficiency of a steam engine operating between the boiling point of water and an ambient environment at about 25°C gives a result of about 0.2. (*Eff.* = (373 − 298 K) / 373 K = 0.2.) This is a fairly low efficiency. Do we really need to waste 80 percent of the heat provided by our fuel to operate a steam engine? The answer, according to Carnot's principle, is, no, we don't. If we can raise the upper temperature of operation we can obtain a higher efficiency, and there are many ways to do this.

First of all, water under pressure boils at a higher temperature and produces high-pressure steam. High-pressure steam engines were developed before the second law of thermodynamics was discovered, and in fact heat engine efficiencies were improved dramatically over time with very little input from the laboratory. (This fact led one historian of science to remark that science owes more to the steam engine than the steam engine owes to science.) Most electricity-generating plants running on fossil fuels use high-pressure steam at 500°C (773 K) and reach a real, as opposed to a reversible, efficiency of about 40 percent (the ideal efficiency would be 61 percent for these temperatures). While other working substances might permit higher temperatures than those possible with inexpensive and convenient water, we run into limitations on the ability of available materials that can be used for the moving parts of heat engines—such as pistons, cylinders, or turbines—to withstand higher temperatures. Proposals have been made for a process that would convert thermal energy directly to electricity from hot, electrically charged gases flowing in magnetic fields (first studied in the atmospheres of stars; the field is called *magnetohydrodynamics*). Estimates have been made that gases at temperatures of 2,000–2,500°C could be used and that real (not reversible) efficiencies of 50–60 percent could be achieved.

The conclusion that all working substances have the same *reversible* efficiency can be a misleading guide to the practical details of heat engine design. No *real* heat engine or heat pump is run reversibly: we cannot wait inordinately long periods of time to get jobs done. Real efficiencies fall below the reversible limit and depend on such properties of the working substance as the time it takes heat to flow into it (its "thermal conductivity"), its boiling point and melting point in relation to the temperatures of operation, and various properties of the materials of which the engine itself is made. It is no accident that we use water in steam engines but not in refrigerators, where ammonia and other low-boiling substances which evaporate faster and do not freeze under operating conditions perform better.

It should also be remembered that the Carnot limit applies only to the conversion of thermal energy to other forms. There is no comparable restriction on the conversion of mechanical energy to electrical, as in a hydroelectric station, or electrical to chemical, as when we charge a battery, or chemical to electrical to mechanical, as when we use the charged battery to operate the starter of an automobile. In a later chapter we will examine how living organisms do work by a direct conversion of the chemical energy of foods to motion. There is no heat engine, and no Carnot limit. None of these processes involves cyclical operation, however; if they did, the second law assures us that they would do no net work.

Entropy: A New Property of Matter

The importance of the second law is not so much what it tells us about heat engines, but rather what it tells us about the properties of matter. Any body, whether it is a single pure substance or a complicated combination of substances, can be imagined to be the working substance of a heat engine. When the body is subjected to a series of changes in temperature and in the forces between it and other systems, there will be heat and work inputs and outputs predictable from the properties of the body. We can then calculate what its reversible efficiency in some cyclical process would be from the calculated inputs and outputs. This efficiency will depend on one hand on the specific properties of the body, but on the other hand it must always be given by the general formula

$$Eff. = \frac{T_H - T_L}{T_H}$$

This in turn implies that the physical properties from which the efficiency is calculated must be interrelated in just such a way as to always give us that general formula. Thus we have been able to learn something useful and important about those physical properties, even though we may not have discovered a better heat engine. This approach was used, as we will see in Chapter 6, by Kelvin and his brother early in the history of thermodynamics to explain some baffling properties of water and ice.

There is, however, an alternative and more powerful way to explore the consequences of the second law. We have already shown that the first law implies that matter has a previously unsuspected property, thermal energy, and prescribes procedures for determining this property. The statements of the second law given earlier can be shown by some logical and mathematical analysis to imply that matter has an additional unsuspected property, called *Entropy* by its discoverer Rudolf Clausius, and to prescribe procedures for determining this new property.

The first law requires that when any process takes place, the total energy change for all bodies involved must be zero: what one body loses another must gain. The second law requires that when any process takes place, the total entropy change for all bodies involved cannot be negative. This means that the total entropy cannot ever decrease, but it may increase, and in real (as opposed to reversible) processes, where friction and other forms of inefficiency always operate, the total entropy will *always* increase. Unlike energy, entropy is not conserved. The amount of it in the universe gets greater all the time.

The reasoning that led Clausius to the concept of entropy as a property of matter involves some difficult logical and mathematical steps and requires the use of calculus. In the following summary some of these mathematical steps are omitted; we hope that even without them the reader can get some insight into the argument.

First, what do we mean when we speak of a "property of matter"? Let us think of definite samples of matter—a kilogram of water, 2,000 kilograms of a mixture of 20% oxygen and 80% nitrogen, 1 gram of platinum—and consider what properties these samples might have: density, electrical resistance, speed of sound propagation, hardness, color, and so on. Some instrument or combination of instruments will measure each of these properties on our sample of matter. We can easily verify by simple experiments that changes in the "state" of the matter— a change in its temperature or in the pressure or in any electrical fields acting on it—will produce changes in each of its properties, though the changes need not always be large. And we expect also that if, after having changed the state, we then return to the original state (that is, to the original temperature, pressure, electric field) so that the process has been a *cyclical* one, the properties will return to their original values. If they did not, we would hesitate to call them "properties."

For example, measurements performed on any sample of matter undergoing rapid and violent chemical reaction—the contents of a blast furnace, or a sample of rubber decomposing at 1,000°C—would give unreliable and rapidly changing values for anything we try to measure. Our concept of "property" therefore seems to require two things: first, that we should be able to perform a direct measurement of the property, and, second, that the value should be stable: there should be no net change during a cyclical process. This stability implies that the property has a definite value associated with each definite state of the sample of matter. If we change the state, the property takes on a new value, and we can measure it again in the new state.

Properties That Cannot Be Directly Measured

Now, although we have said energy and entropy are also properties of matter, they lack one of the crucial defining features: there are no instruments to measure them, no energy-meters or entropy-scopes we can apply to our kilogram of water or our gram of platinum that will give us numerical values for the properties. Why do we claim that they are properties, then? While we cannot measure either of them directly when our sample of matter is in a definite state, we can measure the *changes* in energy or entropy when the sample changes from one state to another. Because we can measure these changes, we can show in the laboratory that the net changes of either energy or entropy in cyclical processes always add up to zero. So although we can't measure the value of the energy or entropy of a kilogram of water at, say, 20°C, 10 atmospheres pressure, and an electric field of 50 volts per meter, we can measure their net changes when the water is brought to that state from 0°C, 1 atmosphere pressure, and zero electric field. We therefore can measure these particular properties relative to some arbitrarily selected reference state, but otherwise they are properties as definite and as stable as electrical resistance, hardness, and all the others.

This idea that the existence of a property can be proved by showing that changes in all cyclical processes are zero can be made clearer by a more easily visualized analogy. We will consider the property "height-above-sea-level" or, more simply, "elevation." There is a definite numerical value of elevation at each point on the earth's surface. Suppose we are exploring a mountain all trails on which are within sight of a landmark at sea level, and we are equipped with surveyor's instruments, including a telescope. At any point on any trail we are able to make a measurement of its elevation. For example, we can first determine with our instruments that the elevation of a base camp is 250 meters and, after climbing to the summit, determine that *its* elevation is 4,000 meters. We expect, of course, that if we return to any point after wandering anywhere else on the mountain, a repetition of the measurement of elevation will give the same answer as before. This verifies that elevation is a property (barring earthquakes, of course).

Now, however, suppose that our particular mountain is always shrouded in an impenetrable fog that debars us from using our telescope. Is there any way we could test whether elevation is determined by our position on the mountain? Suppose we were to use a meter stick and a carpenter's level to measure accurately in centimeters the *change* in elevation for each step we take along the trail. Thus, for example, we could determine by tediously adding up all the changes of each step that the summit is 3,750 meters higher in elevation than the base camp. Obviously we could also add up all the changes in elevation for any hike that returns us to some particular point on the mountain—the hike is thus cyclical. We would find that the net change on returning to any starting point is zero. We would conclude that elevation really is a property. We could therefore determine the elevation of any point on the mountain relative to some arbitrary reference point, such as the base camp, even though we do not know the elevation above sea level of either base camp or summit.

How Entropy Is Defined

Clausius was aware that the net heat inputs and outputs of a heat engine do not add up to zero. The first law forbids it, since if work is done by the engine, it must take in more heat than it gives out. But he did realize that there was a quantity related to the heat inputs and outputs that did add up to zero for any *reversible* cycle whatever. Let us follow his reasoning:

The first law requires that the work, W, must equal the difference between the heat received by the engine, Q_H, from the hot reservoir and the heat, Q_L, discharged to the cold reservoir:

$$W = Q_H - Q_L$$

This permits the equation for the reversible efficiency of a heat engine operating between two temperatures to be written:

$$Eff. = \frac{W}{Q_H} = \frac{Q_H - Q_L}{Q_H} = \frac{T_H - T_L}{T_H}$$

A little algebraic manipulation of the above equation gives us

$$\frac{Q_H}{Q_L} = \frac{T_H}{T_L}$$

In words, the ratio of the heat extracted from the hot reservoir to the heat discharged to the cold one is equal to the ratio of the absolute temperatures of the two reservoirs.

A little further algebraic manipulation leads to:

$$\frac{Q_L}{T_L} = \frac{Q_H}{T_H}$$

Q_L is the quantity of heat lost by the engine to the cold reservoir and Q_H the quantity of heat gained from the hot one. We paraphrase the above equation by saying that the loss of Q_L/T_L is equal to the gain of Q_H/T_H. This doesn't by itself make Q/T a property, as the heat engine operating reversibly with only two temperatures is a very restricted kind of cycle. But Clausius was able to prove, by using calculus, that any completely arbitrary reversible cycle, no matter at how many different temperatures it gains or loses heat, can be divided up into a large number of two-temperature reversible cycles. For each of them the gain of Q/T at the upper temperature is balanced by a loss of Q/T that just cancels it at the lower temperature. It follows that for any arbitrary reversible cycle the net changes of Q/T add up to zero. The conclusion: since the net change around any arbitrary reversible cycle is zero, there must be a property of matter associated with those changes. He chose the name *entropy* [in German, *Entropie*] from a Greek root meaning "turn" or "change"; the same root is present in other English words, such as *tropism* and *troposphere*.

Entropy Changes in Real Processes

What about heat engines operating under real conditions, which means nonreversibly, and with lower efficiencies? First of all, once we are convinced that entropy, determined by measurements performed during reversible changes, is a property of matter, then we must concede that the change in entropy of the working substance of a heat engine operating in a cycle must be zero, even if the cycle was not carried out reversibly. This may sound contradictory, but it isn't (we recognize that this point is one of the more difficult for the reader not trained in science or mathematics). Entropy is a property whose numerical value must be determined by a reversible process, but once determined it is a property and depends only on the state of the system. Hence if the working substance in the heat engine has been returned to the same state, it must have again the same entropy as before.

Let us turn instead to the two reservoirs, which are also samples of matter and which therefore also have the property entropy. The reservoirs do not operate cyclically: in reversible operation the hot reservoir gives out the heat Q_H, and thus loses an amount of entropy Q_H/T_H, while the cold reservoir gains an amount of entropy Q_L/T_L. So, like the engine, there has been no net entropy change in the reservoirs. Now let us consider the opposite extreme from reversible operation: an engine so inefficient that no work is done at all. Under such conditions we can ignore the heat engine entirely and imagine that the quantity of heat Q_H flows directly from the hot reservoir to the cold one, so that $Q_L = Q_H$. The entropy decrease of the hot reservoir is still Q_H/T_H, but now the entropy increase of the cold one is Q_H/T_L. Since the temperature T_L is smaller than T_H, the entropy increase of the cold reservoir is now greater than the decrease of the hot reservoir. There is therefore a net entropy increase of the two reservoirs.

If the real engine, although not operating reversibly, is not this ridiculously inefficient and does some work, it is easy to show that there is a net entropy increase, although this increase is less than if the heat were just to flow unimpeded from the hot to the cold reservoir.

Once this net increase of entropy in real operation of a two-temperature heat engine is established, further detailed logical analysis shows that it is also true for all processes in nature, not just those of conventional heat engines, and not just those involving heat reservoirs at just two temperatures. The conclusion: there can be no net decrease of entropy in any natural process. In reversible processes, an unattainable ideal, there is no net change in entropy. In all real processes there must be an increase.

What Is Entropy, Anyway?

Attempts to find an intuitively simple meaning for entropy without reference to a molecular picture have not been particularly successful.

The Carnot limit on the efficiency of even reversible heat engines implies that not all the thermal energy of a heat source is available for conversion to mechanical energy, and there is a relation between the entropy of the heat source and the unavailable portion of the thermal energy. Unfortunately, the amount of thermal energy that is unavailable depends also on the temperature of the ambient environment that we are using as our cold reservoir, and that in turn depends on where we happen to site the engine. So entropy is related to the concept of unavailable thermal energy, but not in a simple way.

We have come to the end of a long story. Let us summarize it before we go on in the next chapter to describe one of the first experimental tests of the second law and discuss some of its implications.

Starting with the conviction that a heat engine operating cyclically at a single temperature should be unable to do any work, Carnot and Clausius were able to derive a remarkable consequence: that the maximum (or ideal) efficiency of a heat engine should depend only on the temperatures of operation, not on the specific properties of the working substance used. Once this was established, an even more remarkable consequence followed that extended beyond the operation of heat engines to describe the properties of all matter and the course of all natural processes. This was the discovery of entropy. Entropy is a property of matter, readily measured in the laboratory. The total amount of entropy, unlike that of energy, is not conserved, but rather must increase in all real processes.

Appendix: Entropy Changes and How They Are Determined

When we described how the thermal energy in matter is determined in the laboratory, we stated that we were required by the nature of energy to determine it relative to an arbitrarily selected reference state. The energy relative to this state was measured by keeping track of the total heat flowing into the substance and the total work done on it during a process going from the reference state to the state we are studying. It doesn't matter what kind of process is used or whether it is carried out reversibly or not.

Entropy is also determined relative to a reference state, and its determination also requires us to have knowledge both of the heat absorbed and the work done. The process by which the reference state is changed to the new state we are studying, in contrast to changes in energy, must be a reversible one: it must be a process of negligible frictional loss, or anything equivalent to friction that would reduce the work done below its maximum possible value for the path of change. Such a process cannot be carried out exactly in practice, but it can be closely approximated in the laboratory, with corrections made for any frictional losses. Under reversible conditions, and if the process takes place at a single temperature, the change in entropy of the process, ΔS, is given by the heat absorbed divided by the absolute temperature at which the process is carried out:

$$\Delta S = \frac{Q}{T}$$

As an example, consider the following experiment, which approximates a reversible process. When water is boiled slowly at 100°C under normal atmospheric pressure and is thus converted to steam also at normal atmospheric pressure, the heat absorbed per kilogram of water evaporated (the latent heat) is 2,254,000 joules. The absolute temperature corresponding to 100°C is 373 K. Dividing the latent heat by this temperature, we find the entropy change on evaporation to be 6,043 joules per kelvin per kilogram. In other words, the entropy of 1 kilogram of steam at this temperature and pressure is 6,043 joules per kelvin greater than that of 1 kilogram of liquid water.

More generally, we need to determine entropy changes for processes in which the temperature changes. The formula for these cases has to be expressed in the language of calculus, but the calculation of the entropy change is still quite straightforward. In Table 5.2 we give the energy and entropy of 1 kilogram of water as measured in the laboratory at various temperatures. We have chosen the temperature 0°C as the reference state, so we arbitrarily call the energy and entropy zero in that state.

Now let us calculate the energy and entropy changes when 1 kilogram of cold water at 0°C is brought into contact with 1 kilogram of hot water at 100°C. We know, first, the energy cannot change. We note that the energy of 2 kilograms of 50°C water is just equal to the sum of the energies of 1 kilogram of 0°C water and 1 kilogram of 100°C water (418,400 joules). It follows that the final temperature is 50°C. (This is not a particularly surprising result, but accurate measurements show that it is not strictly true because the specific heat of water depends slightly on temperature between 0°C and 100°C.) The entropy at the start is the sum of the entropies of the two kilograms of water, one at 0°C and one at 100°C. Thus $S = 0 + 1,305 = 1,305$ joules per kelvin.

The entropy at the end is twice the entropy of one kilogram of water at 50°C, or $S = 2 \times 703 = 1{,}406$ joules per kelvin.

There has therefore been a net increase in entropy of 101 joules per kelvin. The hot water has *decreased* in entropy in the process, but the cold water has *increased* by a greater amount, so that the net change is an increase. The requirement of a *net* increase clearly does not mean that some of the participants in the process cannot undergo an entropy decrease (as did the hot water). This numerical example illustrates the process considered earlier: a flow of a quantity of heat Q_H from a body at a higher temperature T_H to another body at a lower temperature T_L, leading always to a net increase of the total entropy.

The entropy of liquid water, as can be seen from the table, increases when the temperature increases. So does the entropy of everything else,

Table 5.2
The Energy and Entropy of 1 Kilogram of Water

Temperature (°C)	E (joules)	S (joules per kelvin)
0	0	0
10	41,840	151
20	83,680	297
30	125,500	435
40	167,400	573
50	209,200	703
60	251,000	833
70	292,900	954
80	334,700	1,075
90	376,600	1,192
100	418,400	1,305

provided pressure and other relevant variables remain the same; it is one of the most general rules about entropy.

There are other ways to increase entropy: expanding a gas to a larger volume while holding its temperature constant always increases its entropy. Compressing liquids and solids to smaller volumes at constant temperature usually results in decreases of their entropies; but there are exceptions to this rule: very cold water (between 0° and 4°C) is one.

When soluble substances like sugar dissolve in water there is a net entropy increase. The process of dissolving something in a large volume of a liquid solvent is in some ways like expanding a gas.

When substances undergo "changes of phase"—a solid melts, a liquid vaporizes—the phase formed by heating has the higher entropy, so invariably liquids have higher entropies than the solids they are formed from, and gases have higher entropies than liquids. Since water always has a higher entropy than ice at the same temperature, one might think that the freezing of water on a cold day leads to a decrease in entropy and a violation of the second law. But it is the *net* entropy change that must be an increase. When water freezes outdoors, it gives out heat to the cold surroundings, increasing the entropy of those surroundings by more than enough to cause a net entropy increase.

6

Implications of the Second Law

~~~~~~

We have stated that the profound significance of the second law lies not in what it tells us about heat engines but in what it tells us about the properties of matter. We will illustrate this by an example of some historical interest: an application of the second law that was made before it was recognized as a new law of nature. Then we will describe how the second law applies to perpetual-motion machines and, finally, what it tells us about the direction of time.

The Ice Heat Engine

We are aware of the tremendous forces developed in nature by the freezing of water. Given enough time, rocks are split and ultimately mountains are eroded away. On a more humble scale, potholes appear in the paved streets of cities, and in unheated houses the water in pipes freezes and bursts the pipes. Could we use the freezing and thawing process as a source of useful energy?

The expansion of water as it freezes (ice occupies a volume about 9 percent larger than the water it is formed from) is not typical of liquids when they freeze. It is why ice floats on water, whereas most other solids sink in the liquid they are formed from on freezing. If water were confined in a cylinder with a movable piston, the piston would be pushed outward as freezing takes place. It would not move far, but the forces developed on solidification are, as we are aware, large enough to split rocks. Since work is the product of distance and force, it is possible that the piston could do a significant amount of work.

So an ice heat engine can be conceived as operating as follows. Water cooled to its freezing point of 0°C is confined in a closed cylinder. The temperature outside the cylinder is allowed to fall very slightly to make the water freeze, and the ice exerts a large force on the piston, which then performs some desired work, such as raising a weight or setting a flywheel in motion. After the freezing is completed the cylinder is then placed in an environment very slightly above the freezing point, the ice melts, and we are ready to repeat the process so as to raise the weight still further or increase the speed of the flywheel (Figure 6.1). The difference between a temperature cold enough to freeze water and one warm enough to melt ice can be made negligible if we are willing to

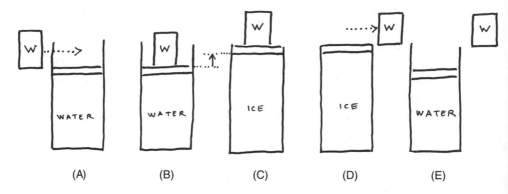

Figure 6.1 The Ice Heat Engine
(A) Initial state: water is confined in a cylinder equipped with a piston and is under atmospheric pressure. *(B)* A weight is placed on the piston. The water is now under a higher pressure than atmospheric. *(C)* The system is cooled and the water freezes. The expansion on freezing raises the weight, doing work. *(D)* The weight is moved laterally, returning the contents of the cylinder to atmospheric pressure. No work is done in moving the weight laterally. *(E)* The temperature is returned to the initial value and the ice melts. The state of the system is now identical to the initial state.

The energy bookkeeping is as follows: there is a heat input Q_H, the heat necessary to melt the ice in step *(E)*. There is a heat output Q_L, the heat given out by the water when it freezes in step *(C)*, and a work output *W*, the work done when the weight is raised during the freezing in step *(C)*. The first law requires that the total energy input (Q_H) equals the total energy output (Q_L plus *W*). If the freezing and melting steps both took place at the same temperature, the second law would be violated: there would be a net absorption of heat from the environment, equal to $Q_H - Q_L$, converted completely to work. Only if the freezing temperature under excess pressure were lower than under atmospheric pressure, and by the right amount, could this violation be avoided.

operate the engine very slowly. Ice melts, eventually, at −0.001°C, and water freezes, eventually, at +0.001°C. To all intents and purposes, this engine can operate at a single temperature. Now this possibility would not only violate the second law as we know it today, but it also violates Carnot's inference (from the erroneous caloric theory and the irrelevant impossibility of perpetual motion) that a heat engine requires at least two temperatures to operate and does no work if there is no temperature difference. Work has been done: there has been a net transfer of thermal energy from the environment at 0°C, but no compensating transfer of a smaller quantity of thermal energy to a colder environment. Our engine obviously violates the second law. It therefore should not work, but where does it fail?

An answer was suggested by Kelvin's brother, James Thomson, Professor of Mechanical Engineering at Glasgow University (where Kelvin was a Professor of Physics), who realized that if the freezing water is to lift a weight, there must be a back-pressure due to the weight, easily calculated by dividing the weight (a force) by the area of the piston. He then made the inspired conjecture that if the melting temperature of water was lowered by pressure, the contradiction would be resolved. There would be the requisite two temperatures in a cycle of operation: water would freeze under excess pressure at, say, −10°C, and the ice would melt under ordinary atmospheric pressure at 0°C. The water on freezing gives out a quantity of heat to the surroundings at −10°C and simultaneously raises the weight; later, when we melt the ice under no extra pressure at 0°C, heat is absorbed from the surroundings. Our engine now operates with two temperatures, as Carnot said it must. When the first law was discovered (after the Thomson brothers did their work on ice) it became possible to perform the bookkeeping on the energy inputs and outputs and show that energy is conserved in the cycle: the heat given out to the surroundings when water freezes under excess pressure is *less* than the heat absorbed when ice melts under normal atmospheric pressure at 0°C, by an amount equal to the work done.

The steps in the operation of the ice engine are thus as follows:

1. Water at 0°C is placed under pressure by a weight and cooled down to −10°C.
2. Water freezes under pressure at −10°C, giving out a quantity of heat Q_L to the environment at −10°C (263 K: this is T_L).
3. The weight is raised because the water expanded when it froze, so work is done.

4. The weight is removed from the ice, returning the ice to atmospheric pressure.
5. The ice is warmed to 0°C.
6. The ice melts under no excess pressure at 0°C, absorbing a quantity of heat Q_H (larger than Q_L) from the environment at 0°C (273 K: this is T_H).

The water and the piston have now been returned to the same state from which they started. The ice engine is now clearly a cyclical heat engine, and it violates neither law of thermodynamics when it does the work of bursting pipes or making potholes when cold nights (at −10°C) alternate with moderate days (slightly above 0°C).

The study of the effect of pressure on the melting point of ice gave Kelvin so much confidence in the correctness of Carnot's reasoning that he was reluctant to believe Joule's experimental results until the apparent contradiction between Joule's ideas and Carnot's could be resolved.

Efficiency of the Ice Engine and Other Solar Engines

We can use the equations derived in Chapter 5 to calculate the reversible efficiency of the ice engine. If we assume that the heat is supplied by day at about 0°C (= 273 K) and given out to the colder environment at night at −10°C (= 263 K), we find:

$$Eff. = \frac{T_H - T_L}{T_H} = \frac{10}{273}$$
$$= 0.037$$

An efficiency of 4 percent is not very good, but the fuel, the fusion of hydrogen in the sun, is free. Solar energy can be collected cheaply under water in a device called a "solar pond," which heats water nearly to its boiling point of 100°C or 373 K. If the ambient air, at a temperature of 20°C (293 K), is used to condense the stream, the reversible efficiency is much higher (about 21 percent):

$$Eff. = \frac{373 - 293}{373} = \frac{80}{373} = 0.214$$

Optical focusing of the sun's rays with lenses or mirrors allows much higher temperatures and thus greater efficiencies to be achieved, but then we were faced with an economic question: how much do the lenses or mirrors cost, and how long does it take for the free fuel to compensate for the initial investment?

Solar energy may also be converted directly to electrical energy in solar cells and to chemical energy by growing plants. None of these methods has yet been developed to be economically competitive with fossil fuels for most energy needs, though this leaves out of the economic calculation the relative damage done to the environment by different energy sources. One form of environmental damage, the "Greenhouse Effect," will be discussed in Chapter 10.

A Quantitative Prediction

The reversible efficiency of the ice heat engine was calculated from the Clausius formula, which is independent of the properties of ice, water, or any other substance. But the heat and work inputs and outputs in a cycle of operation, and thus the reversible efficiency, can also be calculated without the use of this formula from four easily measured physical properties of ice and water (or of any other pure substance): the ordinary melting temperature, the latent heat of melting, the expansion on melting, and the pressure generated by freezing water in a cylinder in which the piston is prevented from moving. Since these two ways of calculating the efficiency must give the same answer, we can obtain an equation relating the change of the melting temperature produced by a given increase in pressure to those four properties of the solid and liquid forms of the substance. The equation enables us to predict, for example, that a pressure of 1,000 atm lowers the melting point of ice to about −7.5°C. We do not give the equation here: its exact form is less important than the facts that the relation exists and that we would never have conjectured a relation among these properties without the second law.

This episode encapsulates a whole treatise on how scientific theories are tested and why they are believed. They are tested by their ability to predict something unknown—ice melts under pressure—or something known but not yet explained—as Newton's laws of motion and his law of gravity explained Kepler's laws of planetary motion. Theories earn belief to the extent that their predictions are surprising and specific. Among other factors that make a prediction surprising and specific is quantitative precision. The equation for melting points under pressure predicts not only that ice will melt under pressure, but also the precise temperature of melting under a given pressure. A prediction that hits the answer on the head like this surprises us and makes more plausible the theory that led to it.

The same kind of analysis that gives us this equation also enables us to predict the effect of pressure on the boiling points of liquids, the effect of a magnetic field on the temperature at which an ordinary metal becomes superconducting, a relation between the cooling of a gas when it expands through a nozzle and the extent of its deviation from "ideal" behavior, and much else.

What other surprises does the second law hold for us? What else does it imply about the properties of matter, about what is possible and what is not? There are many other consequences: new ones are discovered all the time and reported in the scientific journals by the dozens each year.

One-Way Membranes

Kelvin's forbidden engine, as we have noted, does not violate the first law and thus is not a perpetual-motion machine; it does not create energy out of nothing. It does, however, offer the possibility of supplying work in unlimited quantities from the thermal energy that will always be available in the environment. Such an engine has been called, clumsily, a "perpetual-motion machine of the second kind," requiring us to call the kind of machine that creates energy from nothing a "perpetual-motion machine of the first kind." This terminology permits the first and second laws to be stated in the parallel forms:

> First law: It is impossible to build a perpetual-motion machine of the first kind.
> Second law: It is impossible to build a perpetual-motion machine of the second kind.

Suppose some kind of device or membrane could be fabricated that without any input of energy would allow molecules, say those in a gas, to pass through it in one direction and not the other. Such a device, placed between two containers of gas in which the concentrations of the molecules are initially equal, would lead to a buildup of the molecular concentration in one container and a depletion in the other. Figure 6.2 suggests a possible simple design for a membrane that might accomplish such a one-way transmission: a barrier with conical holes, wider on one side than the other. The excess pressure built up by such a device could be used to drive a compressed air engine. The work done

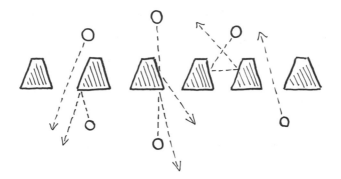

Figure 6.2 A Perpetual-Motion Machine of the Second Kind
Intuitively we might expect that if a membrane with tapered holes, the smaller openings all on one side, were placed in a container of gas molecules, the molecules would have a greater tendency to pass through the membrane from the side where the openings are larger *(top)*. If this were the case, pressure would build up on the other side *(bottom)* and work could be performed by the high-pressure gas. When the high-pressure gas does work it would tend to cool, but if it is in good thermal contact with the outside environment heat would flow in from that environment and the temperature would not fall. As a result, an unlimited quantity of work could be done at the expense of the thermal energy of the environment.

would be at the expense of the thermal energy of the gas molecules, so that the gas would cool as it expands, but its temperature could be restored by its contact with the surrounding environment and the energy could thus be replenished indefinitely from the thermal energy of that environment.

This is a perpetual-motion machine of the second kind, and we can confidently predict that it will not work: there will be no one-way transmission and no pressure build-up. In spite of the ingenious shape of the holes, the rate of molecular flow in both directions must be the same when the molecular concentrations on the two sides are the same.

Entropy and Time

The devices we use to keep track of time sometimes operate cyclically, as clocks do, and sometimes give progressively increasing readings without ever repeating themselves, such as the numbers we give calendar years. There is a corresponding conflict in conceptions of time itself: is it cyclical or not? The earth undergoes repeated revolutions around the

A Perpetual-Motion Machine with a U.S. Patent

The U.S. Patent Office, as a matter of policy, rejects out of hand patent applications for perpetual-motion machines of either kind (it is not true that they request the submission of working models with such applications). The office did, however, inadvertently grant a patent in 1972 to a Mr. Arthur Schultz of Ohio for a "thermodynamic power system" using a one-way membrane to create compressed air (U.S. Patent No. 3,670,500). Mr. Schultz's membrane is similar to but a little more complicated than the one illustrated in Figure 6.2. It includes a "unidirectionally permeable membrane" *(24)*, a compressed air engine *(34)*, and a heat exchanger *(38)* ensuring good contact with the ambient environment.

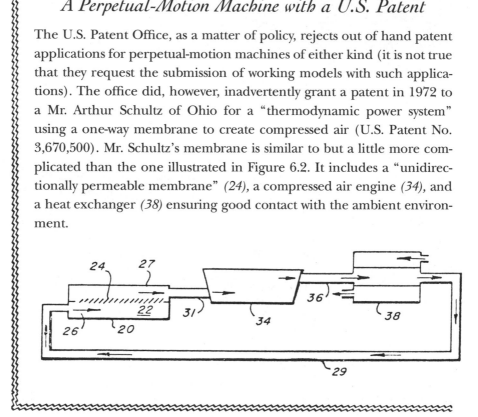

sun, seasons follow one another in order, the tides ebb and flow, crops are planted and harvested again and again. "One generation passeth away, and another generation cometh, but the earth abideth forever." Is there any reason to think time can't go on repeating itself "forever"? The second law tells us it can't: there *is* something that indicates that time flows in a single direction only, the increasing total entropy of everything. There is nothing cyclical about entropy, only an inexorable increase.

That the total entropy can never decrease provides us with a powerful tool for predicting the future, though one with limits. Using it, we can tell what will *not* happen. We can rule out as possibilities all conceivable events that violate this rule, and need consider only the conceivable events that do not. But when we ask which of these conceivable events—those for which the total entropy is larger than it is now—will happen, and how fast, we discover the limits of the second law. It has no answer for questions of "whether" or "when." We must look elsewhere.

There is one particular prediction of the future the second law seems to imply, and it is a depressing one. We know enough about the nature of entropy as a property of matter to be able to say that for any system of finite extent, in isolation from all others and thus unable to gain or lose energy, there is a maximum possible entropy it can have. Once that maximum is reached, nothing further can happen anymore. The capacity for change will have been exhausted, the system will have run down, like a clock whose spring has unwound.

When Clausius first discovered the concept of entropy, he reasoned that because the universe itself is an isolated system, unable to gain or lose energy to others (there are no others), this was its destiny also. Clausius did not presume to tell us the future history of the universe in detail, nor how long it will take to reach its final state, only that according to the second law such a final state would be inevitable. This dismal end has been called "the heat death of the universe."

The modern view of the universe, provided by the theory of relativity and the "big bang" hypothesis of its origin, offers a very slim chance of avoiding this grim prognosis, as we will discuss later on. First, however, we need to know more about entropy, to have some sort of intuitive understanding of what it is and why it tells us such bad news. For that we must turn to a molecular picture.

7

The Molecular View of Entropy

〜〜〜

Our goal in this chapter is to lead the reader to an intuitively satisfying meaning of entropy, which can be achieved only with an appreciation for the meaning of *probability*. We will therefore begin by applying the theory of probability to some elementary problems—tossing coins or throwing dice—and then to a slightly more complicated problem—shuffling a deck of cards. It is a matter of common experience that when a deck of cards ordered by number and suit is shuffled, it becomes disordered, but when a disordered deck is shuffled, it does not usually become ordered. While the experience is common, the reason for it is less familiar: it is that there are so many more disordered arrangements of the cards than ordered ones. To go further, we will need a more precise meaning for the terms *ordered* and *disordered*.

The same reasoning applies to collections of large numbers of molecules, which, like decks of cards, can have many arrangements, some orderly, but most not. Like cards, they also have a tendency for orderly (and less probable) arrangements to become disordered as time goes on, and very little tendency for disordered arrangements to become ordered. This behavior of large collections of molecules, a consequence of the overwhelmingly greater probability of disordered arrangements, seems to have features in common with the so far purely macroscopic principle that the entropy of large collections of molecules always increases. In particular, it suggests that entropy and probability are closely related. There is a significant difference between the absolute prohibition by the second law against a decrease of entropy, however, and the more qualified view that spontaneous increases in molecular order are only highly improbable, not impossible. Is a spontaneous decrease in entropy impossible or only extremely rare? We must turn to the laboratory for an answer.

In the laboratory we observe a phenomenon called *Brownian movement,* a spontaneous, erratic motion by small but microscopically visible particles suspended in a liquid or a gas. First discovered early in the nineteenth century, Brownian movement provided experimental evidence a century later that spontaneous decreases in entropy can indeed be observed if the system studied is sufficiently small. This conclusion served two purposes: it provided strong evidence for the real existence of atoms and molecules, and it made clear the molecular basis of the second law.

Finally, we will use the principles discussed in this chapter to establish a precise relation between entropy and probability and give a molecular interpretation of temperature. Before we begin, however, we want to acknowledge a conceptual difficulty with the proposed endeavor: what right have we to apply the laws of probability to real molecules, as opposed to games? The short answer is that in the course of over a century of research, speculation, and controversy, doing so has given the correct answers, though we are not sure why. A more careful discussion will be given in the next chapter.

Probability Applied to Coins, Dice, and Cards

To deduce the macroscopic properties of matter from the microscopic perspective of the kinetic-molecular theory, we must take into account the motions of enormous numbers of molecules. As was noted in Chapter 4, this seemingly intractable problem was made not only tractable but easy, almost, by the use of the theory of probability. We now want to understand the molecular meaning of the important but so far obscure concept of entropy, and for this, too, the theory of probability is the key.

We will need only a few elementary results from the theory for our analysis, which we review briefly in this chapter. Let us begin with some simple experiments—tossing a coin, or six coins in a row, or throwing two dice, or fifty. For example, a trial of tossing six coins, or of tossing one coin six times, might have the outcome THTTTH (T denoting a tail, H a head). This completely specified outcome is a "simple outcome," one that occurs in only one way. A 7 on the throw of two dice is not a simple outcome; instead, it corresponds to several possible simple outcomes: a 6 and a 1 *or* a 1 and a 6; a 2 and 5 or 5 and 2; and 3, 4 or 4, 3.

The theory defines the *probability* of any event by a number between 0 and 1, 0 denoting an event that has no chance whatever of occurring and 1 an event that is certain to occur. *The probability of any outcome of one trial is defined as the fraction of times in a very long series of trials that the outcome occurs.* The outcome "heads" when a coin is tossed occurs half the time, and therefore has a probability 0.5; a 7 with two dice occurs less often, 0.167 of the time, and has a probability 0.167. An example of an event with a probability 1 is the outcome "*Either* a head or a tail" on the toss of a coin. (We assume that if the coin lands on its edge or rolls out of sight the toss doesn't count as a trial.)

Two outcomes of a single trial are said to be *mutually exclusive* if only one of them can occur but never both. "Head" and "tail" are thus mutually exclusive outcomes of a coin toss. A 7 and an 11 with two dice are also mutually exclusive. Examples of outcomes of the throw of two dice that are not mutually exclusive are "At least one of the dice shows a 3" and "The sum is 6 or more." Here there is partial overlap between these two outcomes. Sometimes simple outcomes are *equally likely,* as are "head" and "tail" in one toss of a coin; sometimes they are not, as when we draw a marble from a bag in which two-thirds of the marbles are white and one-third are black.

As already indicated by some of the examples we have given, there are ways of defining outcomes more broadly, outcomes that can be achieved by several of the mutually exclusive simple outcomes. For example, rolling a 7 with two dice corresponds to six simple outcomes. The outcome "Greater than 7" with two dice can be achieved by a number of simple outcomes corresponding to totals of 8, 9, 10, 11, or 12. These are called *compound outcomes.*

We will be able to develop most of the probability theory we will need here by considering only trials in which the possible simple outcomes are *both* mutually exclusive *and* equally likely, though we must caution that this is an incomplete version of that theory.

A Simple Example

When a coin is tossed six times in a row, or when six coins are tossed at once and laid out in a row on a table, the theory of probability shows that there are 64 possible mutually exclusive, equally likely, simple outcomes, a small enough number for us to list them. Table 7.1 iden-

Table 7.1
Possible Outcomes of Tossing Six Coins

Compound outcomes	Simple outcomes	Number *(W)* of simple outcomes corresponding to the compound outcome
6 heads	HHHHHH	1 outcome (There is also 1 with 6 tails.)
5 heads	HHHHHT HHHHTH HHHTHH HHTHHH HTHHHH THHHHH	6 outcomes (There are also 6 with 5 tails.)
4 heads	HHHHTT HHHTHT HHTHHT HTHHHT THHHHT HHHTTH HHTHTH HTHHTH THHHTH HHTTHH HTHTHH THHTHH HTTHHH THTHHH TTHHHH	15 outcomes (There are also 15 with 4 tails.)
3 heads	HTTTHH HTTHTH HTTHHT THTTHH THTHTH THTHHT TTHTHH TTHHTH TTHHHT TTTHHH	20 outcomes (These 10 are outcomes with at least 2 tails in the first 3 tosses; another 10, not shown here, have at least 2 heads in the first 3 tosses.)

tifies 32; the other 32 can be obtained easily by changing each head (H) to a tail (T) and each tail to a head in the 32 we list. We found these outcomes by a careful and exhaustive enumeration of all the possibilities. When the numbers are larger we will have to use algebraic formulas.

Examples of compound outcomes are "3 heads and 3 tails in any order" or "5 heads and 1 tail in any order." Most of us would expect intuitively that "3 heads and 3 tails" is more probable than any of the others of this kind, but we might not be able to justify our expectation. An examination of Table 7.1 suggests the reason. The theory of probability shows that if the simple outcomes are mutually exclusive and equally likely, the probabilities of compound outcomes are equal to the number of simple outcomes corresponding to the compound outcome divided by the total number of all simple outcomes possible.

We will use the symbol W for the number of simple outcomes corresponding to some compound outcome of interest. When a coin is tossed 6 times, each of the 64 simple outcomes tabulated above is equally likely. But if we lump together outcomes under broader categories such as "3 heads in any order" or even "4 or more heads"—so that they are compound outcomes—they need not be equally likely, as the summary of outcomes in Table 7.2 shows.

If we were to put six coins in a cup, shake it well, throw the coins onto a table, record the numbers of heads (a compound outcome), and

Table 7.2
Probability of Outcomes of Tossing Six Coins

Compound outcome	W	Probability of compound outcome
6 heads	1	1/64 = 0.016
5 heads	6	6/64 = 0.094
4 heads	15	15/64 = 0.234
3 heads	20	20/64 = 0.312
2 heads	15	15/64 = 0.234
1 head	6	6/64 = 0.094
0 heads	1	1/64 = 0.016
Total	64	

Table 7.3
Possible Outcomes of Tossing Ten Coins

Compound outcome	W	Probability of compound outcome
10 heads	1	0.0010
9 heads	10	0.0098
8 heads	45	0.0439
7 heads	120	0.1172
6 heads	210	0.2051
5 heads	252	0.2461
4 heads	210	0.2051
3 heads	120	0.1172
2 heads	45	0.0439
1 head	10	0.0098
0 heads	1	0.0010
Total	1,024	

repeat this procedure a large enough number of times, we will find that the frequencies of each of the possible compound outcomes in our series of trials will be equal to the probabilities in Table 7.2. Put another way, the frequencies will be *proportional* to the numbers *W*. Three heads will, in the long run, occur 20 times as often as 6 heads; in fact, 3 heads will occur more often than any of the other outcomes but still less than half the time: 20/64 of the time in fact. The *odds* of getting 3 heads is the ratio of *W* for 3 heads. (20) to the sum of all the *W*'s for the other outcomes (44)—the odds are 5 to 11 against getting 3 heads.

For ten tosses of a coin there are 1,024 equally simple outcomes. This is too many to list, but we can use formulas from algebra to produce a table of probabilities (Table 7.3). Note that the probability of 3 heads when 6 coins are tossed is 20/64 (= 0.312), but the probability of 5 heads in 10 tosses is 252/1024 (= 0.246). When a coin is tossed a very large number of times, we expect the result to be an equal, or almost equal, number of heads and tails. It is interesting to note that the probability of getting 5 heads in 10 tosses (0.246) is *less*, not more, than the probability of getting 3 heads in 6 tosses (0.312). That phrase *almost equal* is the important difference: it represents a broader, more inclusive, compound outcome than the category "Exactly half heads."

For example, the probability of getting exactly 50,000 heads in 100,000 tosses is very small, about 0.0025. But if by "half" we would be satisfied with getting within 1 percent of half heads, we note that 1 percent of 50,000 is 500, so we must add together the probabilities of getting 49,500 heads, 49,501 heads, 49,502 heads . . . on up to 50,500 heads. When all these probabilities are added up they give 0.9999999993, implying virtual certainty.

The basic difference between the problems of probability theory that we have considered so far and the problems that arise with molecules is the numbers involved. We may have to imagine experiments equivalent to tossing 10^{23} coins at a time, but this is an advantage, as the predictions of the theory of probability are more precise under such conditions. Numbers as large as 10^{23} are hard to conceive of (the U.S. national debt in 1992 is about one trillion dollars, or 10^{14} pennies; this is exactly one-billionth of 10^{23}), but the numbers of simple outcomes possible when 10^{23} coins are tossed at one time are inconceivably greater than 10^{23}. To get some feeling for how enormously large are the numbers of simple outcomes that are possible even with small collections, we will shuffle some cards before we turn to molecules.

The Shuffling Paradox

Let us imagine ourselves to be the audience for a performance of card tricks. The magician shuffles the deck ostentatiously, then hands it to a member of the audience for further shuffling. Then he takes the deck and begins to turn the cards face up one by one from the top of the deck.

Since we are watching what purports to be a magic show we expect that the cards will turn up in some surprising way: the spades first and in numerical order, followed by the other suits, also in numerical order. This would satisfy us that we have seen a "magic" trick, and we would have only to wonder how he did it.

But this time we are in for a real surprise: the magician lays out the cards in the following order:

Four of clubs
Nine of diamonds
Jack of diamonds
Seven of hearts
Ace of clubs . . .

and so on, with no discernible order, just what we expect from a well-shuffled deck. We object indignantly that this is no trick; we could have shuffled the deck ourselves and got the same result.

But the magician answers, "Could you? Could you produce the precise order I did, four of clubs first, then nine of diamonds, then jack of diamonds and so on to the end? Is it any easier than getting the cards in perfect order, by number and by suit? If you think so take the deck and try it!"

The magician is of course right, yet something is wrong. No single, precisely specified order of the cards is any more or less probable than any other, so why should a perfectly orderly outcome be more surprising than the one the magician got?

Let us put it another way. There are about 10^{68} possible arrangements of a deck of 52 cards. (See the section on "factorial numbers" in the appendix for the calculation of this number.) The odds of getting any specified outcome is 1 chance in 10^{68}, corresponding to a probability of 10^{-68}, an inconceivably small number. If we were to define a "miracle" as anything that has less than one in a trillion (10^{12}) chance of happening, but happens anyway, then *any* outcome of shuffling a deck of cards is a miracle, and that is the magician's point.

But intuitively we know that the magician has not worked a miracle. When a deck of cards is shuffled, there must be *some* outcome of the shuffle. It will of course be one of the equally likely (or equally unlikely) 10^{68} possible simple outcomes. It is clear that the perfectly ordered arrangement beginning with the ace of spades has only a 10^{-68} probability of occurring after shuffling. While it is true that the magician's result is just as unlikely, the magician did not tell us in advance what order he would shuffle the cards into. He could call "miraculous" any result he got.

The great majority of the 10^{68} simple outcomes we lump together under the rubric "disordered." The probability of getting one of this enormous number of outcomes is so close to 1 as to be a near certainty. The number of simple outcomes corresponding to the result "Having enough order in the arrangement to be surprising" is very small. They include such possibilities as having the suits come in various orders, or getting four aces first, followed by four twos, on up to four kings. Maybe a hundred or so outcomes would surprise an audience. Even a thousand (10^3) or ten thousand (10^4) is negligible compared with the 10^{68} possible outcomes. Thus the first category, "disordered," is overwhelmingly more probable.

The categories "ordered" and "disordered" are clearly compound outcomes. Let us represent the number of simple outcomes corresponding to the "disordered" category by W_D and the number in the "ordered" category by W_O. Our conclusion can be summarized as

$$W_D \gg W_O$$

In other words, it is far more likely to find the deck in one of the multitudinous disordered outcomes than in one of the much fewer ordered ones, even though each simple outcome is neither more nor less probable than any other.

Order and Disorder

We have used the terms *order* and *disorder* so far without defining them, relying on the reader's understanding from ordinary usage. But they are so significant here that they deserve more careful analysis. In particular *disorder* can be deceptive. Our dictionary defines *order* as "a condition of logical or comprehensible arrangement among the separate elements of a group." The word *comprehensible* is crucial: comprehensible to whom? Some forms of order are comprehensible at a glance, others look like disorder unless the underlying structure is known: for example, a coded message. The letters RIAPS EDDNA ERUTP ARFOIE VAHS EVOLO WT look pretty disorderly, but they are in fact a line from Shakespeare written backward, with spaces placed irregularly for added obscurity.

A criterion for distinguishing ordered from disordered arrangements has been proposed that came to us from computer science but is intelligible without it. The arrangement of a deck of cards or any other group of objects can be specified by a list of the objects in the order they appear. Such a list is 52 symbols long for a deck of 52 cards, 1,000 symbols long for 1,000 objects, and so on. Can a set of directions be given for determining the arrangement in question that is shorter, that requires fewer symbols, than the complete list itself? If so the arrangement is ordered.

Let us forget about hearts and aces and imagine a group of N numbered objects. Then any list giving the N numbers in any order whatever corresponds to an arrangement of the objects. But what if we gave *rules* for ordering the list? We could say, "Arrange the objects in inverse numerical order," or "Put the even-numbered objects first, in ascending order, then the odd-numbered objects in descending order." Each rule enables us to determine the position of each object with far fewer symbols than was required to list them all.

If this definition of *order* is accepted, then it can be shown by mathematical arguments that the number of ordered arrangements of any large group will be a small fraction of all possible arrangements; furthermore, the larger the group, the smaller the fraction.

In brief, there are always more ways to produce nonsense than sense. What drives decks of cards—and, as we shall see, much else—toward disorder is not some attractive force favoring disorder, but simply the fact that there are so many more ways of getting it.

Shuffling Simpler Decks

Let us return to our deck of cards, but this time we will use a greatly simplified deck, where suit and number value of the cards are erased and only the categories red or black are considered. We will begin by arranging the deck so that the 26 red cards are on top and the 26 black cards are on the bottom. What will happen when we shuffle the cards? This is not a magic trick anymore: the cards gradually become randomly distributed, with about half the red cards in the top half, and the remainder in the bottom half.

Shuffling cards well is not a simple matter. Different people do it differently, and some take longer than others to produce a well-shuffled deck. Let us devise a process equivalent to shuffling that will accomplish the same end result but be easier to perform and monitor, especially if we have a microcomputer or programmable calculator.

Our model "shuffle" will proceed as follows. We select a card at random from the top half of the deck (by having the computer pick a number at random from 1 to 26) and a card at random from the bottom half of the deck (by picking a number at random from 27 to 52). Computers have programs for picking what are called "pseudo-random" numbers. They will serve here: we will discuss how programs like these operate in a later chapter. Now we exchange the chosen cards. We repeat this operation a sufficiently large number of times and we have shuffled the deck.

Note that on the second and subsequent exchanges we continue to select the two cards at random, so that the card in position number 17 may be selected twice, even though the card in position number 4 has not yet been chosen.

Let us observe the course of events by recording, after each exchange, or even after each group of five or ten exchanges, the number of red cards in the bottom (originally all black) half. When we begin it is 0. After the first exchange it must be 1. After that its course is not precisely determined. On the second exchange it is most likely for a red card to be selected from the top half because now 25 of the 26 cards are red, but there is a small chance the 1 black card there could be chosen. So while 2 red cards in the bottom is most likely after the second exchange, it is possible that we may end with 1 or, even less probably, 0. As the number of exchanges increases, the numbers of red cards in the bottom half tends to increase, toward the expected number of 13 for a well-shuffled deck.

In Figure 7.1 we have plotted the results of two separate trials of our

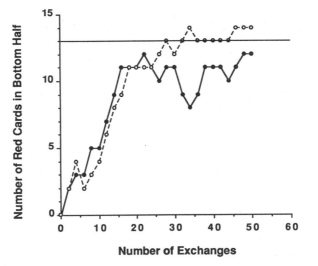

52 Card Deck

Figure 7.1 Card Shuffles

This graph shows the results of two simulated shuffles of a deck of 52 cards, of which half are red and half black. Initially all the red cards were in the top half of the deck. Cards from top and bottom halves were interchanged one pair at a time by a "random" process, until 52 interchanges had been performed.

100 Card Deck

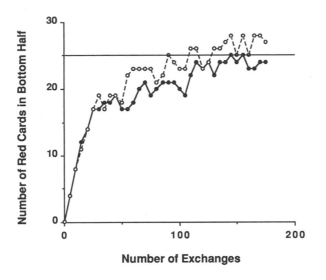

Figure 7.2 Simulated Shuffles of a 100-Card Deck

shuffling procedure. Note first that the two trials do not duplicate each other in every detail but that both do approach the well-shuffled result. Note further that neither approaches the result 13 and then stops there, but rather tends to fluctuate around this value, sometimes increasing beyond it and sometimes falling below. We can infer that if we played this game long enough, the number could occasionally fall back to the starting value of 0, or overshoot to 26, but even if these values were reached, continued exchanges will bring us back to the neighborhood of 13.

We can do the same procedure with a deck of, say, 100 cards, of which 50 are red and 50 are black. We expect the behavior to be generally similar. It will take more exchanges to approach the well-shuffled value of 25 red cards in each half. There will still be excursions from this number, but the tendency to return to it remains (Figure 7.2).

As we increase the number of cards it becomes more time-consuming to run the experiment and more difficult to plot the graph: we would need too large a sheet of paper. The graph can be drawn more easily if we focus not on the *number* of red cards in the bottom half of the deck but on their *fraction,* which varies between 0 and 1 and fluctuates around 0.5 whatever the size of the deck. Figure 7.3 gives results for a deck of 200 cards. To make the results more general we have changed

200 Card Deck

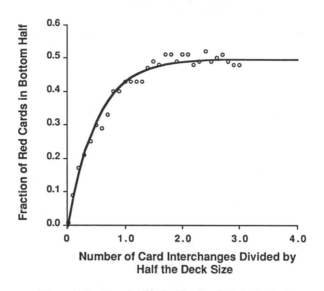

Figure 7.3 Simulated Shuffle of a 200-Card Deck
The vertical scale now shows the *fraction* of red cards in the bottom half of the deck instead of the *number* of red cards, and the horizontal scale the number of card interchanges divided by half the deck size (100 in this case) rather than the number of interchanges. The data points, shown as small circles, are given only for every fifth interchange.

The solid line in the graph shows the results calculated theoretically for a deck of "infinite" size. The larger the deck, the smoother the behavior appears, and the fluctuations about the expected average value of 0.5 can no longer be seen.

the horizontal scale (which previously represented the number of exchanges performed) to represent the number of exchanges *divided by* half the number of cards in the deck. So the point labeled 1.0 on the horizontal scale represents 100 exchanges for this deck.

Comparing this result with those for the smaller decks, we can see that the larger the deck the smoother the behavior. The fluctuations, on a fractional or percentage basis, are smaller the larger the deck, and we can imagine that for a sufficiently large deck the fluctuations would no longer be visible. We can in fact use the theory of probability to calculate what would happen in a deck of infinite size. The results are shown as the smooth curve in Figure 7.3.

Our conclusion: The theory of probability predicts an approach to a final or "equilibrium" value, about which there are fluctuations that on a percentage basis are less and less noticeable as the system under study grows larger.

The Significance of W

It should be clear by now that the tendency of our shuffling procedures to lead to an "equilibrium state" is just a consequence of the enormously large value of W—the number of equally probable simple outcomes associated with the equilibrium state—compared with the much smaller values of W associated with states that differ noticeably from equilibrium.

For a deck of C cards, the number W of arrangements near equilibrium (nearly equal numbers of red cards in each half of the deck) is about 2^C, or 2^{52} for a 52-card deck. This equals about 10^{15}. Our handheld calculator does not handle numbers larger than 10^{99}, the number of arrangements of a deck of only 330 cards.

When we apply the theory of probability to large collections of molecules, as we will do shortly, we find that they act much like decks of cards, except that the number of molecules involved in any real macroscopic system is enormously greater than the numbers of cards we have considered: 10^{20} or greater. But that is a gain, not a loss: it gives us greater assurance in our results, not less.

In explaining scientific concepts to the nonscientist it is often necessary to use metaphors: to compare something new and strange to something familiar. So amino acids are described as the "building blocks" of proteins, and the curved and expanding but finite universe of relativity theory is compared to a balloon being continuously inflated. For the second law of thermodynamics, we will not need metaphors. The behavior of a deck of cards on being shuffled is not a sort of analogy to the second law: the second law itself will be seen to be a consequence of applying the theory of probability to molecules, just as we apply it to cards.

Molecular Probability

There are practical difficulties in applying probability to molecules that we did not have with cards. In our example, each card had only two features, red or black, whereas each atom or molecule has many more: how fast it is moving, in what direction, where it is in the containing vessel, how near it is to other molecules, and others. We require a more complicated card game than we have played so far, but a card game nevertheless.

Before we go further we will make one change in our terminology. For the simpler problems of probability, we have used the term *simple outcomes* for the completely specified and mutually exclusive results of a single trial. When we consider molecular probability problems, we will use instead the term *microscopic state* or *microstate,* for short. This term implies everything *simple outcome* does: it is a complete description of the position (and of the energy) of each molecule in the system. It represents far more detailed information about the system than we really care to know. The averaging over microstates that give us the macroscopic properties does not require all that detail. Let us summarize by giving a molecular definition of W: the number of equally likely microstates of a large collection of molecules consistent with some particular macroscopic state of the collection. The *macroscopic state* is specified by the number of molecules N, the total energy E, and the volume V only: no other information need be given.

The molecular process analogous to shuffling cards is one that occurs spontaneously. The driving mechanism is just the motions all molecules engage in: they move from place to place in the container; as they collide, energy is transferred from one to the other. Molecules move at about the speed of sound, and each one undergoes collisions with others at rates of about 10^8 to 10^9 times per second under ordinary conditions of temperature and pressure. The system therefore changes rapidly from one microstate to another, though it may take an extremely long time for all the possible microstates to be explored.

Let us begin with one simple molecular example to which the theory of probability can easily be applied. Let us first describe the example from a macroscopic perspective. A container is divided into two equal parts by a partition, with a gas on one side of the partition and the other side empty. We make a hole in the partition so that the gas can flow over to the previously empty side and observe how the pressure on each side changes. Clearly, the two pressures will tend to equalize, and each side will end up with half the initial value in the side that originally had all the gas. Because the entropy of a gas increases when it expands into a larger volume, the entropy is now higher than it was when the gas was all on one side of the container.

Let us now view the gas microscopically, as a collection of a larger number of moving molecules in a container. The molecules might occur in a very large number of spatial arrangements. Among the many arrangements are some in which the molecules are all on one side of the perforated partition, and others in which they are evenly divided between the two sides. Let us compare the number of equally likely arrangements W_1, when all the molecules are on one side of the partition, with the number W_2, when they are equally divided between the left and right halves of the container.

We will make a number of simplifying assumptions:

1. We will ignore the contribution to each W of the different kinetic energies the molecules may have and the different directions in which the molecules may be traveling. This omission does not affect the ratio of the W's, which is what we need here.

2. We will assume that the average distances between the molecules are large, compared with their diameters, so that most of the space in the container is empty. This means that we are considering a dilute gas (as in our model of an "ideal" gas; see Chapter 4.)

3. We will ignore the attractive forces between the molecules, although we know such forces are present: they are what cause gases to condense to liquids on cooling. (The absence of attractive forces is another criterion of "ideal" gas behavior.)

4. There are also repulsive forces between molecules that prevent them from getting too close together. Strictly speaking, "ideal gases" don't have repulsive forces but their presence here simplifies the calculation slightly. To visualize the repulsive forces, we may imagine that the molecules are hard, like marbles or ball bearings, which obviously cannot occupy the same space at the same time and would "bounce off" each other if they collided.

5. Finally, we will imagine the space in the container divided up into little cubical boxes, each large enough to hold only one molecule at a time. As we assumed most of the space is empty, we must conclude that most of the boxes or "cells" are empty and that only a few contain a molecule (Figure 7.4).

A reasonable choice for the size of the cell is the size of a molecule. If each half of the container has a volume of 1 liter, and the gas is initially (before opening the hole in the partition) at a pressure of 1 atm (the average pressure at sea level) and a temperature of 20°C, there will be about 2.5×10^{22} molecules in the liter. This number is almost independent of which gas we use. The volume to be assigned to the little cells, however, does depend on the gas: for oxygen a reasonable figure for the volume of a molecule (which can be estimated in a number of ways) is 1×10^{-26} liters. This gives us 10^{26} cells per liter, and there is 1 liter of space on each side of the partition. Initially, the oxygen molecules occupy 2.5×10^{22} of the 10^{26} cells on the left side. This means that one cell out of 4,000 contains a molecule: the rest are empty.

The *macroscopic* states before and after we open a hole in the partition are, respectively:

1. The molecules are all on one side of the partition, occupying half the volume of the container, and have a total energy E.
2. The molecules are equally divided between the two sides of the partition, occupying the whole volume of the container, and have the same total energy E. (When a real gas expands into a vacuum its energy usually changes, but the energy of an ideal gas does not.)

Now let us introduce some symbols, which will make our formulas more useful for other problems as well.

Let C = the number of cells in 1 liter (= 1×10^{26} in this problem)
Then $2C$ = the number of cells in the whole container, which has a volume of 2 liters (= 2×10^{26})

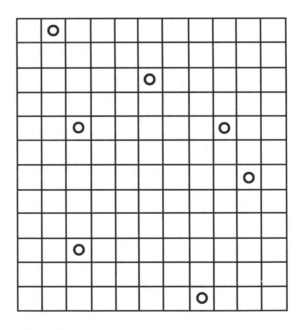

Figure 7.4 An Arrangement of Molecules in Space
Here we show one of the approximately 12 million ways of placing 7 molecules on a "checkerboard" of 132 squares.

Let $N =$ the number of molecules $= 2.5 \times 10^{22}$

Let $W_1 =$ the number of equally likely arrangements of the molecules when they are confined to the left side of the container; *this is the number of different arrangements of a deck of C cards, of which N are red and the remainder (C − N) are black*

Let $W_2 =$ the number of equally likely arrangements when the hole in the partition is opened and the molecules are free to move throughout the container; *this is the number of different arrangements of a deck of 2C cards, of which N are red and the remainder (2C − N) are black*

Our claim is that $W_2 \gg W_1$. If so, this would make the probability of finding the molecules distributed throughout both halves of the container overwhelmingly greater than finding them all on one side; one may conclude that if the hole is opened, the gas flows to produce approximately equal numbers of molecules on both sides.

To calculate the individual values of W_1 and W_2 and then their ratio, it is necessary to use algebraic formulas for what are called "combinations" and also methods of approximating the results to simplify the calculation (these are discussed in the mathematical appendix). We will not go through the details of the calculation, but only give the results for the desired ratio of W_2 to W_1, the ratio of the probability of finding the molecules distributed throughout the container to the probability of finding them all on the left side. The result is, to a very good approximation,

$$\frac{W_2}{W_1} \cong 2^N = 2 \text{ raised to the power } 2.5 \times 10^{22}$$

Note that the result depends on the number of molecules N and the fact that the volume was doubled (which is where the "2" in 2^N comes from) but *not* on C, the total number of cells or, in more physical terms, not on the actual volume in liters occupied by the gas before its expansion.

Astronomical Numbers and Beyond

The number 2^N, with N equal to 2.5×10^{22}, is large beyond belief and beyond comprehension. We conventionally refer to huge numbers as "astronomical." A literally astronomical number is the distance to the furthest visible stars, ten billion (10^{10}) light-years. Converting this distance to centimeters raises the number to 10^{28}, and measuring the distance in hydrogen atoms laid down one next to the other raises it to 10^{36}. This is hopelessly negligible compared with the number W for one liter of gaseous matter. The values for W are the largest that arise in any scientific context, swamping astronomical numbers to insignificance. The tendency of a compressed gas to flow into a vacuum, as well as much else that happens in the universe, is a reflection of the inconceivably great increase of W that takes place when it happens.

Numbers so large are not only hard to comprehend, they are also something of a nuisance to write down and do arithmetic with. For convenience we choose to work instead with the logarithms of these enormous numbers. In effect, instead of writing numbers like this out in full, which would take much more space than there is in one trillion books this size, we use instead the power to which 10 must be raised to express the number W. Logarithms are described in more detail in the mathematical appendix. For our discussion here we need to know the following properties of logarithms: first, the logarithm of a large number is a smaller number; second, as numbers increase, their logarithms also get larger; and, third, the logarithm of the product of two numbers is the *sum*, not the *product*, of the logarithms of the individual numbers, and the logarithm of the ratio of two numbers is the *difference*, not the *ratio*, of the logarithms.

We obtain for the logarithm of the desired ratio

$$\log \frac{W_2}{W_1} \cong \log 2^N = N \log 2 = 0.3010N$$

The logarithm of the ratio is itself very large in the ordinary sense of "large," being (except for the factor 0.3010) equal to the number of molecules N, but not quite as incomprehensible as the ratio itself. It is only about one hundred million times greater than the U.S. national debt in pennies—a merely astronomical number.

The Distribution of Energy

In the example above, we considered the total number of spatial arrangements of molecules in a container when the speed (translational energy) and direction of motion of each molecule are ignored. We were able to calculate this number—let us call it W_V, for the number of arrangements possible in the volume available—because we could treat it as a card-shuffling problem, with a deck of red cards (cells with molecules) and black cards (empty cells).

We turn now to the calculation of the *total* W for such a collection, where we no longer neglect energies and directions of motion; we want the total number of different microstates W_T possible for a number N of molecules in a container of volume V and with a given total translational kinetic energy E. For simplicity we disregard the possibility of rotational and vibrational motion (see Chapter 4).

Each of the molecules must now be assigned both a position in the container (each put in one of those little cells) and a kinetic energy. We can do this in two independent steps, first putting the molecules in their cells and then dividing up the total translational kinetic energy among them. We have already done the first and labeled the number of arrangements we found W_V. Let us call the number of ways of distributing the total kinetic energy W_E. Then the theory of probability permits us to conclude that the total W is the product of W_V and W_E:

$$W_T = W_V \times W_E$$

Whenever the calculation of W_T for a system can be broken down into two or more independent calculations, W_T is always the product of the separately calculated W's. An important special case of this rule applies when we want to calculate W_T for two physically distinct systems: 4 kilograms of hydrogen gas in one container and 20 kilograms of liquid ethyl alcohol in a second container. If we wish to consider the two systems as forming parts of a combined system, then the number of arrangements of the combined system (W_T) is the product of the W_T for the hydrogen system (W_{T1}) and the W_T for the ethyl alcohol system (W_{T2}). Figure 7.5 illustrates for a simple example why this is reasonable.

Can we find W_E here with the techniques we used for the card-shuffling problem? What kind of deck do we use? We will conclude that finding W_E is like a card-shuffling problem, but one that uses a more complicated deck than before. There are two reasons for the extra complication. One is the need to specify directions of motion as well as energies. The second is that the total energy has a definite value and must remain the same. The first is not a terribly difficult thing to handle, and we will put it aside for the time being.

In this calculation we are imagining that the system is isolated, so that no energy can leave or enter it, a condition that introduces difficulties that we did not encounter with the calculation of W_V. For that problem, the placement of any one molecule was essentially unaffected by the

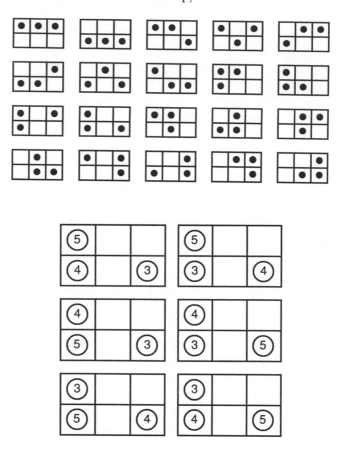

Figure 7.5   Why W_T Equals W_V Times W_E

The blocks at the top show all twenty arrangements W_V of three molecules in a space consisting of six cubical boxes. Below them are the six possible distributions, for *one* of the spatial arrangements given above, by which twelve units of energy (as an example) may be divided up in bundles of 3, 4, and 5 units among the three molecules. Since for each of the twenty spatial arrangements there are six ways to distribute these particular bundles of energy, the total number of arrangements both in space and in energy is the product, 120. All other ways of dividing up the energy (such as 2, 4, and 6 units, or 3, 3, and 6 units) must also be calculated and added up to obtain W_E. For twelve units of energy divided among three molecules, W_E is found by direct enumeration to be 91, of which 72 correspond to all three bundles having different quantities of energy (like 2, 4, 6 rather than 3, 3, 6). W_T is thus 6 × 91, or 546. The tendency for the contribution to W_E to be greatest for those distributions of energy that produce the greatest possible variation in the energies assigned to the individual molecules leads, for large numbers of molecules, to the Maxwell-Boltzmann distribution.

placement of the others, except for the rule that we could put only one molecule in any box. However, the assignment of a particular energy to one molecule is not independent of the energies assigned to the others: the total must add up to E, and if some get more, the others get less. To take an extreme example: if one molecule were assigned all the energy *E,* there would be none left for the others; they would all be standing still.

Extremely uneven energy assignments like this would contribute only a relatively small number of arrangements to W_E. We could avoid giving too much energy to too few molecules and starving the rest by giving exactly the same amount to each. The deck of cards that represents this distribution of the energy is therefore composed of cards all of the same color; it is only one of the enormous number of arrangements given by W_E.

Maxwell found that the largest contribution to the number W_E is made by a distribution of energy among the molecules that has the mathematical form known as a "negative exponential"; it implies that if you look for a molecule having a particular energy, no matter how high or how low, you will always be able to find one, but the higher the energy the fewer such molecules there are (There is a section on exponents in the math appendix, and an illustration of the negative exponential curve in Figure 14.5.)

The number W_E of microstates corresponding to this distribution overwhelms all others to such an extent that the chance, once a large collection of molecules has had time to reach equilibrium, of finding any other distribution of energy is negligible. It is therefore the only distribution we ever observe at equilibrium.

We return now to include the direction of motion of each of the molecules. We can visualize both the energy of a molecule and its direction of travel by assigning to the molecule an arrow, whose length is proportional to the speed v and whose direction is the direction of travel. Such a representation is called a *vector,* a concept we will encounter again when we discuss the theory of relativity later on. An assignment of speed is also an assignment of energy, since $\varepsilon = \frac{1}{2} m v^2$. Part *E* of the figure on page 191 illustates the molecules with attached speed vectors. In the panel on the right, the speeds differ, while in the left panel the speeds of all molecules are the same. We may also interpret the arrows in the left panel as representing only directions of motion, but not speeds, which may differ. It is then easy to visualize why taking directions of motion into account increases the number of microstates W_E: just imagine all the different ways the arrows may be switched around among the molecules,.

As noted earlier, the total number of arrangements when both the translational kinetic energies and the spatial distribution of the molecules are taken into account is given by:

$$W_T = W_V \times W_E$$

The complete calculation, including speeds and directions, can be subjected to an experimental test. Because translational kinetic energy ε and speed v are related by $\varepsilon = \frac{1}{2}mv^2$, we can calculate the fraction of molecules traveling at various speeds. The results, shown in Figure 4.9, do not seem to resemble the negative exponential curve, but the discrepancy is only an apparent one. The difference is a consequence of two factors: taking directions of travel into account and considering speeds rather than energies.

Recapitulation: Probability and Entropy

Let us summarize.

We have applied the theory of probability to some simple problems involving shuffling a deck of cards. We showed that various states of the deck—all red cards in the top half (an ordered deck), or equal numbers of red cards in each half (a disordered deck)—can each be characterized by a number W, the number of equally likely simple outcomes of a shuffle of the deck consistent with the chosen state. If the deck is in a state described by a relatively small value of W (all red cards on the bottom), and we shuffle it, the chances are good that it will go to a state with a larger value of W, and the chances are better the greater the ratio of the larger W to the smaller. The more cards in the deck, the more overwhelming the probability of a disordered arrangement.

We have shown further that if we apply the theory of probability to collections of molecules, the values of W, the numbers of microstates, are very large indeed, and such collections, like shuffled decks of cards, tend strongly to go to states of the maximum possible W. Since there are so many more molecules even in small amounts of matter than there are cards in the usual decks, the numbers W are correspondingly larger, and the tendency to disorder is correspondingly greater.

Now these statements sound something like the second law. According to that law, only changes that lead to an increase of total entropy can occur. When a system reaches a state of maximum entropy, so that no further increase is possible, nothing further can happen.

We have described two ways to predict what large collections of molecules can do. One, macroscopic, is based on the entropy S, a property determined in the laboratory by measuring heat inputs and temperatures, and asserts a tendency for entropy to increase. The second, microscopic, calculates the number of microstates W from the kinetic-molecular theory and asserts a tendency toward larger values of W. When the largest possible value of W is reached, further change is highly unlikely.

This suggests an intimate relation between the microscopic quantity W and the macroscopic quantity S. Let us compare S and W to see if such a relation can be established.

Macroscopic Approach	*Microscopic Approach*
1. S will always increase.	1. W will *almost* always increase.
2. The total entropy of two independent systems is the *sum* of the separate entropies.	2. The total W of two independent systems is the *product* of the separate W's.
3. Entropy is measured in units of energy divided by temperature, so its numerical value depends on the arbitrary choices we make for the units in which to measure energy and temperature.	3. W is a pure number. Its value does not depend on the choice of units of energy or temperature.

An Irreconcilable Difference?

The differences between the macroscopic and microscopic views appear difficult to reconcile. Our attempt to justify the second law by applying considerations of probability to molecular energies led us to a statement that sounds *something* like the second law, but the difference is crucial. When a gas is placed on one side of a divided container and a hole is made in the partition, the gas will flow until the pressure is equalized, and then the flow stops. According to the second law, nothing further can happen: as the pressures equalize, the entropy increases, and when the pressure is uniform it has reached its largest possible value. Any tendency of the gas to concentrate on one side or the other of the partition will lead to a decrease in entropy, which is impossible. From the microscopic view, the state of uniform pressure is overwhelmingly probable, but it is not an absolute certainty. There is some small probability that even after equilibrium is reached the gas could concentrate once again on one side of the partition, and in any event, small differences between the numbers of molecules on each side, and therefore small differences in pressure between the two sides, will come and go.

These predictions cannot both be correct, and we must go to the laboratory to resolve the conflict.

Can Entropy Decrease?

How is the experiment to be done? If we hope to observe all the molecules in a container spontaneously collecting on one side of a partition we are betting on an event at hopelessly small odds.

It is not hard, given a deck of 10^{20} cards with the red cards distributed at random, to calculate how many shuffles would be needed to give us an appreciable chance of finding all the red cards concentrated in only one-half of the deck. From our knowledge of how rapidly molecules change their positions we can calculate how many "shuffles" of the 10^{20} molecules take place per second, and thus how long we would need to wait to observe such an event. It is inconceivably long, many times the estimated age of the universe (10–20 billion years).

In a way the sheer improbability of observing the molecules collecting all on one side goes some distance toward resolving the conflict. If the probability of observing a violation of the second law is too hopelessly small, maybe we can conclude that the two approaches really give the same answer. But somehow we want more. This kinetic-molecular theory, which assumes that matter is composed of particles that are too small to be seen but that are in constant motion, does seem to explain certain phenomena of nature, but one should be wary of letting that settle all doubt. The caloric theory, in its day, also seemed to explain certain phenomena of nature fairly well. There is a touch of the fanciful about the molecular picture, and indeed, in spite of the general acceptance of the two laws of thermodynamics in the mid-nineteenth century, a number of scientists were skeptical about the reality of atoms and molecules.

The leading spokesman for this skeptical view was the physicist and philosopher of science Ernst Mach (after whom the Mach number of supersonic flight was named). He and his followers had two reasons for mistrusting the kinetic-molecular theory.

First, although the theory could be used to predict some properties of matter correctly, it failed in others. Some of the discrepancies could be resolved only by introducing additional hypotheses about the properties of molecules, which had no justification other than that the theory gave the wrong answer without them. The skeptics, appealing to the principle of Occam's razor, rejected these as wild speculations about unobservable entities.

Second, there were serious reservations about applying laws of chance to molecules, which as material bodies were assumed to obey exactly Newton's laws of motion. The theory of probability assumes a degree of ignorance about the systems to which we apply it, and it is not clear how we justify using probability to study systems obeying strictly deterministic laws. These reservations, which trouble scientists to this day, will be discussed at length in Chapter 8.

It was important for those who believed the kinetic-molecular theory to devise some experimental means of demonstrating that molecules were real. Could direct evidence be found that entropy does occasionally decrease, and could the decreases be explained by applying probabilistic reasoning to moving molecules?

We have noted, in our card-shuffling experiments, that fluctuations *do* occur away from the most probable state but that the fluctuations, on a percentage basis, are smaller in larger systems. This suggests that we might more easily observe fluctuations if we look with more sensitive instruments at smaller systems. The kinetic-molecular theory explains the pressure of a gas as the result of the combined impacts of the molecules on the walls of the container. We measure pressures with a barometer, which consists of an evacuated vertical glass tube greater than 76 cm tall, with its open end immersed in a pool of liquid mercury (Figure 7.6). Under the pressure of the earth's atmosphere the mercury rises to a level of close to 76 cm. If the pressure is produced by the bombardment of the surface of the liquid mercury by gas molecules, it is reasonable to expect that from time to time either the number striking the mercury surface or their mean kinetic energy deviate from the expected values. The pressure and thus the height of the column will therefore change. So we may infer that, if the kinetic-molecular theory is correct, the level of the mercury in the tube will vary erratically about its average value, in a way similar to the shuffled decks of cards of Figures 7.1–7.3. If the second law were strictly correct, the pressure would be strictly constant. But how sensitive is a barometer? How large must the fluctuations be for them to register, and will this instrument, in which a few hundred grams of mercury must be moved slightly to show a change in pressure, respond fast enough to fluctuations that may only last a very small fraction of a second? The outlook is not promising unless we can devise a barometer much more sensitive than the conventional one.

Brownian Movement

The invention of the achromatic lens in the early nineteenth century led to a great improvement in the power of optical microscopes. In 1828, a British botanist, Robert Brown, described his microscopic studies of suspensions in water of pollen grains 5×10^{-4} cm in diameter. A French colleague had reported that pollen grains constantly moved about erratically, the more so the higher the temperature of the water. This seemed a clear manifestation of biological activity in the grains; the increase with temperature was interpreted as an increase of metabolic rate. Brown confirmed these observations, not only on pollen grains freshly gathered from living plants but also, to his surprise, on grains that had been stored under dry conditions for up to twenty years. He then examined inorganic powders suspended in water: ground-up rocks, earth, glass, and even a piece of the Sphinx. All the substances, alive or dead, engaged in equally lively motion. It was clear to Brown that the motion had nothing to do with biology. What then was its origin?

In an ingenious experiment he was able to rule out two possible explanations that occurred to him. He shook an aqueous suspension of the tiny particles with oil, thereby breaking up the solution into tiny droplets. In some of the water droplets suspended in the oil only one particle was present. Each of these moved as violently as before, ruling out the possibility that the particles moved due to forces that they exerted on each other. Further, as the rate of evaporation of the water was greatly diminished by the oil, currents created in the water by evaporation could be excluded. This second conclusion was confirmed by others in experiments using covered slides from which evaporation was impossible.

Brownian movement, as it is called today, remained a scientific oddity for the next fifty years. It certainly did not arouse the interest of the scientists who founded the kinetic-molecular theory: Kelvin, Maxwell, Clausius, or Boltzmann. Various attempts were made to explain it, but most could be refuted by simple experiments. Some attributed it to absorption of the intense light used to illuminate the microscope stage by the particles, with consequent increases in temperature and convection currents. But it was shown that neither the color nor the intensity of the light used had any effect on the motion. Electrical forces between particles were invoked by a number of scientists in spite of Brown's observations that particles isolated in droplets continued to move; but other scientists showed that the phenomenon took place to the same extent in liquids whose electrical properties were quite different from those of water.

By the last quarter of the nineteenth century a consensus among

Barometer

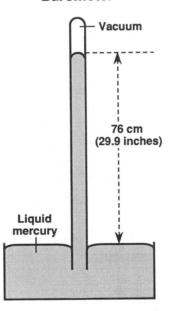

Figure 7.6   The Barometer

Torricelli made his barometer by filling a long glass tube, sealed at one end, with liquid mercury (13.6 times as dense as water), placing his thumb over the open end, and inverting the tube in a pool of mercury. The level of the liquid mercury did not rise above about 760 millimeters (29.9 inches), so a vacuum was left in the space above 760 millimeters. Previously, the tendency of liquids and gases to fill inverted containers was explained by the principle that "Nature abhors a vacuum," but Torricelli's vacuum showed that Nature's "abhorrence" has limits. That the rise of the mercury was due to the weight of the atmosphere was convincingly proved by Pascal, who persuaded his brother-in-law to carry a barometer up a mountain in France and predicted the level of the mercury in the tube would fall as a result of the lower atmospheric pressure at the summit.

those interested in Brownian motion was reached that the phenomenon could be best explained as a consequence of the constant bombardment of the small but still visible particles by the moving molecules of the suspending liquid, thus making molecular motion, or at least its effects, visible to the eye. It was a reasonable conclusion, and no plausible alternative explanation had been offered, but at a time when the kinetic-molecular theory was regarded by many as too hypothetical to be taken seriously it did not carry much weight. As an explanation it was purely qualitative. Could not some quantitative prediction about Brownian movement be made that would make a more convincing argument for that explanation? At the time, the outlook for making such a prediction was not promising. Although Brownian movement could be observed in dust motes floating in air, it was much easier to study in liquids, and the mathematical complexities of a suitable theory were much greater for liquids than for gases. But a solution was found.

First consider why the motion of water molecules should make pollen grains move. We must recognize that the particles we can see with a good microscope, even those at the limit of visibility, are much larger than the molecules of the liquid. Nineteenth-century scientists had already obtained some rough estimates of the sizes of ordinary molecules and it could be shown that Brown's pollen grains, 5×10^{-4} cm in diameter, have the same volume as 10 billion water molecules. As a grain floats in water, it is subject to an intense and rapid bombardment by the water molecules. A crude estimate suggests a rate of about 10^{22} collisions per second, distributed over the surface area of about 10^{-6} square centimeters of the grain. One might expect that in a small time interval the number, or mean kinetic energy, of the impacting molecules striking one side of the grain might differ slightly from the number or mean energy of those striking the other side. During this brief time interval the grain is subjected to a net force tending to move it in a direction opposite to the side receiving the more frequent or forceful impacts. A little while later, another fluctuation of the impacts drives it in another direction.

To sum up: our pollen grain is a very small, sensitive barometer. It detects, by its motion, the fluctuating pressure acting on its surface. Although it is much larger than a molecule, it is enough smaller than a mercury barometer to provide direct evidence for the fluctuations predicted by the kinetic-molecular theory. As the entropy of a liquid varies as its pressure varies, a fluctuating pressure implies a fluctuating entropy; an entropy that falls as well as rises, even though by very small amounts.

Quantitative Study of Brownian Movement: Failure

The movement of pollen grains in water provides qualitative evidence that entropy may sometimes decrease, which connects the second law to the theory of probability and deprives it of the absoluteness it once seemed to have. In view of the skepticism of many scientists, however, it was necessary to provide more convincing evidence. As the fluctuations could now be directly measured, should it not be possible to predict their size quantitatively from the kinetic-molecular theory and compare theory and observation? Clearly the size of the fluctuations depends on how many individual molecules are present in the sample of matter under observation. It would therefore be necessary to determine how many molecules are present in a given small quantity of matter—a cubic millimeter, or a milligram—and then to use the kinetic-molecular theory to predict how large the fluctuations in pressure or entropy should be.

In fact, the reverse of this procedure was followed. The number of molecules in a cubic millimeter was calculated in several different ways from the observed fluctuations (on the basis of the assumption that the kinetic-molecular theory was correct). Then the calculated numbers were compared both with each other and with the number of molecules measured independently by other methods that had just become available toward the end of the nineteenth century. These other methods did not depend on the particular combination of the theory of probability and Newton's laws that constituted the kinetic-molecular theory. The results agreed wonderfully, the skepticism disappeared, and a new era in science began.

The key assumption in testing whether Brownian movement was a consequence of molecular motion was that the Brownian particle—pollen grain or ground-up piece of Sphinx—is, from the viewpoint of the kinetic-molecular theory, just another molecule. It is an enormously large molecule to be sure, 10–100 billion times the mass of a water molecule, but except for its size it should act like any other molecule. It moves, it collides, it has kinetic and potential energy, and it should obey the equipartition theorem by having a kinetic energy of translation equal to $\frac{3}{2}kT$ (see Chapter 4), the same as all other molecules, large or small. Although the constant k was not known with any accuracy in the nineteenth century, a very crude estimate had been made. Since the mass of a Brownian particle could easily be determined, it was only necessary to measure the average speed of motion v of the particles, calculate $\frac{1}{2}mv^2$, and compare it with crude estimates of kT then available. It appeared a simple matter to measure the speeds with the microscope, by observing how far the particles travel in a given time interval.

The results of such experiments were in dramatic disagreement with the expected value: mean speeds were much less than predicted, and they varied erratically, depending inexplicably on the time interval used to measure them. The results were an embarrassment to supporters of the kinetic-molecular theory: they were not explained until 1905, in one of the three revolutionary papers Einstein published that year. He pointed out that Brownian particles in a liquid, unlike gas molecules, never have a period between collisions during which they have a chance to travel in a straight line at a definite and constant speed. The combination of their size and the large number of molecules in the liquid per unit volume ensure that they are under constant bombardment, forcing them to follow a tortuous path whose twists and turns are invisible to microscopic observation. They really do travel with an average speed given by the kinetic molecular theory, but the blurring of the fine details of their paths makes a determination of that speed with a microscope impossible. For all their rushing about they don't get very far very fast, and their apparent speeds are much lower than their true speeds.

Quantitative Study of Brownian Movement: Success

The problem of obtaining an accurate count of molecules from Brownian movement was solved by two men, Einstein and the French physical chemist Jean Perrin. Initially the two were unaware of each other's work, but when Perrin learned of Einstein's treatment of the problem he was able to put it to good use.

Perrin was a fervent believer in the reality of molecules and eager to convince the skeptics in the scientific community. He was aware of the failure of a number of scientists to obtain consistent measurements of Brownian particle speeds by direct observation. He was aware also that the various substances studied had densities that differed, more or less, from that of the suspending liquid. Most, especially the inorganic substances, were denser. By the laws of mechanics, the particles should eventually have settled to the bottom of the liquid, at a rate dependent on their excess density and the frictional resistance of the liquid. But in fact they didn't: whatever caused their erratic motion also kept them from sinking. In fact, if the particles had been placed at the bottom of an empty container, and water carefully poured in above them without stirring them, they would gradually wiggle their way up from the bottom and spread throughout the water. But although the Brownian particles "floated" in a liquid in spite of their greater density, their apparent disregard of the law of gravity was not total: another French scientist, J. Duclaux, had shown that the concentration of Brownian particles in a liquid was not uniform in a vertical direction; an equilibrium was reached in which there was a higher concentration at the bottom of a container than at the top.

Perrin realized that there was a parallel between Brownian particles and ordinary gas molecules subjected to the force of the earth's gravity. The pressure of the earth's atmosphere decreases with increasing elevation. The oxygen pressure at the summit of Mount Everest is one-third of its sea-level value, and jet airplanes, flying at an elevation of several miles, require cabin pressurization to provide enough oxygen for passengers to breathe. The fall-off of pressure with elevation occurs because of a balance between the kinetic energy of the molecules, which tends to spread them out in space, and gravity, which tends to pull them down to the surface of the earth. The rate of decrease of pressure with elevation is determined by the ratio of the gravitational potential energy of a molecule ($= mgh$; see Chapter 2) to its average kinetic energy of translation, $\frac{3}{2}kT$. Indeed, the proportions of the very light gases helium and hydrogen in the atmosphere are greater at greater elevations.

By measuring the way the pressure of each gas falls with elevation, we can find the value of the ratio mgh/kT for that gas. Knowing the ratio unfortunately does not tell us the numerical values of all the five quantities in it. Three of them, g, h, and T, are independently known, which means that the ratio of m/k can be found, but the problem is to find either m or k individually. Knowing either, we could determine the number of molecules in a kilogram or a liter of gas, but not if we know only the ratio.

Perrin realized that the variation of the concentration of suspended Brownian particles with height should obey the same relation, but now the mass of the Brownian particle could be measured directly, from its known size and density. The masses of Brownian particles, even with a correction for their buoyancies, are large enough to produce appreciable decreases of concentration in a few millimeters, unlike air molecules, whose masses are so small that the decreases occur only in miles (Figure 7.7). The experimenter does not have to climb a mountain carrying a barometer (as did Pascal's brother-in-law in the seventeenth

century), but can perform the measurement to determine the ratio of
m to *k* in the comfort and safety of a laboratory. Since the mass of a
Brownian particle could be measured independently, *k* could be determined with reasonable accuracy. Once *k* was determined, the unknown
mass of an oxygen molecule could be found from the decrease of
oxygen pressure with elevation, which gives us the ratio *m/k* for oxygen
molecules. In turn, knowing the mass of one molecule and the easily
measured mass of a cubic centimeter of oxygen gas, the number of
molecules in a cubic centimeter can be found.

Using this simple idea in the laboratory was not so simple, however.
It was first necessary to produce a suspension of Brownian particles all
of the same mass, or as nearly the same mass as possible. The reason is

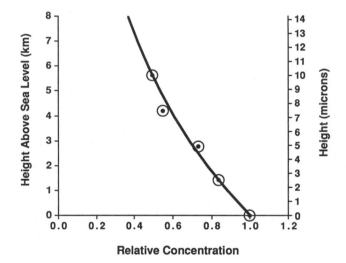

Figure 7.7 Gas Molecules and Dust Particles Obey the Same Laws
This figure compares the decreasing pressure of oxygen gas at higher elevations
above sea level, represented by the solid line, with Perrin's data on the decreasing concentration of gamboge particles suspended in water at greater heights
in the volume of water, represented by circles. It is clear that in spite of the
enormous difference in mass between a gamboge particle and an oxygen molecule, they both obey the same dependence of the fall of concentration at higher
elevations on the gravitational force acting on them and on the absolute temperature. The dependence on elevation is a negative exponential one. (*Note:*
The units of measure for altitude, kilometers, are shown on the left vertical scale;
those for the height of water relative to a reference level are indicated in
microns, millionths of a meter, on the right vertical scale.)

that in a mixture each size of particle will have its own variation of concentration with elevation, and if we observe how the concentrations of particles depend on height without distinguishing them by mass, the results will be in error. Perrin used a powdered pigment called *gamboge*, prepared by grinding a plant resin. By centrifuging a suspension of gamboge he could separate out particles of sufficiently uniform size. He was able to take advantage of the invention of the new "ultra-microscope," which enabled the viewer to keep smaller particles (as small as 5×10^{-7} cm in diameter) under observation, although it produced distorted images of them.

Chemists had for some time been using as a measure of a quantity of molecules something called the *mole*. One mole was defined originally as the number of hydrogen atoms in 1 gram of hydrogen (see Chapter 4). This number, now called *Avogadro's number*, and symbolized by N_A, was not yet known with any accuracy. Perrin's goal was to determine N_A and his first result by this method, published in 1908, was $N_A = 6.7 \times 10^{23}$. From this, k was found to be 1.4×10^{-23} joules/kelvin, and the mass m of an oxygen molecule 4.8×10^{-26} kilograms.

Einstein's Approach

In 1905 Einstein published three papers in the prestigious journal *Annalen der Physik*. One proposed the special theory of relativity; another explained the photoelectric effect by proposing a revolutionary new theory on the nature of light, which was crucial in the development of quantum mechanics; and the third pointed out that as a result of molecular motion, microscopically visible particles suspended in a liquid should engage in observable erratic movement and that this movement could provide a means of counting molecules. It is not known whether this was suggested to Einstein by an awareness of Brownian movement or whether he thought of it independently. His paper makes the statement, "It is possible that the movements to be discussed here are identical with the so-called 'Brownian molecular motion'; however, the information available to me regarding the latter is so lacking in precision that I can form no judgment in the matter."

At the time Einstein wrote this paper, the kinetic-molecular theory had only dealt in depth with gases at low pressure, and applications to gases under conditions where they no longer behaved "ideally"—lower temperatures, higher pressures—were just beginning. No kinetic-molecular theory of liquids existed: the mathematical complexities would have been too much even for Einstein, but he was able to out-flank these difficulties by a remarkable combination of macroscopic and microscopic reasoning. On the one hand, a pollen grain can be thought of as a molecule, large to be sure, but having the expected statistical properties of any other molecule, including $\varepsilon = \frac{3}{2}kT$; on the other hand, the grain is large enough to obey the same macroscopic laws governing the motion of a solid body in a liquid that submarines do.

Einstein's conclusion was that while it was impossible to determine experimentally the mean velocity and thus the kinetic energy of a Brownian particle, measurement of the mean distances the particles travel during a certain time interval permits N_A and hence k to be determined. Note that it is not the total distance along the tortuous path that is referred to—this would be impossible to follow—but only the straight-line distance between the starting point of each particle and its position after a known time interval (Figure 7.8).

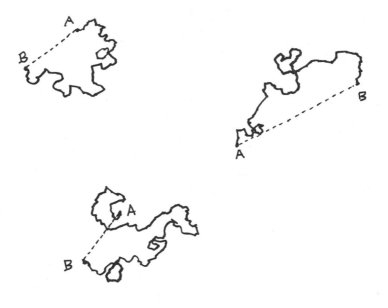

Figure 7.8   The Path of a Brownian Particle
Because it would be impossible to follow the exact path of a Brownian particle, the mean velocity of the particle cannot be determined experimentally. Einstein was able to show, however, that the total straight-line distance traveled by the particle (from *A* to *B*) in a given time interval could be used to determine Avogadro's number.

The Triumph of the Kinetic Theory

Perrin learned of Einstein's paper in the same year (1908) he published the results of his first determination of N_A. Recognizing that Einstein's approach provided another independent means to determine N_A, he put one of his graduate students to work testing it on granules of caoutchouc (natural rubber). N_A was found to be 6.4×10^{23}, in very good agreement with his earlier value. A further study using a new equation of Einstein's for rotatory Brownian movement (changes of orientation of the particle, rather than changes of position, analogous to the distinction made in an earlier chapter between rotation and translation of a single molecule) gave $N_A = 6.5 \times 10^{23}$. Other determinations by Perrin and his associates gave values as high as 7.15×10^{23}, but these are all gratifyingly close together by the standards of earlier attempts to estimate N_A by other means.

At the same time as Perrin was engaged in these researches, other physical scientists were obtaining estimates of N_A by completely different kinds of experiments. At least some of these methods involved direct measurements of events associated with a single atom or a single electron and did not rely on the kinetic-molecular theory but did, of course, assume there really were such things as atoms. For example, the disintegration of one radioactive atom produces an observable flash of light on a fluorescent screen, or a single count on a Geiger counter; it was a simple matter to compare the fraction of the mass of a sample of radium that decays in a second—a macroscopic measurement—to the number of flashes seen on the screen, which counts the number of radium atoms that decayed in that second. Ernest Rutherford and his associates used such observations to obtain values of N_A in the range of 6.0×10^{23} to 7.1×10^{23}. The electric charge of a single electron was measured by Robert Millikan: N_A was calculated from his result to be 6.8×10^{23}.

Perrin summarized his work and the work of others in his 1913 book *Les Atomes.* He gives the results of the measurements of N_A by 13 different methods; they range from 6.0 to 7.5×10^{23}. The modern value is 6.02205×10^{23}: it has been said we now know the number of molecules in a liter of gas more accurately than we know the population of a large city (which, according to arguments about the accuracy of the U.S. 1990 Census figures for large cities, are uncertain to several percent).

Perrin was, in addition to a skillful experimentalist, an excellent publicist for the kinetic-molecular theory. But one may wonder, looking

The Music of the Spheres

Science is unkind to myths. The followers of Pythagoras discovered a relation between musical harmonies and mathematics. Plato, seeing a connection between the order of mathematics and the order of the cosmos, reported the Pythagorean view that the planets are attached to transparent spheres of crystal that turn on adamantine spindles on which sirens sit, each singing a single tone, all together forming a harmony: the music of the spheres. It is a sound unheard by the human ear, perhaps because of our low, earthy character. Plato's music sounded, in the inward ears, at least, of philosophers, poets, and even scientists, through the Renaissance and after. Kepler, in his astronomical researches, tried to relate planetary orbits to musical harmonies.

No scientific evidence for music made by the planets in their orbits has been found, but a radio, tuned between stations and with its volume turned up, will emit a hissing sound. This is the noise of molecular motion, in this case of electrons moving and colliding in the metal wires of the radio, producing fluctuations in voltage that are amplified by the circuits of the radio.

It is the noise usually referred to in the phrase "signal-to-noise ratio." Its loudness is determined by the temperature of the radio and its circuits; it can be quieted by cooling the radio, but this is an unrealistic procedure for radios (though not for sensitive scientific instruments). It should be obvious that we cannot use a radio to pick up a station whose signal is weaker than the noise level: amplification amplifies both, and the signal we want is drowned out by the noise.

Harmony in music associated with tones whose frequencies are in the ratios of simple whole numbers: molecular noise contains all frequencies mixed, and no harmony. It is unpleasant to listen to, and nothing happens but the hissing. There is no progression or development. But it is the only music the universe provides.

at the rich discoveries of the late nineteenth and early twentieth century, whether publicity was really needed. By this time the flood of evidence for the kinetic-molecular theory and its conclusion that entropy fluctuations are real had swept away the doubts of even the most critical scientists—except for Ernst Mach, who published his last book in 1913, shortly before he died, affirming his skepticism. Perrin was awarded the Nobel Prize in physics in 1926; he died in 1942 in New York, in exile from Nazi-occupied France.

Reconciliation

We showed earlier that in a probabilistic treatment of molecular behavior the number W, the number of microstates consistent with a particular macroscopic description of the system, has properties similar to the thermodynamic entropy S. Just as a system has a uniform tendency to an increasing entropy, it has a (nearly) uniform tendency to go to states with larger W. When a greatest possible value of S is reached, no further change can occur; when a greatest possible value of W is reached, no further change is likely to occur.

There are discrepancies between S and W also. The major one, touched on in the preceding paragraph, is that while the entropy principle is absolute, the probability argument allows for fluctuations from the most probable state of affairs. However, we have shown that in the real world the fluctuations occur, and we recognize that the improbability of observing large fluctuations makes the entropy principle seem absolute if we measure only macroscopic properties, those that represent the average behavior of large numbers of molecules.

The second discrepancy we noted was between the total S and the total W. S_T of two independent systems of entropies S_1 and S_2 is the *sum*

$$S_T = S_1 + S_2$$

while the total W of two independent systems is the *product*

$$W_T = W_1 \times W_2$$

Finally, we said that W is a pure number, whose value does not depend on the arbitrary choice of units in which energy and temperature are measured—we call such a number "dimensionless"—but entropy S is measured by dividing a heat absorbed by the body, usually measured in joules, by its absolute temperature T, measured in kelvins. Which units for energy and temperature we use are matters of custom and convenience, and the numerical value of entropy depends on this choice. Entropy is thus "dimensional."

Reconciling the first discrepancy is easier. As noted in the appendix, the logarithms of a product is the *sum* of the logarithms of the terms multiplied. The logarithm of W is thus the *sum* of the logarithms of the Ws of the independent systems:

$$\log W_T = \log W_1 + \log W_2$$

This suggests that S should be proportional to the log of W rather than to W, a relationship expressed in the equation

$$S = C \log W$$

where C is a constant whose numerical value we have yet to determine; we must expect that it will depend on the choice of units for energy and temperature. But at least we can see from this equation why entropy increases when W and therefore $\log W$ increases and why when entropy has its largest possible value, W has its largest possible value also.

The Ideal Gas Again

One system for which we can determine the number of microstates W by actual calculation is the ideal gas. W, the number of detailed arrangements of the molecules in space and by energy, depends on the number of molecules N, on the total energy E the gas is given, and on how large a volume V it happens to occupy. Now we can also determine in the laboratory how the properties of an ideal gas depend on the quantity of gas, its total energy, and its volume. We can therefore confirm, first, that the entropy of an ideal gas is indeed proportional to the logarithm of W and, what is more, determine the constant C, which turns out to be 2.303 k. Thus we have the result for the ideal gas:

$$S = C \log W_T = 2.303k \log W_T$$

Note that the numerical magnitude of the constant k depends on the units used to measure energy and temperature. The usual choice is joules for energy and kelvins for temperature, so the units of k are joules per kelvin, the same as for entropy. The numerical value of k was found by Perrin to be 1.4×10^{-23} joules per kelvin.

For substances other than the ideal gas, it is not so easy to calculate W, and therefore C. We can, instead, appeal to the generality of the second law: nothing can happen that leads to a net decrease of entropy.

This principle applies whether or not we can calculate W from a molecular theory. A careful logical analysis of this statement shows that the constant C must be the same in all systems, ideal or not. Hence we can conclude that what we found for ideal gases must apply to everything else as well, so we have for *all* substances and systems:

$$S = 2.303 \ k \log W$$

The number 2.303 appears because we have used logarithms to base 10; the use of logarithms to other bases will produce different numbers. The usual choice of a base for logarithms made in advanced mathematics and science (the number e, discussed in the mathematical appendix) gives the constant the value 1. The resulting formula is engraved on Boltzmann's tombstone in Vienna:

$$S = k \log W$$

Again: What Is Temperature?

When the concept of temperature came into being with the invention of the thermometer, its most significant property, first discovered by Joseph Black (see Chapter 3), was that it determines which way heat flows when two bodies are in contact. In looking for a molecular interpretation of temperature, we considered, and then were forced to reject, the equipartition theorem's conclusion that temperature is proportional to the average molecular kinetic energy of translation.

A completely general meaning can be inferred if we combine two statements about the flow of heat:

1. Temperature is that property of matter that determines which way heat flows.
2. The flow of heat, like any other spontaneous process in nature, must take place in such a way as to cause a net increase in entropy.

Let us picture two solid bodies initially separate and then brought into close contact, so that heat can flow from one to the other. Initially each body had its own definite energy and, assuming each was individually in equilibrium, the entropy of each body had its largest possible value consistent with its given energy. Now if when the bodies are placed in contact an exchange of energy takes place, one body necessarily gaining what the other loses, each body will undergo changes in entropy also. Since it is an almost invariable property of matter that the entropy

Low Entropy and High

At this point a schematic comparison of various low and high entropy states of different systems may be helpful. The entropy-increasing effects of expansion *(A)*, evaporation of a solid *(B)*, mixing of two different substances *(C)*, molecules moving in various directions *(D)* and with various speeds *(E)* are shown. One important example is not illustrated: a uniform temperature throughout a body represents a higher entropy

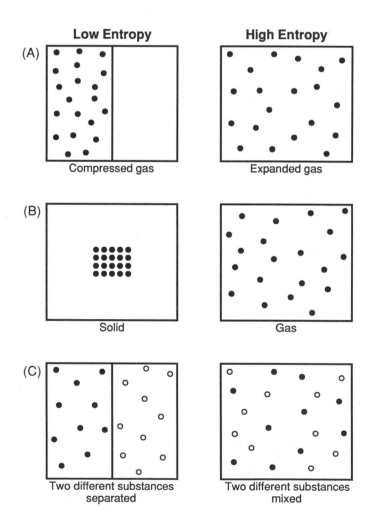

Low Entropy **High Entropy**

(A) Compressed gas / Expanded gas

(B) Solid / Gas

(C) Two different substances separated / Two different substances mixed state than when one part is hot and another is cold. This would be difficult to represent because in both hot and cold bodies molecules are moving at many different speeds (the Maxwell-Boltzmann distribution) and at least some molecules in a cold body are moving faster than the average speed of the molecules in a hotter body.

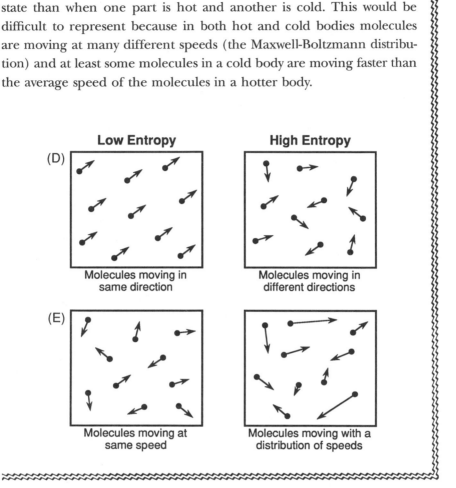

Low Entropy

(D)

Molecules moving in same direction

High Entropy

Molecules moving in different directions

(E)

Molecules moving at same speed

Molecules moving with a distribution of speeds

increases when the energy increases (the exceptions to this rule form too special a case to merit consideration here), it follows that the body that loses energy must decrease in entropy also and that the other, gaining energy, increases in entropy. The actual *change* in entropy per unit of added energy is a *property* of the particular substances involved. Let us call this property "theta." Its definition resembles that of a more familiar property, the thermal expansion, the change of volume per degree rise in temperature. Theta is found invariably to get smaller and smaller as more and more energy is added to the body, as though some sort of law of diminishing returns were operating.

Without using the concept of temperature at all, we may derive from statement 2 the conclusion that heat must flow from that body whose decrease in entropy caused by the loss of a unit quantity of energy is *less than* the increase in entropy caused in the other body by a gain of the same unit quantity of energy. In other words, the property "theta" and the temperature of a body are intimately related. We note a body with a large volume of theta will tend to gain energy from a body with a small theta, since that is what will give a net entropy increase. This implies that the body with the *larger* theta has the *smaller* temperature. A detailed analysis shows that theta is inversely proportional to temperature:

$$\text{theta} \propto \frac{1}{T}$$

This makes temperature a less intuitively graspable concept than if it were always proportional to the average translational energy of molecules. But it is the correct thermodynamic definition of temperature, applicable even when Newton's laws, and with them the equipartition theorem, fail.

8
Why Does Entropy Always Increase?

To explain the molecular meaning of entropy, and why it almost invariably increases, we have used the process of shuffling a deck of cards as an analogy to the rearrangements of a collection of moving molecules. It has been argued that the process of shuffling is a random one, but the motions of molecules obey laws of physics that make their behavior predictable: the laws are said to be "deterministic." The terms *random* and *deterministic* may seem like opposites, but in this chapter we will look more closely at the meaning of each and discover that they need not be as incompatible as they sound.

The Arrow of Time

The apparent contradiction comes out most sharply when we consider the almost invariable tendency of entropy to increase: the irreversibility of all real macroscopic processes. The principle of ever-increasing entropy gives a *direction* to time. Time flows from the less probable to the more probable state of affairs, as exemplified by the common observation of the flow of heat from hot bodies to cold, say, or the burning of wood to form ashes, or the collapse of stars as they radiate their energies away. As noted in Chapter 1, there is a seeming contradiction between these examples and an important feature of Newton's laws, their *time-reversibility*. In brief, if we make a motion picture of a system, no matter how complex, behaving in accord with Newton's laws, and then run the film backwards, the behavior we see will also be in accord with the laws. For example, Newton's laws explain the regularities of planetary motion discovered by Kepler. If a film of the motion of the planets were run backward, the motion would be in the opposite direction from what we see now, but Kepler's regularities, and therefore Newton's laws, would be obeyed as before.

Now, we have concluded that the second law is a consequence of the average behavior of large collections of molecules. If the behavior of molecules is, according to Newton's laws, the same whether time flows forward or backward, why should averaging over large collections of them create a direction to time? This question was first raised by Josef Loschmidt, an Austrian physicist and an early contributor to the kinetic theory. Loschmidt made one of the first estimates of the size of a molecule from data on diffusion in gases, but he is better known today for pointing out the apparent contradiction between time-reversibility and the invariable increase of entropy.

We know something today that Loschmidt didn't: molecules do not obey Newton's laws but those of quantum mechanics. Quantum mechanics will be discussed in Chapter 15, but two things about it need to be said here: it has satisfied every experimental test so far performed, and, like Newton's laws, it is time-reversible. It follows therefore that the problem of irreversibility is present in quantum mechanics no less than in Newton's laws. As Newton's laws are easier to deal with both intuitively and mathematically, and as they are often a good-enough approximation for molecular motion, we will not discuss quantum mechanics further in this chapter. We will instead refer in what follows to the more general "known laws of molecular motion."

Loschmidt's point can be visualized in the following way. Suppose we project a motion picture of burning logs of wood in the correct (forward) direction, but as we do so we gradually increase the magnification of the film. Eventually the logs and the flames dissolve into their constituent molecules, which collide, react chemically, move about, and collide once more, all the time obeying the known laws of molecular motion. Now we reverse the direction of the film. Each molecule instantly reverses its direction of travel. Disorderly motion continues: the molecules still move, collide, and react, and they still obey the same laws of motion just as well as before. Nothing seems wrong, until we reduce the magnification and realize we are watching smoke and flames descending from the chimney to reconstitute the logs from the ashes.

In this example we are following the progress of a system from a low-entropy initial state (wood, oxygen) to a high-entropy final state (combustion gases, ashes, the warmed surroundings). Reversing the film halfway through the process instantaneously reverses the direction of motion of each molecule without changing either its speed or its distances from its neighbors: this implies that the total energy is unaffected by the reversal. It follows, from the hypothesis that equal energy implies equal probability (see Chapter 4), that the state just after the moment of reversal is exactly as probable as the state just before. But the state just before, if allowed to continue, leads to the burning of the rest of the wood, while the equally probable reversed state continues until all the ashes and gases are reconstituted as wood. If the two states are equally probable, why do we see only one happening, never the other?

We might try to hold onto our sanity by arguing that even though the laws of motion are not violated by the reversal of the film, there must be something impossible or at least highly improbable about it. But how do we justify this argument?

The paradox is unresolved today, although attempts to resolve it are still being made. No one doubts that the almost unwavering tendency of entropy to increase somehow follows from what large numbers of molecules do, but there is no scientific consensus yet on the possibility of a chain of reasoning that goes from deterministic, time-reversible molecular motion to the time-irreversible behavior of the macroscopic world.

Card Shuffling Reviewed

On the other hand, there is no paradox in the "irreversibility" of a random process. The "card-shuffling" procedure discussed in Chapter 7 was based on the selection of pseudo-random numbers because a computer can't produce really random ones, but let us imagine that we could somehow come by such numbers. We start with a deck of 52 cards in an ordered state: all red cards in the top half. We select two numbers between 1 and 26 randomly, the first to select a card in the top half, the second to select a card in the bottom half. Then we exchange the cards. After 50 or more exchanges we expect to find about half the red cards in the top half and about half in the bottom. Once the deck has been well shuffled, we expect it to remain disordered on repeated shuffling, most of the time. Occasionally the perfectly ordered state of the 52-card deck will recur, but only very rarely: on the average once in every 10^{15} shuffles.

If we increase the number of cards to 10^{20}, we expect the tendency to approach an "equilibrium" state of nearly equal numbers of red cards in each half of the deck to be all the more overwhelming and excursions to an ordered state all the rarer. Yet in an enormously long sequence of card exchanges, such excursions will inevitably occur.

These features of the behavior of a random process—a tendency to go to an equilibrium state, and once there to tend to stay there, departing only rarely—are just what we mean by the term *irreversibility*. Perhaps that term is too strong. Shuffled cards and collections of molecules *do* revert to states they have passed through at previous times, but they do so only rarely, and more rarely the larger the system. Given enough time, however, they do it again and again.

No one has claimed that there is a paradox in the results of card shuffling: only the molecules, because of the time-reversibility of their equations of motion, seem to behave paradoxically.

Proposed Solutions

A number of solutions to the paradox have been proposed by different scientists. One group suggests that the fundamental laws of nature are not all time-reversible. The laws of molecular motion are only an approximation to some as yet undiscovered combination of quantum mechanics, the general theory of relativity (see Chapter 15), and a theory of the fundamental particles. While both quantum mechanics and general relativity in their present form are, like Newton's laws, time-reversible, the complete theory need not be. There is already evidence for a breakdown of time-reversibility in the decay of a subatomic particle called the K^0 meson. A more radical and not yet accepted view is that of Roger Penrose, a British theoretical physicist, who suggests that the expansion of the universe will require some revision of the general theory of relativity that will introduce time-irreversibility. Even a small degree of irreversibility in the laws of nature might be enough to tip molecular systems in that direction.

A second group points out that in the experiments we do in the laboratory to confirm thermodynamics our systems are never really isolated from outside influences. A molecule of oxygen twenty feet from our apparatus has some effect, though very weak, on the motion of molecules inside. Calculations show that even extremely weak influences would give rise in a surprisingly short time to a noticeable cumulative effect.

Another group of scientists argue that even a system obeying the time-reversible deterministic laws of Newton will still show macroscopic irreversibility. Their view is intuitively plausible, which is hardly a strong reason for believing it. Consider the by now familiar example of a container divided into two parts by a partition, with molecules on one side and a vacuum on the other. If a hole is punched in the partition, the molecules flow until the pressures are equalized, and then, most of the time, nothing discernible happens. This is what experience in the real world leads us to expect, though we are not yet sure if it is a result of some time-irreversibility in the fundamental laws of nature, the expansion of the universe, tiny perturbations coming from outside the container, or none of the above. Would it happen also in an idealized world in which the molecules obey strictly the time-reversible equations of Newton? Billiard balls exemplify Newton's laws, and errors due to any of the factors above are too small for our senses to detect; the effect of friction on their motion is much more important. Our intuitive expectation for the molecules in the container, formed by our experience with real billiard balls, suggests strongly that idealized Newtonian molecules should also spread through a container until they fill it uniformly, just as real ones do.

Newton's Deterministic Laws

Scientists trying to go beyond the plausibility argument and really prove that Newton's laws alone lead to irreversibility have followed a number of approaches, most of which are too complicated to discuss. Ludwig Boltzmann was the first to try, and he spent the major part of his life on the problem. He thought he had succeeded a number of times, but the critical comments of colleagues like Joseph Loschmidt and others forced him to reconsider and try again. In one of his "proofs" he made an assumption about how molecular collisions occur that is not strictly correct, one that had the effect of introducing irreversibility where it would not have otherwise appeared. He had "begged the question" in the original sense of that term: introduced as a starting hypothesis that which the proof was supposed to accomplish. He never completely succeeded: the best he could do was argue that the behavior of systems obeying Newton's laws should be something like a random card-shuffling process, and once considerations of probability were thereby introduced, he was able to show that irreversibility follows.

Let us review briefly how we use Newton's equations to determine the paths of molecules, a subject we discussed in Chapter 4. For two bodies—two molecules or two stars—the solution is simple, and it can be made as exact as we wish or need. For more than two bodies we switch to computers, and the more molecules whose trajectories we want to calculate and the more accurately we want to follow the trajectories, the more computer time it takes. Since we cannot wait an infinitely long time for an answer, we must compromise at some level of accuracy: we want to predict the positions and speeds of N molecules, but only to a limited number of decimal places and only for so many microseconds in the future. How seriously the rounding-off errors affect the results needs to be discussed, but we will come to that later.

Modern computers can follow the trajectories of thousands of molecules colliding with each other in a container for reasonable periods of time. Figure 4.3 shows schematically what the trajectory of one molecule in the collection might look like. If there are 10,000 molecules in the container, there are 10,000 trajectories to describe. We will use *grand trajectory* as a convenient term for the 10,000 individual trajectories.

So we can follow 10,000 molecules, but not 10^{23}, and we do so with limited accuracy and for a limited time. It doesn't sound promising with all these caveats, but there is one thing that helps. For thermodynamic purposes we are mainly interested in macroscopic properties, such as pressure, which is the cumulative effect of all molecular collisions with the walls of the container averaged over the short time interval it takes to measure pressure. The values calculated might be expected to be relatively insensitive to the details of the grand trajectory, and rounding-off errors should tend to cancel.

Pressures and temperatures calculated by computer simulation of grand trajectories are found to be in good agreement both with laboratory measurements on real molecules and with the predictions of the kinetic-molecular theory. Most important, when we start off with an assembly of molecules in a nonequilibrium, low-entropy state—all the molecules with the same speed initially, or all on one side of the container—the molecules redistribute their positions or their speeds to reach the greatest possible entropy, and once they do, they hardly ever depart from it by much. They spread uniformly through the container, and their speeds come to follow the Maxwell-Boltzmann distribution (Chapter 4). Basically, they behave in this respect as the card-shuffling analogy would suggest.

This is reassuring, but it doesn't carry the conviction of a mathematical proof. The dichotomy between *deterministic* and *random* remains a challenge, and we need to examine the terms more carefully.

Randomness

The term *random* has been applied here to such processes as tossing a coin or shuffling a deck of cards: it could as well be exemplified by rolling dice or spinning a roulette wheel. What it implies is a considerable but not total unpredictability. Although we never know what the next result will be, we can be sure that in the long run heads come up as often as tails, the six on a die as often as the two. From this very limited knowledge, a significant degree of quantitative prediction can be obtained, if we confine our predictions to the average behavior in long series of trials rather than to specific individual outcomes.

Whether any particular series of real events to which we apply the term *random* really *is* random is a difficult question. There are, after all, loaded dice, unbalanced roulette wheels, and card sharps. We have not actually shuffled decks of cards by hand to illustrate the concept of entropy—rather, our deck was "shuffled" with the aid of a series of "random numbers" between 1 and 26. Where did we get them? We could have spun a roulette wheel, discarding outcomes from 27 to 36 (as well as 0 and 00). But we have already expressed doubts about the trustworthiness of roulette wheels, and besides they are not really random. The spinning ball moves in accord with Newton's laws, though as a practical matter we never have enough information on the initial position and motion of the ball to attempt a calculation of its path and stopping place.

For our example of card shuffling we used a programmable calculator with a "random number" key. Each time the key is pressed a decimal number between 0 and 1 appears on the display. Repeated use of the key gives a series of numbers in no discernible order: there are as many, in the long run, between 0 and 0.5 as between 0.5 and 1, and there is no tendency for the smaller numbers to cluster together. How does the calculator do it? Well if by "do it" we mean "produce a *random* sequence," it doesn't: the numbers are not truly random. The operating manual that comes with the calculator refers to them as "pseudorandom." How are these numbers produced, and why are they enough like truly random numbers to be useful?

Four whole numbers are stored in the subroutine, a "seed" we symbolize by x and three others, a, b, and c. In one subroutine we have used, the numbers were $x = 21$, $a = 24{,}298$, $b = 99{,}991$, and $c = 199{,}017$. The first time the "random number" key is pressed, the program calculates the (very large) number $ax + b$. Then it divides $ax +$

b by *c*, producing a whole number plus a remainder r_1, also a whole number. The remainder, by definition, must be smaller than *c*. The whole number r_1 divided by *c* is thus a number between 0 and 1; it is the first "random number" of the series we are generating. The subroutine then saves the remainder r_1 to use, at the next pressing of the key, as a new seed *x*, to calculate a new *ax* + *b*, which on division by *c* gives a new remainder r_2 and the second "random number," r_2/c.

For the sake of illustration Table 8.1 gives the results of a subroutine that uses numbers too small to be useful. We choose *x* = 3, *a* = 2, *b* = 7, and *c* = 11. The numbers this subroutine generates are clearly not random at all. Once one knows how the subroutine works and what the four numbers *x*, *a*, *b*, and *c* are, the whole sequence is easily determined. One would love to find a roulette wheel in Las Vegas as predictable. Yet if one is not in on the secret, it would be hard, for a while anyway, to realize that anything is wrong. We can illustrate the random

Table 8.1

Subroutine for Generating Pseudorandom Numbers with
a = 2, *b* = 7, *c* = 11 and the Initial *x* = 3

		Dividing by *c* yields:		Pseudorandom
x	*ax* + *b*	Whole number	Remainder (*r*)	number (*r/c*)
3	13	1	2	0.182
2	11	1	0	0.000
0	7	0	7	0.636
7	21	1	10	0.909
10	27	2	5	0.454
5	17	1	6	0.546
6	19	1	8	0.727
8	23	2	1	0.091
1	9	0	9	0.818
9	25	2	3	0.273
3	13	1	2	0.182
2	11	1	0	0.000
0	7	0	7	0.636

appearance of the sequence by converting the numbers to coin tosses: a tail for each number less than 0.5, a head otherwise. The first 10 tosses are T, T, H, H, T, H, H, T, H, T: 5 heads and 5 tails, in no obvious order.

Of course, the absence of real randomness eventually exposes itself. On the eleventh calculation for our example, the same remainder, $r = 2$, is obtained as on the first calculation, and once this happens the entire series repeats itself, and it will do so indefinitely. With $c = 199,017$, the series need not repeat itself until 199,017 numbers have been produced, which is long enough for many purposes.

The irony is that this program for calculating random numbers is exactly what we mean by *deterministic*. Once the program and the numbers x, a, b, and c are known, the outcome of every single step of the calculation is determined. Start the program over and it will give exactly the same sequence of numbers again. Nothing further from "random" could be imagined, yet we use the numbers produced as though they were random.

Aren't there ways to test if a sequence is random? For example, the average of a long sequence of truly random numbers between 0 and 1 should be 0.5. There should be as many below 0.5 as above, and the numbers should be uniformly distributed in their ranges. Certainly if the average were not 0.5 we would infer that the series is not random. But our pseudorandom series would give the correct average. A more stringent test is to count the numbers in the ranges 0 to 0.1, 0.1 to 0.2, 0.2 to 0.3, and so on. Our pseudorandom series would pass this test also. Of course, we have already pointed out a test our pseudorandom series will fail: after a very large number of steps, it begins to repeat itself. But there are other pseudorandom sequences that never repeat themselves.

The number π, the ratio of the circumference of a circle to its diameter, is a number with an unending sequence of digits after the decimal point, and the sequence of these digits does not ever repeat. It has been calculated by modern computers to billions of decimal places. The first thirty digits are as follows:

$$\pi = 3.141\ 592\ 653\ 589\ 793\ 238\ 462\ 643\ 383\ 279\ldots$$

If one were to break up the digits after the decimal point into groups of three (as we did), they would form a sequence of three-digit pseudorandom numbers as useful for shuffling decks of cards as the sequence produced by our hand calculator. A mathematician unaware of the source of these numbers would not likely be able to find anything nonrandom about them.

The sad truth is that there is no single test or finite number of tests that can assure that a given sequence of numbers is truly random. It is a property that can be disproved, if you can figure out the rule by which the sequence has been produced, but not proved. A sequence would have to pass an infinite number of different tests, and we would never know the outcome of those we haven't yet tried. All we can ask is: are the numbers in a particular sequence enough like real random numbers to do the job we wanted the random numbers for?

Errors and Their Consequences

We have stated that Newton's laws are deterministic. They are expressed in mathematical equations that can be solved to any desired degree of numerical accuracy. Solving a problem to a high degree of numerical accuracy has its costs in time and money, so we limit the accuracy to what is needed for practical purposes. We often simplify the problem we wish to solve by leaving out factors whose effects we have reason to think are small: if a heavy body falls a short distance, both the earth's gravitational force and air resistance influence its motion. The latter force, however, has only a small effect, and for many purposes we can afford to neglect it. There is no point in carrying out the calculations to a degree of accuracy greater than the contribution of small errors arising from factors we have chosen to neglect.

One of the first problems Newton applied his laws to was the calculation of the orbits of the planets around the sun. To keep the problem tractable, given the mathematical tools then available, he chose to ignore the gravitational forces between planets, only taking into account the much stronger force between each planet and the sun. The astronomer Kepler had found three quantitative regularities in the planetary motion but could offer no reason for them. Newton showed that they followed directly from his equations, once the small forces between planets were omitted. More accurate calculations, taking these forces into account, predict slight deviations from the simple pattern, but more accurate observations than Kepler could make at the time are required to detect them. Such calculations are not difficult with modern computers, but for several hundred years after Newton they were carried out tediously with paper and pencil and human ingenuity. That is how Neptune and Uranus were discovered.

Newton's laws gave us these triumphs because the physical and mathematical approximations worked: they represented small errors in the basic equations or in the calculations, and such small errors should and did lead to only small consequences in the predicted outcomes. We don't ask of science that it give us perfect answers, only good enough answers. Indeed, if small errors have large consequences, we could hardly do science at all: no prediction of anything, and thus no experimental test of any theory, would be possible. The real world is too complicated to deal with in full detail, and we are able to get by because of what we can leave out.

The Flow of Fluids

There are some problems Newton's laws apply to, however, which are so complicated—they need so many variables to describe them and so many equations to find the values of the variables—that little progress toward a comprehensive treatment of them had been made prior to the invention of the computer. An example is the flow of fluids: water rushing in a torrent over a rocky streambed or, even more difficult, the motion of the earth's atmosphere and the weather it brings. Even with powerful computers the complexity of these problems requires approximations to simplify the mathematics.

A greatly simplified model of the atmosphere was developed by Edward Lorenz, a meteorologist at MIT in the 1960s, using only three variables instead of the enormous number the real atmosphere would require. The equations of the model could be solved on a simple computer Lorenz built himself, and the way the variables changed with time printed out on a graph. On one occasion Lorenz decided to repeat a calculation the computer had already made, starting in the middle, so he read off from the graph the numerical values the computer had produced at one point in the previous calculation, punched them in again as a starting point, and restarted the calculation. The graph began at first to retrace the results of the first calculation, but after a short time it began to diverge erratically from the earlier one. At first Lorenz assumed his computer was malfunctioning, but after ruling this out he realized that the numbers read off the graph had been rounded off to three decimal places, while in the computer's memory they were stored to six decimal places (for example, a number stored in the memory as 0.691573 would have been rounded off to 0.692 for graphing purposes). This difference in the fourth decimal place eventually made an extraordinary difference in the calculated values.

It was Lorenz's good fortune to have identified the cause correctly and his unusual physical insight to have recognized its profound significance. To predict the weather, we use as input to the equations of motion of the atmosphere such data as the initial wind pattern, the temporal and spatial variations in energy received from the sun, the moisture content of the atmosphere, and much more. None of these variables is ever known to three decimal places, much less to six. If differences in the fourth or seventh decimal place in the inputs produce extraordinarily different outputs, how can we ever predict the weather? If a butterfly flapping its wings in Brazil today can cause a tornado in Texas a month from now, how can we predict tornadoes?

In other words, small errors sometimes do have large consequences.

Chaos

The equations that arise in science can be classified as either *linear* or *nonlinear:* for our purposes it is not necessary to explain the technical meaning of the terms, or even how to tell whether a particular equation is one or the other. Linear equations are much easier to solve and often do not require the use of a computer. Newton's laws sometimes lead to linear equations, but more often to nonlinear ones. The earth-sun problem is nonlinear, but the equation happens to be of an easily solved form. Because of the difficulty of solving most nonlinear equations without computers, most of the applications of Newton's laws for the first few hundred years of their use were to the linear and to the simplest nonlinear problems, and the intuitions of physicists, their concepts of how nature behaved, were shaped by the solutions to such problems and the excellent agreement found with experiment.

When modern calculators and digital computers came along, it became possible to handle many more nonlinear problems and under a wider range of conditions, one of which was Lorenz's model of the atmosphere. What often emerged, as Lorenz was the first to observe, was a startling kind of behavior, with features dramatically different from those seen in linear problems and for which the scientific community was not psychologically prepared. The behavior is called *chaotic,* and its study is a new field of science called *deterministic chaos* or, more formally, *nonlinear dynamics.*

Scientists had long been used to the behavior of bodies moving on smoothly curved paths, without sudden, erratic changes in speed or direction: the elliptical orbits of the planets, the regular swings of a pendulum, the parabolic arc of a projectile fired from the earth's surface. One doesn't need Newton's laws to make a fair guess as to what will happen next from what has already been seen. Astronomers, beginning in Babylonian times, had managed for thousands of years, without Newton's help, to predict the motions of the planets from the regularities of the recent past.

In chaotic behavior, however, one hardly knows what will happen next. The variable we study jumps erratically from one value to another, resembling in some respects a sequence of random numbers. Also, the behavior shows an extraordinary sensitivity to the numerical precision of the calculation and to the original numerical value of the variables at the start of the calculation. We no longer dare ignore rounding-off errors.

Not all nonlinear equations show chaotic behavior, and those that do do not show it under all conditions. Further, chaotic behavior, though occurring often for bodies moving according to Newton's laws, is not limited to such bodies. Phenomena that are suggestive of chaos include heart arrhythmias, the spread of some epidemics, fluctuations in the stock market, and the growth and decline of animal populations. We will choose as an illustration of chaotic behavior a nonmechanical problem from the field of ecology: a nonlinear equation used to describe annual changes in the population of a species in a particular ecological niche—rabbits on a small island, let us say.

The number of individual animals of a particular species present in any one year is a reflection of many factors, such as the population in the preceding year, the rate of mating and number of births, the number of predators, the food supply, the weather, and so on. Let us disregard—or assume constancy for—most of the factors and consider only the interaction among three: (1) the number of individuals in the preceding year; (2) the birth rate; and (3) the negative effect of overpopulation and competition for the food supply.

We would expect that either too few or too many individuals are deleterious for the survival of the community, and we might guess that there is an optimal figure for the population that would guarantee survival, so that the population would remain stable from one year to the next. This so far purely verbal and qualitative reasoning can be expressed in an equation, which predicts the population next year from its value this year and which is surprisingly simple but nonlinear. It contains a parameter whose numerical value depends on the balance between the birth rate and the negative effect of overpopulation. Different kinds of behavior are seen for different values of this parameter, as illustrated in Figures 8.1–8.2.

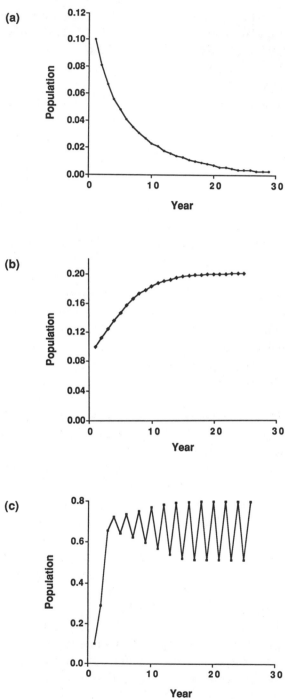

If the birth rate is too low, for example, the population diminishes from year to year and the species soon dies off (Figure 8.1a). When the birth rate is large enough so that the species does not disappear, but not so large as to quickly saturate the available resources, the population, as we guessed, gradually approaches a stable value, which then remains the same from one year to the next (Figure 8.1b). (One should not forget this is a greatly simplified model assuming no influences of weather, fluctuations in the number of predators, epidemics, and so on.) With an even higher birth rate and a greater tendency to strain the food supply (a kind of negative feedback), something happens that we might not have expected. The population does not settle down to a stable value but alternates between high and low values from one year to the next. Every second year the population size repeats itself, but there is no longer a single stable population size (Figure 8.1c).

So far these different types of behavior are fairly regular. Different patterns result from different values of the parameter that measures the balance between birth rate and overpopulation, but the patterns can be understood as consequences of the simplified model. On examining the figures, one can make a pretty good guess as to what will happen next.

But when the birth rate is so high that overpopulation is likely, a new and strange pattern begins (Figure 8.2). Wide swings in the predicted population occur from one year to the next, but there is no tendency for population to alternate in a regular pattern (as there is in Figure 8.1c). The successive yearly values begin to resemble in some respects a sequence of random numbers. What is worse, the system shows an unexpected and extraordinary sensitivity both to the numerical value of the birth rate–overpopulation parameter and to the size assumed for the initial population. When the parameter is small, so that the system tends to settle down to a stable size, a change of a percent or so in the initial population changes the population sizes in successive years by a

Figure 8.1 Variations in the Rabbit Population
The number of rabbits (in thousands) on an island changes from year to year as a result of the reproductive rate and the deleterious effects of overpopulation (we ignore here the effects of weather, predators, and so on). (a) A very low birth rate will not sustain the population. (b) A balance between reproduction and attrition leads to a stable population size. (c) If overpopulation has a more deleterious effect than it has in (b), the population oscillates from year to year between a higher and a lower value.

comparable amount. But when the parameter is large enough for these wild and random-seeming swings to take place, a small change in the initial population produces as time goes on a course utterly unlike the original course; the erratic ups and downs no longer show any similarity (compare the dotted and solid lines in Figure 8.2).

This example of chaotic behavior may make one pessimistic about our ability ever to achieve a scientific understanding of complex systems. How can we ever identify, much less measure, all the variables that may affect a system as large as the earth's atmosphere? Since chaos was discovered, it has become clear that not all the goals of science as we have understood them since the Newtonian revolution in the seventeenth century can be achieved. Scientists have taken for granted that if they can express a problem in terms of an equation, they can solve the problem. Some equations are difficult to solve: numerical calculations even on computers cost money. Investigators may choose not to

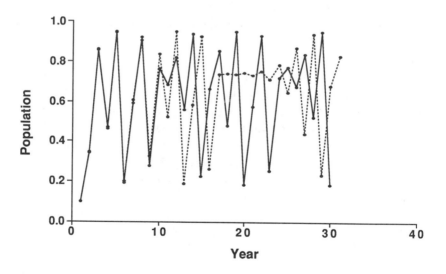

Figure 8.2 Chaotic Variations in the Rabbit Growth
If overpopulation is assumed to be even more deleterious than it is shown to be in Figure 8.1(c), the population varies chaotically. It increases and decreases in what appears to be a random way, and there is no obvious pattern that enables one to predict what it will do next. Furthermore, a very slight change in the conditions assumed—one rabbit more in an initial population of 100, for example—produces a course that eventually becomes utterly unlike the previous course. The dotted graph shows the results for just such a trivial change.

solve an equation in particular cases, but the feeling that they could if they wanted to has given the scientific endeavor its conviction and its brio. The existence of chaos shows that there are important scientific problems governed by easily written and often simple-seeming equations that can be "solved" in the strict mathematical sense but whose solutions are almost useless for predicting what will happen next.

This has not daunted those scientists who, confronted with chaos, have turned their attention from precise prediction of the next event to studying the structure of chaos: when it appears, what it looks like, and what patterns can be found in its seeming randomness. The effort, like the rest of science, is both intellectually challenging and useful.

Probability and Molecules

To summarize, Newton's equations for the motions of even a few molecules are nonlinear, and the motion is often chaotic. The trajectories predicted are extremely sensitive to the initial conditions; a tiny difference in the speed of one molecule at the start leads quickly to a completely different trajectory, not only for that molecule but for all others that collide with it, and for all others that collide with those that have collided with it, and so on. The behavior of molecules therefore takes on a pseudorandom character. This in turn implies that there is much less of a contradiction in assuming randomness and using the theory of probability for molecular motion than was once thought. The *formal* justification is still lacking. Chaotic behavior is not perfectly random: there are still patterns and regularities under the turbulent appearances. But many of those who have studied the problem of irreversibility feel that the recognition of chaos brings us finally close to a solution.

As we indicated earlier, once we are convinced that the behavior of collections of molecules can be described by the laws of probability, we can then conclude that the behavior of a deck of cards on shuffling is not just a metaphor for the collections of molecules but instead an example of the same kind of thing. And just as we expect an ordered deck of cards to disorder on shuffling, so also should we expect heat to flow from hot bodies to cold ones, gases to rush into a vacuum, sugar to dissolve in hot coffee, and wood to burn to ashes.

Chaos among the Rabbits

The equation describing the changes in population levels of rabbits inhabiting a small island is called the *logistic* equation. It predicts the population in the next year from its value this year and from a parameter whose value is determined by a balance between the advantage of a high birth rate and the disadvantage of overpopulation. The parameter is symbolized by A, and the larger its numerical value, the greater the harmful effects of overpopulation. The population is symbolized by x, and a subscript is attached to x to denote which year it is, so x_1 is the population in the first year and x_n in a generic year. The population in the year following year n is x_{n+1}. For simplicity of calculation x is assumed to lie between 0 and 1; in our example, multiplying x by, say, 1,000 gives an actual number of rabbits.

The equation is as follows:

$$x_{n+1} = A[x_n - (x_n)^2]$$

We begin by assuming an initial population (that is, an initial value for x_1) and calculate the population in year 2 from the equation:

$$x_2 = A[x_1 - (x_1)^2]$$

The Direction of Time

Let us close this discussion by asking, if Boltzmann was right and molecules do obey the theory of probability, where does the direction of time come from? If we were to observe an ordered deck of cards gradually becoming disordered by shuffling, we would not ordinarily ask the question, "How did the deck get ordered in the first place?" We would assume that someone originally chose to arrange the cards and think no more about it. But the order in the universe is another matter.

The entropy of the universe is not yet at its maximum possible value, and it seems to be increasing all the time. Looking forward to the

Then we calculate the population in year 3 from:

$$x_3 = A(x_2 - (x_2)^2]$$

and so on. A programmable scientific calculator can quickly produce a long sequence of yearly populations for various choices of x_1 and A. Chaos begins when A exceeds about 3.6. The calculations illustrated in Figures 8.1 and 8.2 assumed the following values for the initial populations and the parameter A:

Figure	x_1	A
8.1(*a*)	0.1	0.9
8.1(*b*)	0.1	1.25
8.1(*c*)	0.1	3.2
8.2 *(solid line)*	0.100	3.8
8.2 *(dotted line)*	0.101	3.8

Experimenting with various values of A will show how sensitive the results can be. We have chosen a simple mathematical model; most examples from the real world will be much more complex. Our illustrations of chaotic versus nonchaotic behavior, however, are very similar to many real-world patterns, be they changes in the weather, fluctuations of the stock market, or variations in the number of measles cases from one year to the next. These examples *resemble* chaotic behavior, but it is often not easy to prove that they *are* chaotic, even when there is no plausible alternative explanation.

future, Clausius and Kelvin foresaw a time when the maximum possible entropy would be reached and the universe would be at equilibrium forever afterward; at this point, a state called the "heat death" of the universe, nothing would happen forever after. Looking backward to the past, we infer a greater and greater degree of order the further we go and conclude that there must have been a state of maximum order, some kind of beginning. Why was there so much order in the past? This is a question for cosmology to answer, and we will come to it later on. We note only that because our originally orderly universe is on the way to disorder, we exist and, existing, feel the passage of time.

9
Entropy and/or Information

〜〜〜〜〜〜〜〜〜〜〜〜〜〜〜〜〜〜〜〜〜〜〜〜〜〜〜〜〜〜〜〜〜〜〜

Early in the history of telegraphy, it occurred to people sending telegrams, aware that the cost increases with the length of the message, that coding saves money. For example, a codebook might inform us that the message "Please send invoice for merchandise received" can be coded as PXUYQ; this will be understood by anyone using the same codebook. Coding is advantageous not only for the customer but also for the telegraph company, which can more than make up for the loss of some potential profit on each message by sending more messages in a day.

It is not surprising that research on the efficiency of transmitting messages should have been conducted by the research arm of a communications company (the Bell Telephone Laboratories, now the AT&T Bell Laboratories) or that the research led to a quantitative measure of the information a message can carry. It is surprising, however, that the measure of information, proposed by Claude Shannon in 1949, resembles closely the formula for entropy, $S = k \log W$. This resemblance, as we will show, is not fortuitous; it reflects an intimate connection between the concepts of entropy and information. Recognition of this connection has led some to ask the question: is the entropy of a particular physical system we are studying in the laboratory an objective property of that system, or does its value depend on how much information about the system the human observer happens to have?

We will try to answer that question here, and in the process we will also deal with a challenge to the second law of thermodynamics posed by a figure in the history of thermodynamics who is perhaps more well known than its creator: Maxwell's demon.

Shannon's Measure of Information

Before we define *information* we must issue the usual warning: the scientific and ordinary meanings of the term overlap to some extent but are not identical. In ordinary discourse we are concerned with the importance and the usefulness of information. We do not regard as equally informative the following two statements:

1. There are 37 cars in the parking lot this morning.
2. Shakespeare wrote 37 plays.

They would require about the same number of symbols for transmission, however, and the quantitative measure of information proposed by Shannon was concerned primarily with that aspect of the message. This does not imply that science is not concerned with the importance or usefulness of statements, but only that this narrow focus on the number of symbols needed was a good way to begin.

Shannon looked for a definition of the amount of information in a message that did not depend on accidental features, such as the language (or alphabet) the message was written in or the coding (if any). Instead, Shannon took the point of view of the recipient who, prior to its arrival, was not only expecting a message but had some expectation of what messages were likely to arrive. To take a simple example, the message expected might have been the answer to a question asked earlier by the recipient: "Are you coming for dinner Tuesday?" There are two possible messages, "Yes" or "No," and receipt of the message confirms one of them. In more complicated situations, such as those in which PXUYQ might have been the message or part of it, the number of possibilities prior to its arrival is much larger. Let us represent the number of possibilities by W. The information content of the message once received was taken by Shannon to be greater the larger the value of the number W. Why this should be so is not immediately obvious, but it can be justified.

Why W?

We must first distinguish between the length of a message and the information it conveys. This is a point readily appreciated by anyone who has ever had to listen to a long and boring speech, whose content is easily summarized as "Honesty is the best policy" or some similar platitude. We must also avoid confusing the information content of a message with its importance. The choice between "yes" and "no" may also be the choice between life or death, but Shannon's approach attributes to such a message only a small amount of information.

We may get at the concept better if we think of *ignorance* rather than *information* first. Let us see if we can quantify ignorance. If we know nothing at all, our ignorance is infinite. If, however, we know of W possible (and equally probable) "facts" or "truths" or messages, but not which is the right one, we know *something* anyway. Obviously our ignorance is greater the greater W is. If we could eliminate half the possibilities, our ignorance is less, and once we discover the one right one, our ignorance is reduced to zero. Our loss of ignorance is a gain of information, and the greater W was originally, the greater the gain.

Another way to think about the quantification of information is through the economics of sending messages. A communications company must be willing to transmit any message for its customers, but it will charge according to the length of the message. From the point of view of the customer who pays for sending it, the shorter the message the better, so the customer's goal is to phrase the message in the minimum length necessary to transmit the desired information. What is that minimum length?

If it is agreed that there are W possible messages the recipient could expect, the sender and recipient can do no better than make a list of the messages in advance together and number them from 1 to W. The longer the list, the larger the particular number the sender might have to send, and therefore the longer the message he must be prepared to pay for.

One might object that an infinite number of answers are always possible even to a simple question: "No, I am not coming to dinner." "Yes, I will come unless my cold gets worse." "Tuesday is bad, could we change it to Friday?" Not all are equally probable, but is there really only a finite number? If we limit consideration to messages of finite length, we can show that the number of possibilities is indeed finite.

Let us avoid coding and consider only messages in the English language. If there is no limit on the length of a message, then of course there is an infinite number of possibilities, but suppose we limit the length of the message to no more than a book of 1,000 pages: we are permitting as a possible message, therefore, anything from a single symbol to any book of 1,000 pages or less that has been written or could be written in English. Is there not an infinite number of possibilities?

The answer, it is simple to show, is no, there isn't. A typical line of

English text has about 10–15 words, of average length 5–8 letters, separated by spaces and other punctuation marks. This implies about 100 or so characters per line of text. There are 40–50 lines per page, and in 1,000 pages we can show therefore that there are less than five million characters (letters, spaces, and other symbols). If we assume about 50 characters—26 letters of the alphabet, 10 digits, and the rest spaces and punctuation marks—there are $50^{5,000,000}$ (or $10^{8,500,000}$) possible arrangements of them. All the books yet written or to be written in the Roman alphabet, and therefore in the English language, are included in this number, the overwhelming majority of them complete nonsense, consisting of lines such as "Asplf.gasme! -wlfgl plsmo., qglzf." Obviously, the number of books making sense in English is much less than $10^{8,500,000}$. The point is that it is not infinite. Of making many books there is an end. W is a finite number.

The Binary System

We will bring the concept of information and the world of the computer closer together by using a binary alphabet for our message, an alphabet with only two letters, which can be thought of as dots and dashes, heads and tails, or the digits 0 and 1 (namely the binary number system). There is no loss of generality in changing to the binary system from the 26 letters of our alphabet and the digits 0 through 9 we used before. Binary numbers are as good for counting as the familiar numbers based on the number system 10, as shown in Table 9.1, where numbers up to 32 are represented in binary notation. The first 64 binary numbers can be used to represent the 26 letters of the alphabet, the 10 digits from 0 to 9, and 29 other symbols, such as punctuation marks, we might need.

Computers are built up of enormous numbers of small elements, each of which can be in one of two states. One may think of the states as those of an electrical switch that may be either "on" or "off," or a small magnet pointing either "north" or "south." The elements may be switched from one state to the other as the computer operates. If we identify each magnet pointing north as a 1 and each magnet pointing south as a 0, we have coded the state of the computer as a binary number. Table 9.2 shows the number of messages (W) that can be made from different numbers of binary symbols.

It is apparent that if there are N symbols in the message, the number of messages $W = 2^N$. If we take W as the measure of information in the message, then a 4-symbol message, which can be thought of as two

Table 9.1
The First Thirty-Two Binary Numbers

Binary notation	Decimal notation	Binary notation	Decimal notation
0	0	10001	17
1	1	10010	18
10	2	10011	19
11	3	10100	20
100	4	10101	21
101	5	10110	22
110	6	10111	23
111	7	11000	24
1000	8	11001	25
1001	9	11010	26
1010	10	11011	27
1011	11	11100	28
1100	12	11101	29
1101	13	11110	30
1110	14	11111	31
1111	15	100000	32
10000	16		

Let us work through an example of a number converted from one notation to the other. In the familiar decimal notation (a system based on the powers of ten), 31 is equal to 3 time ten plus 1 unit, or $(3 \times 10^1) + (1 \times 10^0)$. (Any number raised to the zeroth power is equal to 1.) The binary notation is based on powers of two:

$$11111 = (1 \times 2^4) + (1 \times 2^3) + (1 \times 2^2) + (1 \times 2^1) + (1 \times 2^0)$$
$$= \quad 16 \quad + \quad 8 \quad + \quad 4 \quad + \quad 2 \quad + \quad 1$$
$$= 31$$
$$11010 = (1 \times 2^4) + (1 \times 2^3) + (0 \times 2^2) + (1 \times 2^1) + (0 \times 2^0)$$
$$= \quad 16 \quad + \quad 8 \quad + \quad 0 \quad + \quad 2 \quad + \quad 0$$
$$= 26$$

Table 9.2
Number of Messages *(W)* That Could Be Sent with *N* Symbols

N	Possible messages				W	Information content $(I = \log_2 W)$
1	0	1			2	1
2	00	10			4	2
	01	11				
3	000	001			8	3
	010	011				
	100	101				
	110	111				
4	0000	0001	0010	0011	16	4
	0100	0101	0110	0111		
	1000	1001	1010	1011		
	1100	1101	1110	1111		

2-symbol messages, will carry 4 times the information that a single 2-symbol message does (16 vs. 4). This doesn't make intuitive sense. Two 2-symbol messages should, we feel, be capable of only twice the information content of one such message. Our sense that something is wrong is even greater with a message of one million binary digits. *W* is enormous (about $10^{300,000}$). If we lengthen the message to one million and one digits, *W* is exactly doubled. Whatever "information" is, we do not feel comfortable with the idea that a message of 1,000,001 symbols should be capable of conveying twice as much of it as one with 1,000,000. Shannon, reasoning from this and other considerations, suggested that the proper relation between the information content (symbolized by *I*) and the number *W* should be logarithmic rather than direct. He proposed the formula

$$I = \log_2 W$$

If $W = 2^N$, then $I = N$ (see the appendix, "Math Tools"). *The information content of a message is thus equal to the logarithm (to the base 2) of the number of binary symbols needed to express it.* As noted earlier, this assumes that all binary messages are equally probable, an unrealistic assumption, but one we can get by with here.

Information and Entropy

The close resemblance of Shannon's formula to $S = k\log_e W$ is obvious, but there are also two differences. First, the presence of the factor k, which relates temperature to the average kinetic energy of a molecule, makes the numerical value of entropy depend on the units we choose for measuring energy and temperature. The information content I, however, is a pure number, whose magnitude does not depend on any physical units. Second, we use 2 as a base for the logarithm defining I, but we use $e = 2.718\ldots$ as the base for S. $\log_e W$ is a little smaller than $\log_2 W$, but it is always smaller by the same factor, regardless of how large W is:

$$\log_e W = 0.69315 \log_2 W$$

The important difference is conceptual. In the entropy formula we know only that our system can be in W possible states but we don't know which of them it is actually in; in the formula for information we know exactly which one among W possibilities is our message. Entropy measures ignorance, information measures knowledge.

We can have information about the stock market, English poetry, or telephone numbers in Chicago, but we would not apply the entropy concept to such subjects, as the concepts of energy and temperature are not applicable to them. We can have information about the distribution of energies among the molecules in a tank of compressed gas or in a bottle of liquid. Here information and entropy are closely related. The larger the number W of microstates the system could be in the larger is its entropy, and the less information we have about it. When anything happens that increases entropy, our ignorance is increased. We can express this approximately in an equation:

$$\text{Change in information} = -\frac{(\text{Change in entropy})}{k}$$

where k is Boltzmann's constant (See Chapter 7).

The intimate but inverse relation between the entropy of a system and information we have about its molecular state raises an intriguing question. We have described how the entropy of a system is measured in the laboratory, by measuring heat inputs and outputs and the system temperature while the system undergoes a change from one state to another. Our ability to measure it suggests that entropy is like any other physical property we determine in the laboratory: it has a definite numerical value when the system is in a definite state, and once it is determined we expect it to keep the same value as long as the system is not subjected to changes. Suddenly it appears that knowledge we have or may acquire about the system can change its entropy. Is entropy objective or subjective? Does it have its value only in the eye of the beholder? This is a question scientists have argued about passionately almost since entropy was discovered. Maxwell began the arguments by introducing his demon, to be described shortly, and they are not yet resolved.

When we determine entropy in the laboratory, we do so on the assumption that the only knowledge we have about the system is macroscopic. We know how much of each kind of substance is present, we know the temperature and the pressure and the strengths of any electric, magnetic, or gravitational fields acting on the system (if there are any). On the other hand, we know nothing about where each individual molecule is, or how much energy it possesses, or how fast its position and energy change from one moment to the next. If we agree to use the term *entropy* only when these conditions of ignorance are met, then there is no problem. Entropy is objective: you will get the same answer every time you measure it.

Suppose, however, we have measured the entropy of a system according to the above procedures, and by an additional measurement we acquire some more detailed knowledge about the system on the molecular level: knowledge, for example, about the speeds and directions of motion of some small group of the molecules. Does this knowledge imply that the entropy is decreased? Would such a decrease violate the second law? If we could use the new knowledge to get the system to do work for us, this would be evidence for such a violation. Suppose, for example, that the detailed information we have gained about the molecular motion enables us to predict that at a certain moment in the near future there will temporarily be an excess of molecules on the righthand side of a container and thus a slightly higher pressure there. We can then quickly insert a partition blocking the transfer of molecules and trap the system in this state. In this state the system has slightly less entropy than it had originally. The pressure difference could be used to do (a small amount of) work. It would appear that the second law has been violated, unless we can show that the measurement process itself, which gave us the information needed to produce this entropy decrease, itself had to be paid for with a compensating entropy increase great enough to save the second law.

Maxwell's Demon

Getting around the second law this way is just what Maxwell invented his Demon for. When the second law was discovered, it was assumed to apply universally, without any exception. Maxwell from the beginning recognized its basis in the enormous numbers of molecules in any ordinary physical system, the near impossibility of predicting their motion, and the necessity of using a probabilistic approach. He also realized that the smaller the number of molecules we look at, the more their instantaneous behavior is likely to differ from their expected average behavior. He suggested the possibility, if one's perceptions were refined enough, of taking advantage of these deviations from average behavior to produce a violation of the second law.

To make his point graphically, he visualized "a finite being who knows the paths and velocities of all the molecules by inspection" and uses this information to open or close a frictionless door between two volumes of a gas allowing certain selected molecules to pass through. Maxwell described several ways this "being" might decrease entropy. In his first writing on the subject he showed how it could cause a gas originally at a uniform temperature to be separated into a warmer portion and a cooler portion. For our purposes a simpler mode of operation, also suggested by Maxwell, which produces only a pressure difference, will serve: Maxwell imagined that the being opens the door only when a molecule on the left side is approaching it. When one from the right side approaches, he keeps the door shut. The upshot is that the gas, originally distributed equally between right and left sides, becomes concentrated on the righthand side. The entropy is decreased, violating the second law, and a pressure difference is created that can be exploited to do work.

We can try to save the second law by saying that there are no "demons" or other such "finite beings." In Maxwell's time not enough was known about living organisms to rule out the possibility that they operated with energy extracted from the environment by such processes. Today we know that they don't (see Chapter 12). But an intelligent being need not be necessary: a purely mechanical device may serve.

"A Very Observant and Neat-Fingered Being"

Maxwell did not call his being a "demon": the name was given to it later, by Kelvin, and has stuck. Its first appearance was in a letter Maxwell wrote to a colleague, P. G. Tait, in 1867. (In the following excerpt, we have altered the sense slightly in order to shorten the text.)

Now let *A* and *B* be two vessels divided by a diaphragm [*CD*] and let them contain . . . molecules in a state of agitation which strike each other and the sides.

Let the number of particles be equal in *A* and *B* . . . Then even if all the molecules in *A* have equal velocities, if . . . collisions occur between them their velocities will become unequal, and I have shown that there will be velocities of all magnitudes in *A* and the same in *B* . . .

When a molecule is reflected from the fixed diaphragm *CD* no work is lost or gained.

If the molecule instead of being reflected were allowed to go through a hole in *CD* no work would be lost or gained, only its energy would be transferred from the one vessel to the other.

Now conceive a finite being who knows the paths and velocities of all the molecules by simple inspection but who can do no work except open and close a hole in the diaphragm by means of a slide . . .

Let him first observe the molecules in *A* and when he sees one coming . . . whose velocity is less than the mean . . . velocity . . . let him open the hole and let it go into *B*. Next let him watch for a molecule of *B*, . . . whose velocity is greater than the mean . . . , and when it comes to the hole let him draw the slide and let it go into *A*, keeping the slide shut for all other molecules.

Then the . . . energy in *A* is increased and that in *B* diminished, that is, [one system has got hotter and the other colder], yet no work has been done, only the intelligence of a very observant and neat-fingered being has been employed.

Or in short if heat is the motion of finite portions of matter and if we can apply tools to such portions of matter so as to deal with them separately, then we can take advantage of the different motion of different proportions to restore a uniformly hot system to unequal temperatures . . .

Only we can't, not being clever enough.

Let us imagine a small trap door (Figure 9.1) that opens only to the right, with a spring that tends to hold it shut. An energetic molecule from the left side striking the door will force it open and go through to the right side, while a molecule striking the door from the right side simply rebounds. No demon is needed. One molecule going through at a time would give a fairly slow rate of buildup of pressure, and an economically minded person might wonder whether the work we could get from such a device would pay for the capital cost of installing it. But this device, no matter the cost, threatens the second law. Can anything save it?

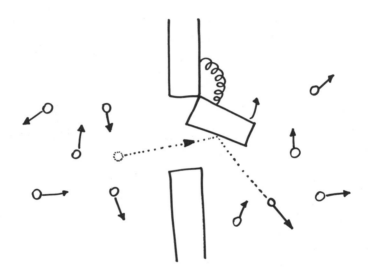

Figure 9.1   A Mechanical "Demon"

A purely mechanical device that would do the job of Maxwell's demon and produce an excess of pressure suitable for operating a compressed air engine can be imagined. It could consist of two compartments separated by a wall containing a spring-driven trap door. A molecule coming from the left and striking the trap door with sufficient force pushes the door open and travels through the door into the righthand compartment. A molecule coming from the right and striking the trap door should only force it to close, and the molecule wouldn't get through to the lefthand compartment. The net effect should be a buildup of pressure on the right side at the expense of the left side. This clever device, however, has never been realized. No barrier permitting molecules to pass through it in one direction and not the other, leading to a net decrease in entropy, has ever been found in nature or devised by human ingenuity.

Let us examine the operation of the device more carefully. A molecule striking the door and getting through imparts some of its kinetic energy to the door. The kinetic energy is transferred to elastic potential energy in the spring as the door swings open, and when the kinetic energy of the door is all transferred to the spring, the spring begins to pull it shut again, giving up its potential energy and increasing the kinetic energy of the slamming door. When the door hits the wall, it has no way of getting rid of its kinetic energy, as the wall does not move. If we neglect friction, the door would then continue to bounce back and forth, alternately stretching the spring and slamming into the wall. It is a kind of pendulum.

This is not good. The effectiveness of the device depended on the door being shut unless a molecule from the left side is butting its way through. If it is open much of the time molecules from the right side get back through to the left and we lose much of what we wanted to achieve. We can imagine adding friction to the device, so that the door doesn't keep swinging back and forth once it has been pushed open. In microscopic systems such as this, however, it is better to avoid using the concept of friction—a macroscopic one—and consider the molecular level only.

Brownian Movement Again

In addition to the impacts of fast-moving molecules from the left that push the door wide open, there are other, less forceful molecular impacts on the door from both sides, ones that might not swing it open widely enough to let a molecule through. Most molecules do not have enough energy to open the door widely, or else they strike it from the wrong direction. But these impacts keep the door constantly rattling back and forth. If the door and spring happened to be without any energy at all, the impacts would tend to get them moving. On the other hand, if the door had been given an extra-large quantity of energy by a fast-moving molecule, the impacts are likely to drain off the excess energy: it would be the impacting molecules rather than the spring-door combination that would tend to gain energy. It can be shown that the door, swinging back and forth on a single arc, would have an average kinetic energy of $\frac{1}{2}kT$. In Chapter 7 we described how Jean Perrin determined Avogadro's number—the number of hydrogen atoms in one gram of hydrogen—by studying Brownian movement. He could do so because he recognized that the floating particles, although composed of enormous numbers of atoms, acted in a sense like single molecules, having an average translational energy of $\frac{3}{2}kT$ as a result of the constant bombardment of the water molecules. The trap door behaves in the same way. After an impact with an especially energetic molecule, its kinetic energy is greater than $\frac{1}{2}kT$, but the subsequent molecular impacts tend to bring it back to the average. The molecular bombardment is basically the origin of what we call "friction" in macroscopic language; the difference from macroscopic friction is that moving bodies are not brought to rest by it: it only looks that way. A 1 kilogram steel ball (2.2 pounds) has an average kinetic energy of $\frac{3}{2}kT$, just as a gas molecule does, and therefore at 300°K (ordinary ambient temperature) has a mean speed of 6.4×10^{-11} meters per second. This speed is unobservable in a large ball, so Perrin used very small particles, barely discernible with a microscope, to observe the motion.

This seems to help. We wanted to introduce friction to keep the trap door closed most of the time, and the constant molecular bombardment provides it. If the door is in some sense only another kind of molecule, however, it is subject to all the other laws of molecular motion as well. One of them is the principle of time-reversibility. Any motion-picture film we make of the action of the door, projected backward, gives an equally good description of its action. Or, put another way, a sufficiently long motion picture will show any event and its exact opposite—the same event apparently run backward in time—equally often.

This is fatal for the mechanical demon. Consider a strip of the film in which a high-energy molecule from the left side strikes the nearly closed door, swings it open, and gets through, losing much of its energy to the door as it does so. The door swings back and forth a few times, until impacts with a number of other molecules drain off its excess energy, and it returns to its nearly closed position, jiggling slightly with its $\frac{1}{2}kT$ of kinetic energy. This is the desired event.

Now play the film backward. The door is jiggling slightly with about $\frac{1}{2}kT$ of kinetic energy, but it is nearly closed. A number of molecular impacts happen to occur in such a way that kinetic energy is transferred from the molecules to the door, causing it to swing back and forth between the wide-open and closed positions. At this point, a molecule from the right side of the container gets in front of the swinging door and is swatted forcefully by it over to the left side. In this collision most of the kinetic energy of the door is transferred to the swatted molecule, which is now on the left side and moving at a higher-than-average speed. The door is back to its nearly closed position and jiggling slightly. The desired event of the preceding paragraph has been undone.

As both events are equally likely, the trap-door demon accomplishes nothing. There is no way to escape this symmetry of equal forward and backward steps. If we try to cut down the amplitude of the Brownian movement of the door by using a stronger spring, we cut down the number of desired events (passage of a molecule from left to right) and cut down the equal number of opposite events (passage of a molecule from right to left) in the same proportion.

The Polish physicist Marian von Smoluchowski, in 1912, explained in this way why such devices won't violate the second law. His analysis would seem to cover any inanimate device, not only the spring-loaded trap door. In a paper published a few years later, he stated his conclusion as follows: "As far as we know today, there is no automatic permanently effective perpetual motion machine [of the second kind] . . . but such a device might, perhaps, function regularly *if it were appropriately operated by intelligent beings*" [italics added].

The intelligent being envisaged by Maxwell could see the molecules, judge their speeds, and accordingly make an intelligent choice about when to open a door. This possibility of a difference between intelligent beings and inanimate mechanisms remained open until the 1920s, when a new understanding of electromagnetic radiation provided by quantum mechanics suggested a resolution. In brief, quantum mechanics showed that "seeing" is not a neutral process that leaves the object seen unperturbed but is instead an interactive process. Light has energy, and the more accurately a body—a moving molecule, for example—is to be *seen,* the more energy is needed and, ultimately, dissipated as thermal energy in the environment, increasing its entropy. Observing a molecule, so that one may decide when to open a door, requires a price to be paid in the form of an entropy increase that more than offsets the entropy decrease the knowledge enables us to produce. If information is not free, not even an intelligent being can get around the second law.

Is Information Free?

Here the matter seemed to rest, until the advent of computers and the beginning of a scientific study of not only how they operate but also of how they could *ideally* operate. This latter question reminds us of Carnot's investigation of how any conceivable heat engine could ideally work.

Computers are, after all, material objects, whose components are made up of various substances with particular chemical and physical properties. In the course of operation they are subjected to various kinds of forces: electrical, magnetic, elastic, and so on. They therefore must obey the laws of physics, including the laws of thermodynamics.

Computers use energy to operate. You have to plug them in or run them on batteries (though the earliest computers were purely mechanical devices using gears, wheels, and belts: the energy to operate them came from human muscles turning cranks). What is the minimum energy an ideal computer would need to run a particular program? This sounds like a question of some economic significance, but it isn't. For what computers give us, the cost of the electric energy to operate them as they are now constructed is trivial. The problem, rather, is one of getting rid of the heat most of the electrical energy is ultimately converted to, which is necessary to avoid damage to the components. The answer to the question is interesting but of more theoretical than practical interest: it is in principle possible to run computers with no energy cost at all. For practical reasons no one would ever design a computer to run this way: one reason, but not the only one, is that the time to run the program is longer the less energy is used to drive it, and so a computer using zero energy would take an infinitely long while to run a program. This result is related to the fact that to perform physical or chemical processes reversibly, with no net entropy increase, we must run them infinitely slowly. In the real world we are willing to tolerate entropy increases to save time.

C. H. Bennet and R. Landauer, working on the problem of the energy requirements of computing, were led to a reexamination of the entropy cost of a measurement. They concluded that although the use of electromagnetic radiation to observe molecules does lead to entropy increases, it is possible to "observe" them by other means that do not. In particular, an observation of the position of a molecule by measuring the pressure it exerts has no entropy cost. Their proposal seemed to bring the demon back to life, but Bennet and Landauer saved the second law by the following reasoning.

The process of making a measurement and using the information to enable the system to do work can be broken down into a series of distinct steps:

1. Observing a molecule's position, speed, direction of travel (or whatever is needed).
2. Recording the results of the observation in some manner: storing it in a "memory."

3. Using the stored information to control some device that performs the work.
4. Resetting the memory so that it is ready to record the results of the next measurement.

The whole process converts molecular (thermal) energy directly to useful work and operates cyclically: the device at the end of its cycle is back in its initial state. The first law shows that if net work has been done, there must have been a compensating absorption of thermal energy from the environment. This implies an entropy *decrease* in the environment. Unless there is a compensating *increase* somewhere, the second law has been violated. Bennet and Landauer showed that the compensating increase is not in the observation step but rather in the last step, when the memory of the previous measurement is erased. It takes work to erase a computer memory, which is dissipated as thermal energy in the environment, and the minimum work needed is the work the device could do as a result of the information.

To demonstrate that the erasure of a memory requires work and leads to an entropy increase takes more physics than we can give here, but it is based, as is the whole approach to computers developed by Bennet and Landauer, on the consequences of the idea that a "memory" is necessarily a material object made up of molecules that obey molecular laws.

Intelligent Beings, Inanimate Devices

This resolution of the paradox of the demon may leave us with the impression that there are two kinds of demons that require two different exorcisms: inanimate demons fail because of Brownian movement and the principle of time-reversibility, and intelligent demons—who make measurements, record them, and make conscious choices based on the recorded information to do some work—fail because erasing the recorded information so as to be ready for a repetition of the cycle requires that work be performed.

This division seems to us to be artificial. After all, what do intelligent beings do when they want to get work done? The answer is that they invent a machine to do it automatically, such as a spring-loaded trap door. That device does everything an intelligent being does.

1. It makes an "observation" to determine whether a molecule is approaching the trap-door from the left with enough energy to

push it open and get through. The observation is the impact with the door.

2. It stores the information in the form of the kinetic energy transferred to the door.

3. The stored information is used to swing the door open, when the kinetic energy of the door is transferred to the potential energy of the spring, at which time the molecule is allowed to get through to the right side.

4. The subsequent impacts with other molecules remove the excess energy of the door-spring combination, erasing the memory of the event and restoring the device to its initial state, ready for the next cycle of operation.

The inanimate device therefore does what the intelligent being (namely, Maxwell) who designed it would want it to do, which is just what "a finite being who knows the paths and velocities of all the molecules" would do. Yet it fails to poke a hole in the second law. There are, instead of two kinds of demons, two languages with which we can explain the demon's failure. The first is a microscopic one, which speaks in terms of molecules, Brownian movement, and time-reversibility on the molecular level; the second is macroscopic, which speaks in terms of observation, computation, and memory. The two are really equivalent, and it is a matter of taste and convenience which to use. It is ironic that von Smoluchowski fully resolved the paradox of the demon in 1912 and then undermined the force of his reasoning by suggesting that it might not apply to intelligent beings. We have been struggling for three-quarters of a century with the idea that there is a paradox, when in fact there is none.

10
Radiant Energy, Black Bodies, and the Greenhouse Effect

The laws of thermodynamics apply not only to matter but also to *light*, a term we will use broadly to include all forms of electromagnetic waves. Since the emission of light by hot bodies is one means of transferring energy and equalizing temperatures, we expect thermodynamics to have something to say about the process.

We see most objects by the light they reflect, light coming from much hotter bodies like the sun. We distinguish one thing from another because each has its own characteristic effect on the light falling on its surface: some reflect green light primarily, as do the leaves of plants; others reflect everything, as does white paper; still others, like coal, absorb everything. When bodies are heated to high temperatures they begin to emit light. This light, too, distinguishes objects from one another; each substance emits a characteristic pattern of colors. To summarize: each of the various substances that make up the familiar objects around us has its own individual properties for the absorption, reflection, and emission of light.

We have said earlier that the importance of Carnot's conclusion about the efficiency of heat engines is not what it tells us about heat engines but what it tells us about the properties of matter. Here we will show that the second law places some extraordinary restrictions on how matter absorbs and emits radiant energy.

That there are such restrictions can be shown by a simple experiment using a furnace, which we can visualize as an enclosed insulated box whose walls are heated by electric currents passing through wires embedded in the walls. Let us imagine a furnace equipped with a small viewing port so that we may look inside, where there are a number of different objects: a block of platinum, a lump of coal, a quartz crystal, a few ceramic tiles of different colors, white, blue, green, and so on. First, before we turn on the current to heat the walls, the furnace is dark and we see nothing. We turn the current on so that the walls begin to heat up, first glowing with a dull red heat and eventually leveling off near a white heat (a temperature of about 1,200°C; this is the lowest temperature at which the furnace emits light seen by the human eye as "white"). We begin to see the objects as the walls begin to glow and can tell them apart by their colors. At first they are still cold and we see them only by the light from the walls. As they warm up they begin to emit light on their own, but we can continue to see them and distinguish them, since each emits its characteristic light.

Now we expect that as the walls come to some definite temperature determined by the electric current we are using to heat them, the objects in the furnace will eventually also come to this same temperature: an "equilibrium of heat" as discussed in Chapter 3. As we watch the objects approach this state of uniform temperature, we find that although there is plenty of light inside the furnace, it becomes harder and harder to distinguish them: the colors of the light coming from each of them become more and more alike, and more like the color of the light coming from the walls. Eventually, when a state of temperature equilibrium is reached, the luminous objects have faded into the luminous background. Blue, copper-colored, white, black, or transparent, all these individual characteristics have disappeared, and there is only a uniform glow.

As we will see, the second law requires that this state of indistinguishability be characteristic of the electromagnetic radiation present when bodies are in true temperature equilibrium. Because it does, the absorbing and emitting properties of each kind of matter must be interrelated in such a way as to ensure that it happens.

As temperature is raised, this universal "color" of bodies at temperature equilibrium changes, as we are already aware in the distinction we make between "red-hot" and "white-hot." The characteristic color we see when we look at hot bodies at temperature equilibrium, which depends on the temperature but not on the properties of the body from which it comes, is called *black-body radiation.*

The sun, whose surface is at about 6,000 K, is not at equilibrium, nor is the earth. Both emit their radiation into a space at a much lower temperature than the surface temperature of either, but it happens that most of the sun's energy is radiated as visible light, the earth's as infrared. The gases of the atmosphere are transparent to visible light, but carbon dioxide, methane, and water absorb infrared and slow down its escape from the earth. The carbon dioxide content of the earth's atmosphere is increasing because of the burning of fossil fuels and the methane content is increasing as more of the earth's surface is devoted to agriculture. As a result, the earth's average temperature may rise: this is "the greenhouse effect," and the climatic changes it may produce are a cause for serious concern.

What Is Light?

Up to now we have considered what the two laws of thermodynamics have to say about matter; now we will see what they have to say about space empty of matter, but not empty of energy.

Hot bodies tend to transfer their thermal energy to cold bodies, leading to an equalization of temperature and an increase of entropy. Thermal energy can be transferred by molecular collisions when the bodies are in contact, but if there is a vacuum between them it may be transferred by "heat radiation." The question of what is radiated played a role in the conflict between the caloric and kinetic theories of heat (Chapter 3). When light was thought to be a substance—Newton believed it to consist of fast-moving material "corpuscles"—it made sense to think of heat radiation also as a substance; when new experimental evidence suggested early in the nineteenth century that light was not a substance but a wave motion, it became more plausible to think of heat as motion, the motion of molecules.

Before this discovery, waves had been studied in other forms. Newton's laws had successfully described the motion of waves on water, whether ripples or ocean breakers, and identified sound as a wave motion in air. Waves show two phenomena that sharply distinguish them from particles flying through space: *diffraction*—an ability to bend around corners instead of always traveling in straight lines—and *interference*—two waves arriving at the same point in space combine to produce either a reinforcement or a cancellation of the motion (see Figures 10.1 and 10.2).

In addition to a definite speed of propagation, waves have a wavelength, a frequency, and an amplitude (Figure 10.3), the latter determining the energy they carry. The wavelengths of visible light are very small (about 10^{-7} meters). This made it more difficult to demonstrate diffraction with light than with sound or water waves, and Newton was not able to produce it with the crude experimental techniques of his

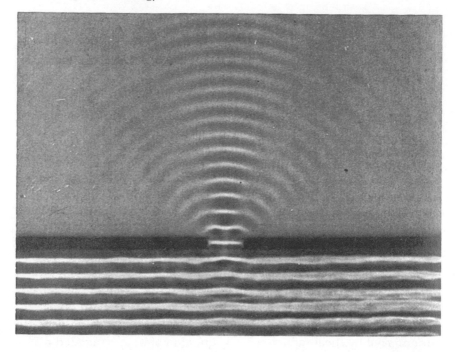

Figure 10.1 Diffraction of Waves
Waves on water, passing through an opening in a barrier, such as an opening
in a harbor breakwater, will bend around the edges of the opening rather than
travel straight past it. The bending is greater the greater the wavelength of the
wave. This is a photograph taken from above the surface of a tank of water.

time. His corpuscular theory of light accounted quite well for the
phenomena he could observe.

He had discovered with his prism that white light, specifically sunlight, was a mixture of light of different colors. Once it was concluded
later that light was a wave, the different colors of the spectrum could
be identified with different wavelengths and the wavelengths measured.
The wavelength of green light, for example, is 5×10^{-7} meters.

In the mid-nineteenth century, following the discovery by Faraday of
electromagnetic induction (Chapter 3), Maxwell developed a set of
equations unifying all electrical and magnetic phenomena then known.
From his theory a surprising new phenomenon could be predicted:
rapidly changing electric and magnetic fields should create, even in
empty space, waves that travel from their source with the same speed
regardless of their wavelength. There was as yet no laboratory demonstration of such waves, but the speed at which they would have to travel

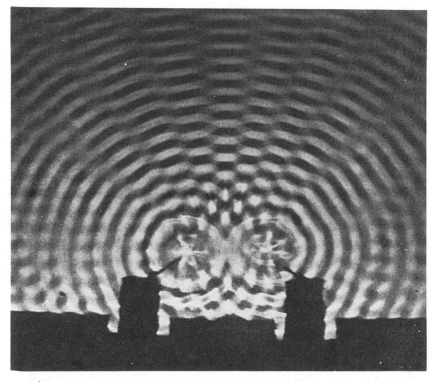

Figure 10.2 Interference between Waves Coming from Two Separate Sources
These waves in a tank of water were photographed from above. The light regions on the photograph represent areas on the surface of the water where there is wave motion, the darker regions are those areas where the surface is momentarily still. When waves from the two sources meet, there will be a large amplitude if the crests of the two waves coincide and stillness if a crest of one arrives at the same time as the trough of the other. The complex pattern that results is called an "interference pattern."

could be calculated from the known strengths of magnetic and electric forces. Maxwell calculated the expected speed; it was exactly the speed at which light was known to travel. The inescapable conclusion was that light is an example of the electromagnetic waves predicted by his equations. Prior to Maxwell's theory there had been no reason to connect light with electricity and magnetism. Most physicists regard Maxwell's electromagnetic theory not only as the single most significant scientific discovery of the nineteenth century but also as one of the most significant in the history of science.

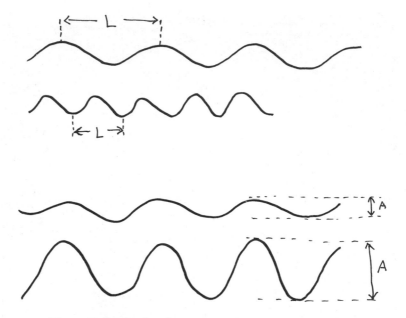

Figure 10.3 Wavelength, Frequency, Speed, and Amplitude
A wave is characterized by its wavelength L (the distance between successive crests), frequency f (the number of crests passing a point in space per second), speed v (the product of L and f), and amplitude A (the vertical distance between the crest and the trough). The *speed of light*, abbreviated c, is the speed of a light wave in a vacuum, about 3.0×10^8 meters per second; in glass or in air, however, the speed of light is somewhat slower. Weather reports warning of "waves four to six feet high" refer to the amplitude of the waves. The energy carried by a wave is proportional to the square of the amplitude.

The range of known electromagnetic wavelengths is enormous. Figure 10.4 gives those commonly encountered, extending from gamma rays, produced by nuclear disintegrations (10^{-10} meters), to the long wavelengths used for AM broadcasting (10^4 meters). The tiny range occupied by visible light is striking. In what follows, the word *light* will refer to all forms of electromagnetic radiation, *visible light* to those forms the human eye responds to.

Measuring the Energy of Light

Light carries energy, as we realize when we stand in the sun. The energy of light (any electromagnetic radiation) can be measured by the warm-

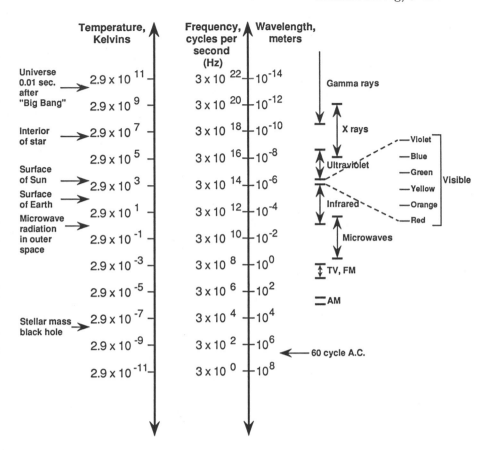

Figure 10.4 The Electromagnetic Spectrum
There is no upper or lower limit on the wavelengths possible for electromagnetic radiation, but this figure shows only those between the wavelengths associated with 60-cycle alternating current (million-meter waves) and gamma rays (10^{-14} meters). The boundaries between different types of radiation are somewhat arbitrary, as is indicated, for example, by the overlap of gamma rays and X rays. As the wavelength L decreases, the frequency f increases according to the relation $f \times L = c$, where c is the constant speed of light in a vacuum. As the frequency increases, so does the energy of a photon (see Chapter 14).

The temperature scale at the left applies to "black-body radiation," to be described later in this chapter. A body heated to any particular temperature shown on the lefthand scale will radiate more energy at the frequency corresponding to that temperature than at any other frequency (Wein's law).

ing it can produce in bodies that absorb it, much as Joule measured the warming effects of electric currents or of friction (Chapter 3). This is not the only way to measure the energy of light, nor is it always the most convenient. The photoelectric light meter on a camera does it a different way. Infrared light is absorbed easily by the skin, and its heating effect is easily felt—for example, the heat from a hot iron. This "heat radiation" was first shown to be a form of light when a thermometer with a blackened bulb was being used to measure the warming effect of the various colors in the spectrum of sunlight. The temperature rose not only when the thermometer was illuminated by the visible light but also when it was placed outside but adjacent to the red end of the spectrum of colors, where no light could be seen. In contrast, ultraviolet light was first discovered photographically. Infrared radiation is misleadingly called "heat rays"; however, *any* wavelength of electromagnetic radiation carries energy, which, when absorbed, increases the energy of the absorbing substance.

Seeing Things

We see objects by the contrast between the visible light coming to us from them and that coming from their immediate surroundings. The objects "seen" don't necessarily have to be sending light to our eyes, as long as the surrounding objects do, which is how we see a black fly on a white tablecloth.

Sometimes the objects we are looking at actually emit the light we see them by: the sun, an electric light, a candle flame. More often they are not emitting light but reflecting it from primary sources like the sun, as does the moon or this book.

Even transparent objects like a pane of glass can be "seen"; some light may be reflected from the surface, or we may see the distortion, due to refraction, of the image of an object behind them. Objects that are transparent to some visible wavelengths but not to others are seen by the contrast between the color of the light transmitted through them and the color of the surrounding light; red cellophane and the lenses of sunglasses, for example.

Light and Matter

In this section we have distinguished several different kinds of interaction between light and matter.

1. *Emission*. Bodies may emit light if they are heated, or if an electric current is passed through them (neon, sodium, and mercury lamps), or if a suitable chemical reaction is taking place in them (fireflies). Also, light of one wavelength falling on certain substances causes the emission of light of a different (usually longer) wavelength (fluorescence under ultraviolet light). The emission of heated bodies will concern us in particular. In this kind of emission, thermal energy in the emitting body is converted to electromagnetic energy traveling away from it. Energy is lost by the body and usually its temperature will fall.
2. *Absorption*. Here the opposite process takes place: electromagnetic energy falling on a body is converted to thermal energy within it, usually raising its temperature.
3. *Reflection*. Light falling on the surface leaves from the surface without any gain or loss of its energy.
4. *Transmission*. Light passes through the body and comes out the other side. Again there is no gain or loss of its energy.

Each of these phenomena are dependent on wavelength. Some bodies absorb blue light and others absorb red. Lead absorbs X rays, but clothing doesn't. Metals are good reflectors for all wavelengths of visible light, but copper reflects red a little more efficiently than it reflects the other wavelengths, which gives it its reddish tint. Mercury lamps emit blue-green light primarily, neon lamps orange light.

The Radiation from Hot Bodies

Heating alone causes substances to emit radiation. The yellow light of a candle is emitted both by hot gases and by small particles of solid carbon (soot). The light from a gas flame can be either blue or yellow, depending on the mix of air and gas that determines both the temperature and the chemical composition of the flame. The surface of the sun is gaseous matter heated to about 6,000 K by radiation coming from the hotter interior, and it in turn radiates heat to the surrounding space. Hot solids also emit radiation, whose wavelength characteristics depend on the substance. For all substances the wavelength characteristics of emitted radiation also depend on the temperature of the radiating body, as we recognize in the distinction between the terms *red-hot* and *white-hot*.

The eye and the skin are not very discriminating guides to the wavelengths of radiation: most wavelengths we neither see nor feel. Instead, to characterize the radiation, we use spectrometers, instruments that first separate the radiation into its component wavelengths and then measure how much energy is carried by each wavelength (the "intensity" of light is equivalent to its energy). The separation can be carried out by a prism, for example, which produces a rainbow of colors from white light, the red light falling on one place on a viewing screen, the orange next to it, and so on through yellow, green, blue, and violet. Any piece of glass with nonparallel faces will produce a visible spectrum. To separate invisible forms of light by wavelength we need either prisms made of substances other than the glass used by Newton or devices of an entirely different sort. An example is a "diffraction grating," made by inscribing a series of parallel lines on a transparent or reflecting surface: the spacing of the lines must be comparable to the wavelengths of the radiation to be separated. A compact disc has a spiral track with spacing between successive turns having the right separation to act as a

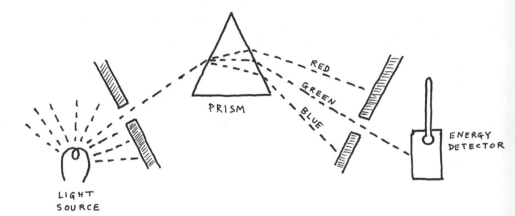

Figure 10.5 Measuring the Energy Spectrum of Light
The energy carried by light can *in principle* be measured by allowing light to be absorbed by a black liquid and measuring the rise in temperature of the liquid. Shown here is a source of white light (a mixture of wavelengths). A beam of this light passing through a prism is separated into its component wavelengths, which fall on different areas of a screen. A slit in the screen permits only a narrow range of wavelengths to reach the energy-measuring detector, represented schematically by a box with a thermometer. In laboratory practice, the rise in temperature produced by radiation is usually much too small to measure, and therefore other methods are used.

diffraction grating, and gives beautiful spectral colors when examined in white light.

Once one color is separated from the other colors, we may collect only this light, so that instead of the broad mixture of many wavelengths in the original white light, we have only a narrow band of wavelengths, all of which produce, say, the response "red" from English-speaking humans. In Figure 10.5 we show how a slit in a screen can select such a narrow band. The energy in this band of wavelengths can be measured, in principle, by the temperature rise in a body that absorbs it. (In practice, this is not sensitive enough; other methods of measuring the energy are used.) Following this measurement, we can move the slit or the prism to select other bands of color and determine the amounts of energy they carry. Once we tabulate all the resulting energies according to the average wavelength (in effect, the color) of each narrow band, we have the energy spectrum of the light.

Energy Spectra

Figure 10.6 gives an example of a spectrometric determination of a radiant energy emission, that of the sun. This quantitative relation between energy and wavelength is called the "energy spectrum" of the radiation. Spectrometers are also used to measure the spectra of *absorbed* radiation. Both emission spectra and absorption spectra may be used to identify an unknown substance, or determine how much of it is present as an impurity, because every substance absorbs and emits radiation in an identifiable pattern. Almost all of modern chemical analysis is based on spectroscopy. It is used to determine the composition of the stars, the amount of mercury in canned tuna fish, and the hemoglobin content of the blood.

Energy spectra may be discrete or continuous or a combination of both. A familiar example of a *discrete* spectrum of radiation is the wavelength range in which the signals of radio stations are transmitted. As we search for a station by tuning the radio, we pick up signals of the various stations, some louder than others, between intervals of near silence. The radio set is a kind of spectrometer, with the energy of the signal proportional to its loudness. The spectrum of a hot solid body, such as the filament of an ordinary electric light, appears continuous; at whatever wavelength (at whatever color) we look, we find some energy, and the amount of energy changes only gradually as the wavelength is changed.

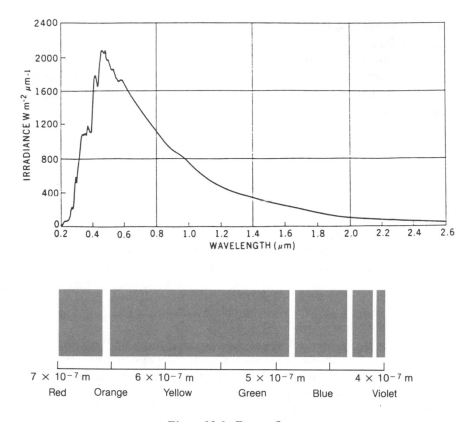

Figure 10.6 Energy Spectra

All substances emit light, more noticeably the hotter they are, and they all absorb at least some of the light falling on them. They emit and absorb some "colors"—wavelengths—more than others. Here we show two examples of the distribution of energy by wavelength of light emitted by heated or electrically excited substances. On top is a graph of the energy emitted by the sun; this is a *continuous* spectrum—all wavelengths are emitted, but some carry more energy than others. (The vertical scale on the left represents watts per square meter per unit wavelength range.) Below it is a drawing of that portion of the emission spectrum of the hydrogen atom that is visible to the human eye. The drawing is similar to what would appear on a photographic film taking the place of the screen in Figure 10.5, if the light source were a gas of very hot hydrogen. The energy in each discrete wavelength could be estimated from the darkness of the line on the photograph and the known sensitivity of the film to light of that wavelength. There are four discrete wavelengths only, corresponding to the colors orange, blue-green, blue-violet, and violet. No other visible light is emitted, but in other regions of the electromagnetic spectrum many other discrete wavelengths are observed. A more familiar example of a discrete spectrum of radiation is the wavelength range of radio signals.

The energy spectrum of radiation can be thought of as a "distribution" that has something in common with the Maxwell distribution of molecular speeds described earlier (Chapter 4). The kinetic energy, $\frac{1}{2}mv^2$, of a molecule can be calculated from its speed v, so that knowing the speed distribution of a collection of molecules is the same as knowing the distribution of molecular kinetic energies. The Maxwell distribution is a kind of energy spectrum also—a distribution of moving molecules rather than of radiation.

To summarize what we have said about the energy spectra of the light emitted by heated bodies:

1. Different substances heated to the same temperature have different energy spectra. Each substance when heated emits its own characteristic energy spectrum, just as each substance has its own characteristic absorption pattern (Figure 10.6).
2. The energy spectrum of the light emitted by any substance depends on the temperature, and the higher the temperature the more energy is radiated.

What the Second Law Tells Us

Normally when we are aware of bodies emitting light, they are emitting light to colder surroundings. The sun, an electric light, a glowing coal, all shine in a space that would otherwise be dark. The system is not in equilibrium: the body is steadily radiating energy and receiving much less from the colder environment. It follows that its temperature must fall unless we supply energy in some other form, such as by an electric heating coil inside the body, to compensate for the steady loss.

We can, however, set up an experimental situation in which an emitting body is in equilibrium with radiation; that is, with as much radiant energy falling on the body from outside as is leaving from the surface. In this case radiant energy is flowing in both directions at the same rate. An example would be a body in a closed container when heat is steadily supplied to the outside of the container so that the interior walls are maintained at *some fixed temperature;* in short, a furnace. Inside the furnace a state of temperature equilibrium should be reached, with whatever body is inside radiating energy to the walls and with the walls in turn sending the same amount of radiation back to the body.

Since we haven't said what substance the walls of the furnace are made of, we might expect that the energy spectrum of the light coming from the walls need not necessarily match that coming from the body inside the furnace. One might guess that all that is necessary for temperature equilibrium to be maintained is that the *total energy* radiated in any direction should be the same as in any other, but the reality is more complicated than this, as we find when we apply the second law.

The View at Temperature Equilibrium

Imagine that the furnace walls are made of some white ceramic material and the body is a black lump of carbon, "black" meaning that it absorbs most of the visible light falling on it. Suppose further that the furnace is kept "white-hot," so that most of the energy radiated from its walls is in the form of visible light and the carbon will glow also.

Now let us introduce into the furnace an (imaginary) observer equipped with a spectrometer. He first turns his instrument toward the glowing carbon and records its energy spectrum, then turns it toward the furnace wall. One might expect that because the two substances, carbon and ceramic, are different, the two spectra would be different. In fact, however, if the furnace and its contents are really in temperature equilibrium, the spectra will be identical. The different emitting properties of the two substances seem to have completely disappeared. The observer inside cannot tell the white ceramic walls from the black carbon. What is going on?

We remind the reader once again of "the equilibrium of heat": bodies at different temperatures tend to transfer heat in such a way as to reduce and ultimately eliminate temperature differences. The bodies reach a uniform temperature, and once there they stay there. Once temperature equilibrium is reached, temperature differences cannot spontaneously arise. Heat cannot flow between bodies at the same temperature so as to raise the temperature of one at the expense of the other: this would cause a spontaneous decrease in total entropy. The entropy of any body with a uniform temperature throughout is higher than its entropy would be when the temperature varies from one point to another, other things, including the total energy, being equal.

Let us consider two different substances, A and B, placed near each other in a furnace, both at the *same* temperature. Radiant energy will be flowing between them in both directions, from A toward B and from B toward A. Certainly for temperature equilibrium to be maintained, the total radiant energy flowing in each direction must be the same. This is necessary, but is it sufficient?

Suppose the light traveling in one direction carried the same *total* energy but had a *different* energy spectrum from that traveling in the other: put simply, suppose the light was of a different color, red in one direction (*A* to *B*), yellow in the other (*B* to *A*). Why should the color matter as long as the total energy flows either way are the same?

If, however, the energy spectrum of the light flowing in the two directions were not identical, it would be possible to introduce a device between the bodies to regulate the flows of energy and produce a spontaneous decrease of entropy. There is a type of light filter, for example, that transmits light of one color only, reflecting all others in the visible range. If the colors traveling in opposite directions differed, a suitable filter could return the red radiation heading from *A* to *B* back to *A*, while leaving the yellow radiation traveling from *B* to *A* unaffected (Figure 10.7). The result: *A* rises in temperature, *B* cools, and the entropy decreases. A filter of this type is not a Maxwell demon: it acts symmetrically, reflecting red radiation coming from *either* direction, and such filters really exist.

Because a spontaneous decrease of entropy is impossible, we conclude that for temperature equilibrium the energy *in each wavelength* must be in balance: the energy spectra of radiation traveling in each direction must be the same. Under these conditions the filter is receiving, and reflecting back, equal intensities of red light from both *A* and *B*, so no temperature difference is created.

The conclusion, that at temperature equilibrium the energy spectrum of the radiation coming from all bodies must be the same, seems to contradict the common observation (illustrated in Figure 10.6) that different bodies when heated emit different colors. How can the contradiction be resolved?

Balancing Energy Losses and Gains

We resolve this problem when we make the distinction between light "emitted" by a body and light "coming from" a body. Light emitted is produced within the body, and its energy comes from the thermal energy of the body. Light coming from a body includes both the light emitted by it and any light transmitted through it that originated elsewhere. Let us imagine two bodies in the furnace: a lump of carbon and a quartz crystal. Carbon absorbs almost all the visible light falling on it. Quartz is transparent, or nearly so.

If the furnace is "white-hot," the carbon absorbs most of the energy

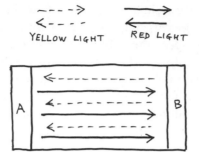

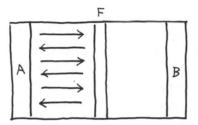

falling on it and rises in temperature fairly rapidly. The quartz, nearly transparent, absorbs only a small fraction of the energy falling on it and rises in temperature relatively slowly; most of the energy passes through it unabsorbed, and does not increase its thermal energy. Since the quartz absorbs so little of the energy, will it ever reach the temperature of the carbon? Should it not remain permanently colder?

Although the carbon continues to absorb most of the radiation falling on it, its temperature does not continue to rise indefinitely, for a simple reason: the hotter it gets, the faster it emits radiant energy. Eventually, a state of balance is reached between emission and absorption. Experiment shows this balance is reached when the temperature of the carbon reaches the temperature of the furnace.

The quartz does the same. It is absorbing less energy per second at

Figure 10.7 Good Emitters Must Also Be Good Absorbers

There is no obvious reason why the colors emitted by different substances need have any particular relation to the colors they absorb—why, for instance, a substance emitting red light and not yellow should not be capable of absorbing yellow light falling on it. The second law shows, however, that this is not possible: *a substance will absorb only those wavelengths it emits.* Here we illustrate a proof of this by contradiction. Two different substances coat opposite walls of a furnace, and the system is heated to a high temperature. The substance on the left *(A)* emits red light only but absorbs yellow (we will call it Red Stuff), the substance on the right *(B)* emits yellow but not red (it is Yellow Stuff). When the system—furnace plus two substances—is at temperature equilibrium, we interpose between the two walls a filter *(F)* that reflects red light but transmits yellow. The red radiation is returned to the Red Stuff on the left, but the yellow emitted by the Yellow Stuff is transmitted and falls on the Red Stuff. The Red Stuff is thus receiving energy from the Yellow Stuff but not giving it any: the result would be a cooling of the Yellow Stuff and a warming of the Red Stuff. A violation of the second law can be avoided if the Red Stuff does not absorb the yellow radiation. The conclusion: to satisfy the second law it is not enough that at equilibrium the rate that the *total* energy is absorbed must equal the rate at which it is emitted, but this relation must be true *at each individual wavelength also.*

each wavelength than the carbon absorbs, and the only way it can reach a balance between incoming and outgoing energy at the same temperature as the carbon is by emitting less energy per second at each wavelength than the carbon. There must be, if the second law is to be satisfied, a relation between the absorbing and emitting power of substances at each wavelength: if they absorb light strongly at any wavelength, they will emit light strongly at that wavelength, and if they are weak absorbers at any wavelength, they will be weak emitters at the same wavelength.

The conclusion is an important one. Let us summarize the logic:

1. if the quartz *absorbs* less energy per second at a particular wavelength than the carbon, and
2. if the carbon and quartz are to reach the *same* final temperature,
3. then it follows that the quartz must *emit* less energy per second at that wavelength than the carbon emits

In other words, the second law requires that all substances be good emitters for wavelengths they absorb readily, and poor emitters for wavelengths they do not absorb readily.

No substance is a good absorber for all wavelengths of light. "Black" substances like carbon are good absorbers for visible light specifically, but they may be poor absorbers for infrared light or microwaves. Quartz is a poor absorber of visible, infrared, and ultraviolet light. Glass, which looks much like quartz to the eye, is also a poor absorber in the visible range but absorbs both infrared and ultraviolet. So carbon is a good emitter for visible light and quartz a poor one. Yet we have argued that at temperature equilibrium, the energy traveling from carbon to quartz is the *same*, wavelength by wavelength, as that traveling from quartz to carbon. Where does the visible light emanating from the quartz come from, if the quartz isn't emitting visible light? The answer is that quartz is transparent, so the visible light coming from the walls of the furnace behind it passes through it and travels on toward the carbon. The result is that the energy of the light "coming from" the quartz, the sum of the light it emits and the light it transmits, is the same, color for color, as that coming from the carbon. The rule that only strong absorbers are strong emitters ensures this compensation. In particular, if the quartz doesn't either absorb or emit green light, green light falling on it from the walls passes through it and comes out the other side, toward the observer measuring the energy spectrum of its radiation. Glass, on the other hand, is a good absorber of infrared light, therefore infrared falling on it is absorbed and not transmitted; however, because it is a good absorber for infrared it must be a good emitter also, so the radiation coming from hot glass has its "proper" mix of both green and infrared light.

This discussion leaves unanswered the question: what if there were only quartz inside the furnace and even the walls were made of quartz too, where would the visible light at equilibrium at a high temperature come from? We have described quartz as though it were perfectly transparent at some wavelengths, so that it emits no radiation at those wavelengths. In fact, however, no substance is *perfectly* transparent at any wavelength. There is always some absorption and therefore some emission. "Transparent" substances will take longer to reach radiation equilibrium than absorbing ones, but sooner or later they get there.

All Cats Look Alike

If the radiation coming from all substances at temperature equilibrium is the same, all substances must look alike at temperature equilibrium. Although our furnace may contain blocks of copper, glass, silver, or amethyst, at temperature equilibrium we cannot tell them apart, nor can we distinguish them from the walls of the container, which are also at temperature equilibrium. The proverb says that all cats look alike in the dark. It is worse than that: if we want to see cats by the radiant energy emitted by them and coming to our eyes, they all look alike in the light also, and they look like dogs or mice or blocks of glass. At temperature equilibrium, nothing can be distinguished from the background.

But what about the room one is sitting in now, reading this book? It is about at temperature equilibrium, is it not? Thermometers in different places in the room would read about the same temperature, wouldn't they? Yet the black print can be distinguished from the white paper, and both can be distinguished from the green plant in the flower pot and the red rug on the floor. The answer is that the room is *not* at temperature equilibrium. The sunlight coming in through the window is emitted by the outer layers of the sun's atmosphere, at a temperature of about 6,000 K. This radiation would be in equilibrium with any matter at 6,000 K, but it is not in equilibrium with the cool matter in the room, and it is this radiation that makes things distinguishable. Because the eye is such a sensitive instrument, the quantity of radiant energy we need to see by is quite small, not enough to alter the temperature distribution in the room by much, so for some purposes we can disregard it and assume temperature equilibrium. However, the small deviation from exact equilibrium is enough to permit us to distinguish one object from another.

Black-Body Radiation

The radiation at temperature equilibrium is called *black-body radiation.* The origin of the name is as follows: an absorbing body is a good emitter only for the wavelengths of light it absorbs well. *If* a body were "black" at all wavelengths, it would also be a good emitter at all wavelengths, and the energy it radiates away even in cold, dark surroundings would be the *same* as that inside a furnace at the same temperature. No such bodies actually exist. In the English-speaking scientific community these hypothetical bodies are called "black bodies" and the radiation "black-body radiation." The term used in the German literature for black-body radiation translates to "hollow-space radiation," which seems to make more sense than naming the radiation after a nonexistent substance.

So far we have come to two conclusions about the radiant energy inside a furnace at temperature equilibrium, the black-body radiation:

1. The energy spectrum of the light coming from all substances at temperature equilibrium at any given temperature is the same.
2. The energy spectrum changes as the temperature changes.

Let us imagine observing the energy spectrum of the radiation inside a furnace at temperature equilibrium and how it changes with temperature. Doing so, we will learn why "white-hot" is hotter than "red-hot" and why we can feel but not see the light from a hot iron.

First, how do we observe the energy spectrum? The observer inside the furnace would also be white-hot at equilibrium, which is why we made him imaginary. If, however, we make a small opening in the furnace and observe the light emitted from it with a spectrometer, the disturbance of temperature equilibrium is small and can be neglected.

In Figure 10.8 we give examples of the energy spectra of the radiation emitted from furnaces at equilibrium at several different temperatures. It is apparent from these curves that:

1. There is more energy in the radiation at higher temperatures.
2. The amount of energy at each wavelength is greater the higher the temperature.
3. The *relative* amount of energy at shorter wavelengths is greater the higher the temperature.
4. There is always a maximum—a peak—in the curve: there is one wavelength that carries more energy than any other. By comparing curves at different temperatures, we can confirm that this wavelength is shorter the higher the temperature.

These qualitative inferences can be expressed more quantitatively with the aid of equations. The total energy E radiated per second per square meter of surface from any body inside the furnace or from the walls increases as the fourth power of the absolute temperature:

$$E = sT^4$$

where s is our symbol for another constant of nature, the same for all substances and all temperatures. Its value has been determined experimentally to be 5.7×10^{-8} watts per square meter per degree kelvin to the fourth power. (A watt, equal to one joule per second, is a unit of power, the rate at which energy is supplied, rather than energy.) We can use this formula to calculate that a person with two square meters of body surface area and a skin temperature of 30°C (303 K) radiates energy at a rate of 1 kilowatt (1,000 joules per second). In a warm room the person is receiving almost as much from the walls as he is radiating, so he doesn't freeze to death. This calculation has assumed that the person, white or black, is emitting "black-body radiation"; this assumption should be nearly correct, since the radiation from human skin at 303 K is in the infrared, and all of us, white or black, have skins that are good absorbers, and therefore good emitters, of infrared radiation.

The wavelength *at the highest point* of the black-body curve, which tells us which wavelength is associated with the largest quantity of

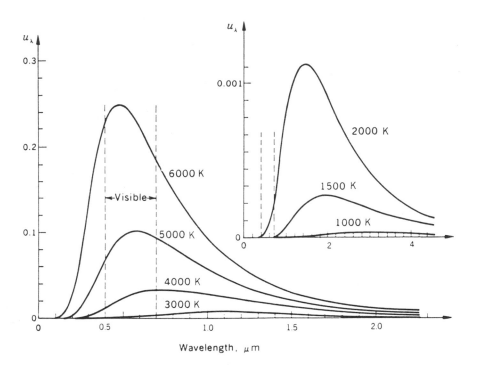

Figure 10.8 The Black-Body Energy Spectrum
These curves show the quantity of energy associated with each wavelength in the radiation emitted from bodies in equilibrium—the black-body radiation—for temperatures from 6,000 K (approximately the surface of the sun) to 1,000 K (red-hot bodies). Because the total energy of the radiation falls so rapidly with temperature (being one million times greater at 6,000 K than at 1,000 K), not all the curves could be drawn on the same set of axes. The curves here are calculated from a theory to be discussed in Chapter 14, but they are in excellent agreement with experimental measurement. The graphs also show qualitatively that the maximum of each curve occurs at a shorter wavelength the higher the temperature.

radiated energy at each temperature, varies *inversely* with the absolute temperature *T:*

$$l_{max} = \frac{w}{T}$$

where *w* is a constant, again for all substances and all temperatures, whose value is found experimentally to be 2.9×10^{-3} meter-kelvins. The letter *w* was chosen to honor Wilhelm Wien, who first discovered this relation. As an example of its use, we can calculate the wavelength of maximum energy in the radiation from the human body to be about 0.011 millimeters (therefore in the infrared), and from the sun's surface at 6,000 K to be about 5×10^{-7} meters (green light). Sunlight is "white" because there is considerable energy in other colors as well.

The two equations show first how strongly the radiant energy increases with temperature: a fourth-power dependence implies a sixteen-fold increase in total radiated energy when the absolute temperature is doubled ($2^4 = 16$). The wavelength associated with the largest quantity of energy can be used as an estimate of the average wavelength at a particular temperature, useful for characterizing the whole broad distribution. Figure 10.4 shows how it shifts with temperature. The human eye also averages, but only over the "visible" wavelengths. "Red-hot" bodies really emit most of the radiant energy in the invisible infrared region; only the tail of the distribution is in the wavelength region we call "red."

The Properties of Space

Space empty of matter may therefore be filled with radiant energy. If the radiation is black-body radiation, the empty space has a temperature also. When the temperature is raised, the quantity of energy at equilibrium in a given volume increases. The increase in energy per kelvin rise in temperature of one gram of a substance is what we mean by its specific heat. It follows that space empty of matter can be thought of as having a specific heat, though instead of basing the definition of specific heat on one kilogram of space we have to think of one cubic meter. It is further a consequence of Maxwell's electromagnetic theory that radiation exerts a pressure. This pressure can be understood as analogous to the pressure of a gas: just as a molecule striking a wall and bouncing back exerts a force on the wall, so also a wave striking a wall and being reflected from it exerts a force. The force of ocean waves is familiar enough. Evidence for the pressure of light is provided by the tails of comets, which always point away from the sun no matter where the comets are in their orbits. Finally, an equation provided by thermodynamics permits us to calculate an entropy for radiation, from the relation experimentally observed between the total radiant energy and the temperature.

All these properties together—energy, entropy, temperature, specific heat, pressure—are also properties of matter. We have discussed the example of an ideal gas, in which molecules move independently, not exerting forces on each other; electromagnetic waves also move independently of each other. Just as Maxwell was able to calculate the energy distribution of gas molecules using the theory of probability, so also it should be possible to calculate the energy distribution of the electromagnetic waves. Once the energy spectrum of radiation had been experimentally determined in the latter half of the nineteenth century, a theoretical calculation was attempted. Maxwell had shown earlier that the average translational kinetic energy of each kind of molecule in a gas should be $\frac{3}{2}kT$. The same kind of reasoning led to the expectation that the average energy of each wavelength of light present in a furnace at equilibrium should be proportional to kT.

The predictions of the theory based on this expectation were in spectacular disagreement with experiment. They were not merely wrong, they were absurd: the theoretical conclusion was that all bodies at equilibrium should be radiating an infinite amount of energy every second. Everything in the universe, hot or cold, small or large, should glow with an infinite radiance. It was one of the worst failures of nineteenth-century science. How the contradiction was resolved will be described in Chapter 14.

Practical Consequences

The Greenhouse Effect

The sun and the earth are not in temperature equilibrium. The radiant energy from the sun's surface is close to that of a black body at 6,000 K. The surface of the earth is warmed partly by the sun and partly by the heat of its interior (see Chapter 13). Surface temperatures on the earth vary daily, seasonally, and with latitude, but the overall average, about 300 K, hardly changes over hundreds and even thousands of years. Since the earth receives large quantities of energy from the sun, its average temperature would not remain so nearly constant were it not also losing energy by radiation, approximately the same amount as it receives. The earth's radiation spectrum can be measured; it is also not too different from the black-body spectrum for its surface temperature.

The radiant energy received from the sun is mainly visible light (the maximum energy is at the wavelength of green light), to which our atmosphere is transparent. The infrared light radiated from the earth's surface, however, does not pass through the atmosphere so easily. Carbon dioxide and water vapor absorb certain infrared wavelengths strongly, though because they are good absorbers they are also good emitters. So the infrared radiation must be absorbed and reemitted many times before it finally escapes from the top of the atmosphere to outer space and is lost to the earth forever.

The atmosphere, with its water and its carbon dioxide, acts therefore as a one-way insulation, letting in the sun's energy readily but slowing down the escape of the earth's. The roughly constant average surface temperature of the earth is determined by the balance between the rates of incoming and outgoing energies.

In recent years it has been observed that the carbon dioxide content of the atmosphere has been increasing, primarily as a result of the combustion of fossil fuels. This buildup inevitably slows down the rate of escape of the infrared radiation from the earth's surface (Figure 10.9). Since the energy radiated by the sun remains the same, the average temperature of the earth's surface is likely to rise, until the increased rate of radiation of energy (rising as the fourth power of the absolute temperature) compensates for the more effective insulation provided by the added carbon dioxide. If the atmospheric content of carbon dioxide continues to increase, the earth's temperature should continue to rise. This hypothesized rise of temperature is called the "greenhouse effect." It could have drastic effects on climate, rainfall, and sea level.

That there is a greenhouse effect large enough to warm the earth appreciably, a matter of international concern, has not yet been confirmed experimentally beyond all doubt. The possibility was inferred and warned against before the gathering of experimental data began.

So far we have applied the thermodynamics of radiation to a system consisting of the sun, the earth, the earth's atmosphere, and outer space. As usual, thermodynamics alone does not answer all the questions raised, and other fields of science must be called upon.

How much will the temperature rise for a given rise in carbon dioxide in the atmosphere? Are we seeing this rise already? Will temperature rise at all? Some have proposed that one consequence of the carbon dioxide increase could be an increased cloud cover and thus a *fall* in surface temperature. Does all the carbon dioxide formed by fossil fuel combustion end up in the atmosphere? Do the oceans take up an appreciable part? Gaseous methane is being added to the atmosphere as a result of increased agricultural activity. Although present in much lower concentrations than carbon dioxide, it absorbs the earth's infrared radiation more strongly. What is its contribution to the greenhouse effect?

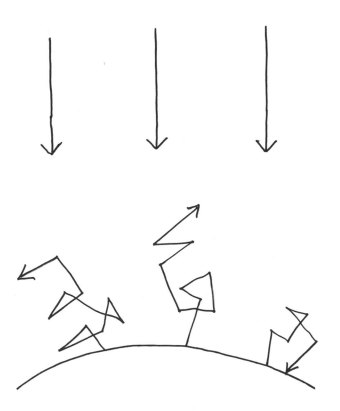

Figure 10.9 The Greenhouse Effect

The carbon dioxide and other industrially or agriculturally produced gases in the atmosphere do not completely prevent the earth's infrared radiation from eventually escaping; this radiation is not trapped but is constantly absorbed and reemitted, which means that its eventual escape takes much longer than it would if there were no such gases in the atmosphere to absorb it. Visible radiation coming from the hot sun *(straight arrows)* reaches the surface of the earth without interference (except on cloudy days), but infrared radiation emitted from the cooler earth *(zigzag arrows)* is absorbed and reemitted by carbon dioxide in the atmosphere many times before it can escape.

What will the climatic and other consequences of a given rise in the average surface temperature be? Which countries will suffer and which benefit? How much will sea level rise, from the melting of ice and snow in cold latitudes and from the expansion of water with a rise in temperature?

Beyond these controversial and as yet unresolved scientific questions there are political and economic ones. Are there feasible alternatives to fossil fuel and at what cost?

Quite a Pandora's box has been opened for us by these two equations, $E = sT^4$ and $l_{max} = w/T$.

How Furnace Temperatures Are Measured

The temperature inside furnaces heated high enough to radiate visible light can be easily measured with an instrument that applies the principles of black-body radiation. The instrument used is called an *optical pyrometer*. It consists of a metal filament enclosed in a transparent envelope (something like an electric light bulb), which is heated to incandescence by an electric current whose strength can be controlled by the observer. The furnace has a viewing port allowing a small portion of its radiation to escape, and the filament is placed so that the observer views it against a background of the furnace radiation coming from the viewing port. As the electric current is increased, the filament begins to glow. At low currents its temperature is lower than that of the furnace and it appears dark against a light background. As the current is increased and the temperature of the filament rises to equal the furnace temperature, the filament ceases to be visible against the background. If the current is increased further and the filament temperature rises above that of the furnace, the filament reappears, this time brighter than the background.

Once the relationship between furnace temperature and that electric current that just makes the filament disappear has been determined, the optical pyrometer is said to be "calibrated" and it then serves as a fast and reliable thermometer.

Selective Absorbers for Solar Energy

Another example of the practical applications of these principles is to absorbers for solar energy. If we want to trap the energy of sunlight as heat, we will of course use some substance that is a good absorber of solar energy. As most of that energy is in the visible region of the spectrum, the absorber must be black, at least to the human eye. However, the conclusion that good absorbers are also good emitters may seem to create difficulties. We want to keep as much of the trapped energy as possible, but if our absorbing substance is a good emitter it will start to re-radiate the energy and we won't accumulate much. We may seem to have run into a serious limitation on solar collectors, but considering the black-body spectra together with what we have said about the properties of different substances, we realize this need not be so. The rule that good absorbers are good emitters is true at *each* wavelength of light. When we collect solar energy, we do not expect our collectors to reach the 6,000 K temperature of the sun's surface. For many applications—home and water heating, for example, and even for heat engines—temperatures of 100° C to 400° C are adequate. At these temperatures the largest portion of the energy of the emitted radiation is in the infrared region, with a much longer wavelength than that of the incoming solar radiation. The second law is not violated if a substance is a good absorber (and emitter) in the visible region and a poor absorber (and emitter) in the infrared. So it is possible to have collectors that absorb solar energy readily and do not re-emit it readily.

This idea was proposed by Dr. H. Tabor of the National Physical Laboratory of Israel, and coatings that have these properties (fine powders of certain oxides of metal, including copper) were easily discovered once it was known what to look for. Solar collectors all over the world use them.

11
Chemistry: Diamonds, Blood, Iron

~~~~~~

The second law states that only processes leading to a net entropy increase are possible, and once a state is reached from which no further increase can occur, nothing more can happen. (Processes involving neither an increase nor a decrease in entropy—reversible processes—are also "possible," but they would require an infinite time.) We have given examples of systems starting out in low-entropy states and ending up at equilibrium: a gas confined to one side of a container by a partition with a small hole in it leaks through the hole until its pressure is the same on both sides; in an iron bar hot at one end, heat flows until the temperature is uniform throughout. The equilibrium states of the gas and the iron bar, when the pressure and temperature are uniform, are states of maximum entropy. No further increase of entropy is possible, unless we do something to change the system: reheat the iron bar, for example, or compress the gas again to one side of the container.

In this chapter we will show how the same principle applies to chemical reactions, in most of which molecules are broken apart and re-formed into different molecules. As before, nothing can happen unless it leads to a net entropy increase. Suppose, for example, that we are interested in the possibility of some particular chemical reaction: can metallic iron catch fire in air? can a catalyst be found to decompose water into hydrogen and oxygen gases? We will find that it is possible to predict the net entropy change that *would* take place *if* the reaction in question occurred, from laboratory determinations of the energies and entropies of the individual chemical species involved. From the principle that the net entropy can only increase and never decrease, we can determine whether that particular reaction can take place at all and, if it can, what chemical species will be present and in what amounts when equilibrium is reached. Reactions that meet this criterion of causing a net entropy increase are said to be "thermodynamically permitted."

Thermodynamics has its power, but it also has limitations. Its laws forbid certain events and permit others, but they do not tell us whether the permitted events *will* happen or how fast they will happen. To tear a molecule apart requires that energy be supplied from somewhere; how much energy is needed, whether that much energy is available in nearby molecules, and how fast it is supplied all determine how fast the reaction will proceed, but these are not questions thermodynamics can answer.

Here we will discuss two examples of chemical reactions that are permitted by thermodynamics but that take place extremely slowly or not at all: the production of diamonds, and the possibility of setting the atmosphere on fire. We will explain why they are so slow, and what can be done to speed them up (assuming we want them speeded up). Then we will discuss the concept of chemical equilibrium: the fact that for chemical reactions involving gases or dissolved substances, the reaction is never complete when equilibrium is reached. The reacting substances are never all used up to form the products desired, but instead both reacting substances and products are simultaneously present. Examples of chemical equilibrium we will consider include the effervescence of carbonated beverages, the transportation of oxygen by the blood, and the operation of a blast furnace.

Physical and Chemical Change

When we increase the temperature of an iron bar, its properties undergo slight changes: it expands a little, its electrical resistance changes a little, and so does its elastic resistance to being bent. The changes are not dramatic, and the bar looks much the same after it has been warmed. Such gradual changes in properties are called "physical" changes. A rise in temperature is one way to produce a physical change; other ways include increasing the pressure on the body, stretching or twisting it, and subjecting it to electrical or magnetic forces.

In contrast, chemical change is dramatic. The properties and appearance of chemically reacted substances are often completely unlike those we started with. Iron when it rusts loses its metallic appearance and properties and becomes a reddish-brown powder. Wood burns, leaving only a small quantity of ash—the rest of it is converted to gases. Acrid and corrosive hydrochloric acid reacts with lye, a substance we use cautiously to clear a stopped drain, to form ordinary table salt.

Processes like the melting of solids, the boiling of liquids, and the formation of solutions of one substance in another are harder to classify as chemical or physical: sometimes they involve breaking molecules apart and sometimes they do not. We will treat them as chemical changes without special apology, because it is convenient to do so.

Chemical changes, as we said earlier, usually involve breaking molecules apart and forming new ones out of the fragments. What is a molecule, and what holds it together?

Even those of us who have never taken a course in chemistry are familiar with some chemical names, like sulfur dioxide and carbon monoxide, and even some chemical formulas, like SO_2 and CO. Both the names and the formulas reflect the atomic makeup of the substances: one sulfur and two oxygen atoms in the former, one carbon and one oxygen atom in the latter. There are about a hundred different kinds of atoms, and they combine to form molecules, which are more or less stable entities held together by fairly strong forces of electrical origin known as *chemical bonds*. In chemically pure substances, all the molecules are identical—they are composed of exactly the same numbers of atoms of each kind, connected together in the same way. We identify these by chemical names or formulas, like those given above, or else by common names: alcohol, benzene, sugar, and water.

Some molecules, such as proteins or nucleic acids, have tens or hundreds of thousands of atoms. For the purposes of this chapter we need consider mainly simple ones, composed of a few atoms, like H_2O, CO_2 (carbon dioxide), or CO (carbon monoxide). Most of the substances we commonly encounter, like these examples, are composed of two or more *different* kinds of atoms, but some substances are composed of only one kind of atom. These we know as the *elements,* such as the metals iron, copper, and gold, the gases hydrogen, helium, oxygen, and nitrogen, and the solids carbon and sulfur. In some elemental gases, the atoms do not join together by chemical bonding: helium (He) is one. For these there is no distinction between the *atom* and the *molecule.* Most elemental gases, however, exist as molecules composed of two identical, chemically bonded atoms, as indicated by their chemical symbols: H_2, O_2, and N_2.

Must "Thermodynamically Permitted" Reactions Occur?

We have made a distinction between reactions that are thermodynamically permitted (those that would lead to a net entropy increase if they occurred) and those that actually occur. All chemical reactions that actually occur and have been examined thermodynamically have been shown to lead to a net entropy increase, which is one of the many kinds of experimental evidence that confirms the second law. There are, however, many examples of reactions which are permitted but which do *not* occur, or at least they do not occur readily.

The active ingredient in dynamite is nitroglycerine, a chemical compound of hydrogen, oxygen, carbon, and nitrogen. An explosion is an extremely rapid reaction that results in the formation of relatively large quantities of gaseous products: in the case of dynamite, steam, carbon dioxide, and oxides of nitrogen are produced. Thermodynamics tells us how to measure the energies and entropies both of nitroglycerine and of the products of its decomposition, and thus we can calculate, even without exploding nitroglycerine, that the total entropy, both of the chemicals involved and of the environment where the explosion takes place, increases because of the reaction. Hence the reaction is thermodynamically *permitted*. But knowing that the reaction is permitted doesn't answer the question, "Will the reaction take place slowly, explosively, or not at all?"

It is hardly surprising that thermodynamics "permits" the decomposition of nitroglycerine. After all, we knew dynamite was an "explosive" before we measured any entropy changes. We need a more convincing example.

Can the Atmosphere Catch Fire?

Let us consider the possibility of a reaction between the nitrogen and oxygen of the air and the waters of the oceans to produce a dilute solution of nitric acid. The chemical equation for this reaction is:

$$2H_2O + 2N_2 + 5O_2 \rightarrow 4HNO_3 \text{ (aq.)}$$

HNO_3 is the symbol for nitric acid; the abbreviation *aq.* is for "aqueous," to indicate that in the reaction we are considering the nitric acid is dissolved in water, the water of the oceans. About 20 percent of the molecules in air are oxygen (O_2) and most of the remainder nitrogen (N_2). There is more than enough nitrogen and ocean water to use up

Chemical Kinetics

Almost all chemical reactions go faster at higher temperatures. Typically, reaction rates double or more than double for each 10°C rise in temperature. Why should this be so?

To break a molecule apart, the first step in most chemical reactions, the strong forces that bind the atoms together in the molecule must be overcome. This means that work must be done, and the potential energy of the atoms is thereby increased. We cannot reach into molecules with grappling irons to pull the atoms apart. Instead, we must rely on the kinetic energy from colliding molecules to knock them apart. Molecular collisions are frequent, but the energy needed to do the job is often very high, and only a tiny fraction of the moving molecules have enough. The higher the temperature, the greater is the proportion of fast-moving molecules, and a collision with a molecule having enough kinetic energy to disrupt the critical chemical bond becomes more likely. This is why chemical reactions take place faster at higher temperatures.

Almost all chemical reactions begin with an input of energy to the molecule, but the process does not end there: when the molecular fragments recombine to form the new molecule, there will usually be an energy output that is transferred to neighboring molecules by collisions, and this energy output may exceed the input needed to make the molecules react in the first place. Under these conditions, the energy needed to make nearby molecules also react may come from the net energy output of the just-reacted molecules. If there is enough energy output to make at least one nearby molecule react, we have the conditions for a chemical "chain reaction" to occur. If the energy

all the oxygen of the atmosphere in this reaction. If it took place, it would mean an end to most if not all life on earth, so the question of its thermodynamic possibility is of more than passing interest. The fact that it has not taken place during the 5 billion years of the earth's history may suggest that it is not possible, but unfortunately direct determination of the entropies and energies of the various chemical species involved shows that the reaction would result in a large net output is enough to make more than one neighbor react, we may get an explosion, as with nitroglycerine.

The fact that reactions go faster at higher temperatures suggests a simple way to make things happen fast when we want them to: warm them up. When this is possible and practical, it is very effective, but there are times when high temperatures have undesirable effects. Many of the chemical reactions that take place in living organisms and that are necessary for their survival and functioning would not take place rapidly enough at normal body temperatures to maintain life. But we do not have the option of raising body temperatures by 50 or 100 degrees to speed them up.

However, there are substances called *catalysts* that enable chemical reactions to take place faster. The "catalytic afterburner" used on cars to reduce air pollution contains a solid catalyst that promotes the reaction of oxygen with gaseous products of incomplete gasoline combustion coming from the engine to produce carbon dioxide and water. The reactions that make life possible are catalyzed by specific protein molecules called *enzymes*, which do their job at normal body temperatures. An example is pepsin, a digestive enzyme that speeds up the breakdown of large protein molecules into their constituent amino acids.

An important property of catalysts is that while they shorten the time to reach the equilibrium state, they do not otherwise affect it. The properties of any system at equilibrium are the same whether it got there with the aid of a catalyst or not. And catalysts are not used up in a reaction, so they may be recovered and used again and again.

The study of how fast chemical reactions go, and how to speed them up or slow them down, is called *chemical kinetics*.

entropy increase. It is thus thermodynamically permitted. So why hasn't it happened? What could make it happen? These are questions thermodynamics doesn't answer. We must look elsewhere.

Chemical and spectroscopic evidence shows that the force that holds the two nitrogen atoms together in the N_2 molecule is very strong. To separate them so that they may react with other atoms requires an energy input about double that needed to separate the oxygen atoms in the O_2 molecule. The chemical "fixation" of nitrogen from the air—combining it with other elements so that it may be used for such human needs as fertilizers and explosives—has proved difficult. The German physical chemist Fritz Haber received the Nobel Prize in 1918 for developing an industrial process that combines nitrogen with hydrogen to form ammonia (NH_3). The reaction requires elevated temperatures and pressures if it is to take place at all, though it too is thermodynamically permitted without them.

Electric sparks (and therefore lightning) and nuclear explosions produce small quantities of oxides of nitrogen, but none of these has yet set the atmosphere on fire. Before we relax in the confident assurance that we are safe, we should be aware that bacteria living symbiotically in nodules on the roots of certain leguminous plants (peas, beans, lentils, alfalfa) fix nitrogen without the aid of high temperatures and pressures. The bacteria produce enzymes that catalyze the reaction: we must hope that no catalyst is inadvertently created that causes the reaction to proceed in an uncontrolled way.

The reaction of hydrogen and oxygen to form water is thermodynamically permitted, which is no surprise. The two gases may be mixed in a container and allowed to stand for long periods of time without any detectable reaction, just as the nitrogen and oxygen of the air do. But the hydrogen-oxygen mixture is more sensitive than the earth's atmosphere. A spark, a flame, or a catalyst, such as a piece of platinum, can bring about an explosive reaction that produces H_2O.

Making Diamonds

Diamond and graphite (familiar as the "lead" in pencils) are both forms of the element carbon. It is possible to show by measuring the thermodynamic properties of each that total entropy is larger at ordinary temperatures and pressures when all the carbon is in the form of graphite. Graphite is said to be the "thermodynamically stable" form under ordinary conditions.

According to the second law, diamonds therefore can spontaneously turn into graphite; fortunately, they do not do so except at very high temperatures. The same factor that makes nitrogen unreactive operates here: very strong forces hold together the carbon atoms in the diamond, the same strong forces that make diamonds the standard for hardness. A thermodynamic analysis—much like the one used by Kelvin for the effect of pressure on melting ice—shows that a high pressure affects the thermodynamic properties of graphite and diamond in such a way as to make the total entropy larger when carbon is in the diamond form, which means that a high pressure should be able to change graphite into diamond. In particular, a pressure of about 15,000 atm should do the job at ordinary temperatures. This is a high pressure, but well within the capability of modern laboratories: unfortunately, it doesn't work. The trouble is that the forces holding carbon atoms together even in the softer substance graphite are too strong to allow them to be rearranged into diamond at an appreciable rate at room temperature (the softness of graphite is a misleading criterion for the strength of the forces between the carbon atoms, for reasons we will not give here). Higher temperatures should permit those forces to be overcome by the greater kinetic energy of the atoms, but a thermodynamic analysis shows that the higher the temperature, the higher the pressure needed to reverse the stability. Nevertheless, the attempt was made with the aid of a new technology combining high pressure and high temperature. Success was achieved at a General Electric Company laboratory in 1960, at pressures of 50–60,000 atm and temperatures of about 1,500°C. The high temperatures gave the atoms of carbon in graphite enough kinetic energy to overcome the strong forces holding them together. Once diamonds were formed, rapid cooling prevented them from reverting to graphite when the pressure was released. The diamonds first produced were not of gem quality, but they were useful for industrial purposes, such as grinding and polishing, and they are as "eternal" as natural diamonds. Refinement of the procedure has since enabled gem-quality diamonds to be made. Even more extreme conditions of temperature and pressure must have been reached underground during the geological histories of the strata in which natural diamonds are found.

Equilibrium

Solids and Liquids

The reaction graphite-to-diamond and the reaction ice-to-water have features in common. In both a pure substance in one form (graphite or ice) changes into another form (diamond or water) of the same pure substance. In both reactions, which of the two forms is thermodynamically stable depends on the temperature and the external pressure. Under atmospheric pressure and below the melting point of 0°C, the

The Thermodynamics of Diamonds

The graphite-diamond reaction illustrates in general how we tell from thermodynamics what to expect at equilibrium. The entropies of both diamond and graphite are determined in the laboratory by measuring the specific heat (the ratio of the heat input to the rise in temperature produced; see Chapters 3 and 5) of each substance. From our knowledge of how the specific heat varies from absolute zero to room temperature, we can calculate the entropy of the substances. It turns out that the entropy of diamond per kilogram (203.25 joules/kelvin at 298 K and 1 atm) is considerably less than that of graphite (474.50 joules/kelvin). This of itself might lead one to expect that diamond is less thermodynamically stable at this temperature and pressure, which happens to be true, but the calculation is incomplete so far.

There is another contribution to the net entropy change: a change in the entropy of the surroundings resulting from heat given off or absorbed in the conversion from one form to the other. If heat is given out, the entropy of the surroundings increases; if heat is absorbed, it decreases. From the first law it can be shown that the quantity of heat absorbed when graphite is converted to diamond can be measured accurately by an indirect method, one that doesn't actually require us to convert graphite to diamond. In brief, both graphite and diamond can be burned in oxygen to form carbon dioxide, and diamond gives

total entropy is higher when all the water freezes; ice therefore is the stable form below 0°C, and at equilibrium only ice is present. While it is possible to cool liquid water well below 0°C and keep it for long periods of time in this supercooled nonequilibrium state, its instability is revealed when we drop in it a small crystal of ice (or certain other substances, such as silver iodide, that act as centers of crystallization); rapid crystallization of ice follows. Above the melting point the total entropy is higher with liquid water; hence at equilibrium only liquid water is present. It is only exactly at the melting temperature that both forms can be present simultaneously at equilibrium.

If carbon always reached equilibrium quickly after changes in temperature or pressure, the transition between graphite and diamond out slightly more heat per kilogram than graphite, showing that its energy is slightly higher. It is not necessary to burn large quantities of diamonds to measure this energy difference accurately.

From such energy measurements it is found that 158,000 joules is absorbed from the environment for each kilogram of graphite converted to diamond. The entropy decrease of the environment is therefore

$$\frac{158,000\ \text{joules}}{298\ \text{kelvins}} = 530.20\ \text{joules/kelvin}$$

Since the entropy of carbon in the graphite form is greater than the entropy of the carbon in the diamond form, there has been a decrease of entropy in the carbon as well. To be precise, there has been a decrease of $474.50 - 203.25 = 271.25$ joules/kelvin. The total decrease of entropy is therefore $530.20 + 271.25 = 801.45$ joules/kelvin per kilogram of graphite converted. This tells us that the reaction of graphite to form diamond is impossible at this temperature (298 K) and pressure (1 atm). It follows that the reaction of diamond to form graphite is thermodynamically *permitted*. The calculation of the entropy and energy changes under high pressures and temperatures is a straightforward extension of this procedure, though a little more complicated.

would act much the same. Unless we happened to be exactly at the transition temperature and pressure, only one of the two forms would be present. Since carbon does not reach equilibrium quickly, a graphite pencil and a diamond ring can exist simultaneously in an ordinary room. This is not a violation of the second law, only an example of its limitations.

Thermodynamic Analysis of Soda Pop

We now turn away from considerations of why chemical equilibrium is sometimes reached slowly or not at all, to an examination of the conditions that prevail once chemical equilibrium is reached.

The sharpness of the transitions discussed above—from 100 percent ice to 100 percent water when the temperature is raised a tiny fraction of a degree—is not typical of most chemical reactions. Consider a bottle of soda pop at room temperature. We see two "phases" in the bottle, a liquid phase, mainly water, and a gas phase in the space above the liquid. The gas phase is mostly carbon dioxide, with some air and some water vapor. A simple experiment—prying off the cap—leads immediately to a (slight) explosive expansion of the gas (the "pop"), followed by an effervescence of gas bubbles from within the liquid. We can conclude, first, that the pressure of the gas was higher than that of the outside atmosphere and, second, that the gas that effervesces must have been dissolved in the liquid. Before the removal of the cap, the contents of the bottle were at equilibrium. The carbon dioxide was distributed between gaseous and dissolved states in such a way that the total entropy (including both the contents of the bottle and of the outside environment) was as large as it could be.

Prying off the cap has the initial effect of reducing the pressure of the carbon dioxide gas to about 1 atmosphere: typical pressures in sealed soda bottles at room temperature (25°C) are about 3–4 atmospheres. Once the pressure has been reduced, the carbon dioxide dissolved in the liquid is no longer in equilibrium with the gaseous phase and tends to form bubbles of gas. Equilibrium is not reached immediately. We enjoy drinking soda pop because more of the dissolved carbon dioxide effervesces with each mouthful, making tickling bubbles. Eventually the opened soda goes flat; when a new equilibrium is reached between dissolved carbon dioxide and the very small pressure of carbon dioxide gas in ambient air, no more bubbles form in our mouths.

Equilibria with Gases or Dissolved Substances

Whether the carbon dioxide is in the gaseous or the dissolved state, its concentration—the number of CO_2 molecules in a given volume, such as a liter—can vary as the reaction takes place, and the entropy of the dissolved substance or of the gas is affected by its concentration. This introduces a new variable in the problem. The entropies per gram of pure substances like ice or graphite are affected by temperature and pressure; the entropies per gram of gases and dissolved substances are *additionally* changed when their concentrations change. A full thermodynamic treatment shows that changes of temperature or of the outside pressure acting on a system in which concentrations can vary produces only *gradual* changes in its state, unlike the dramatic changes of melting, vaporization, or the conversion of graphite to diamond.

The effervescence of carbon dioxide is just such a gradual change. As the reaction takes place

$$CO_2 \text{ (aq.)} \rightarrow CO_2 \text{ (gas)}$$

the concentration of CO_2 in the liquid phase decreases and its pressure in the gas phase increases. When does the total entropy have its largest possible value? This will depend on both the concentration of CO_2 in the liquid and the pressure of the gas above the liquid. In general, the entropy of any substance is greater in the gas form (at the same temperature) than its entropy when it is dissolved in a liquid, but when the dissolved carbon dioxide effervesces to form gas, heat is absorbed from the surrounding environment, decreasing the entropy of the environment. There is thus a kind of tug-of-war between these two contributions to the net entropy change, and a thermodynamic analysis of the soda pop bottling process shows that at equilibrium the concentration of dissolved carbon dioxide is greater at higher pressures of the gas above the liquid. The relation is one of simple proportionality:

$$\frac{\text{Concentration of } CO_2 \text{ (aq.)}}{\text{Pressure of } CO_2 \text{ (gas)}} = \text{Some "constant"}$$

Because this ratio depends on temperature, the "constant" is constant only if the temperature remains the same. In Figure 11.1 we show how the equilibrium CO_2 concentration, in grams per liter of water, varies with the pressure of the gas, in atmospheres, at several different temperatures. We can have equilibrium at *any* pressure of CO_2 provided the concentration in the water is that given in the figure for the pressure in question.

Raising the temperature decreases the amount of dissolved CO_2 for any given CO_2 gas pressure—in other words, the numerical value of the "constant" decreases when the temperature increases. Warming therefore makes CO_2 less soluble in the water and thus increases the pressure of the gas. This property of the CO_2-water system can be put to use, as James Joyce described in his short story "Ivy Day in the Committee Room." In this story a group of men with corked bottles of stout and no corkscrews get the corks out by placing the bottles near a fire. Thermodynamics not only shows that this diminishing solubility with increased temperature is a consequence of the fact that heat is absorbed from the environment when the gas effervesces, but it also provides an

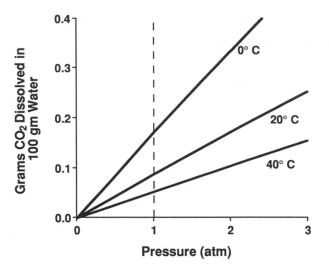

Figure 11.1 Solubility of Carbon Dioxide in Water

The solubility of carbon dioxide in water varies with the pressure of the gas above the water. The variation is shown at three different temperatures; the lower the temperature, the more carbon dioxide is dissolved for any given pressure. The graph also shows why carbonated drinks foam or spray when a warm bottle is opened. At higher temperatures, CO_2 is less soluble in water at any given pressure. In the closed bottle, therefore, warming causes some of the dissolved gas to transfer to the gas phase, increasing the pressure, but the quantity dissolved is still greater than would be in equilibrium with cold soda. Once the cap is removed, the amount effervescing is correspondingly greater.

equation predicting quantitatively the solubility decrease from the amount of heat absorbed, an equation as useful and as applicable to a wide variety of problems as the equation predicting the effect of pressure on the melting point of solids (see Chapter 6).

Unlike the ice-water equilibrium, raising temperature does not suddenly, at one particular temperature, cause all the dissolved CO_2 to go to the gas form. Such an event in this system would involve a net entropy decrease. The largest possible entropy, at any given temperature and with any given amount of CO_2 in the bottle, is always achieved when *some* CO_2 is dissolved in the water and *some* is in the gaseous form. Changing temperature only changes the relative amounts in each form, and does so gradually. Figure 11.2 compares this gradual reaction with the sudden change when ice melts.

To summarize: these gradual changes with temperature in the concentration of dissolved CO_2 and the pressure of gaseous CO_2 in the bottle are typical of chemical equilibria involving gaseous or dissolved substances.

Blood and Entropy

Gases other than CO_2 also dissolve in liquids, and their concentrations similarly rise when the pressure of the gas above the liquid is increased. Oxygen is absorbed by the blood in the lungs and then transported to the tissues where it is used. As blood is mostly water, we might speculate that oxygen dissolves in this water when the oxygen gas pressure is high (in the lungs). The venous blood returning to the lungs, having been depleted of oxygen and now carrying carbon dioxide, is ready to dissolve more. But this arrangement wouldn't work: oxygen gas is much less soluble in water than is carbon dioxide—less than 0.01 grams O_2 will dissolve in a liter of water at the oxygen pressure of ambient air.

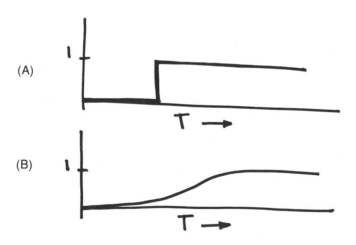

(A)

$T \longrightarrow$

(B)

$T \longrightarrow$

Figure 11.2 Gradual Shift of Chemical Equilibrium
Ice does not melt gradually over a range of temperatures; instead, it melts completely at a single temperature, as in (A). In contrast, the carbon dioxide in a sealed bottle is present both as a gas and as a dissolved substance, and its solubility in the water diminishes gradually as the temperature is raised. It does not suddenly effervesce completely and explosively at a single temperature. The result, shown in (B), is that the gas pressure in the bottle increases gradually rather than suddenly as the temperature is raised. (In both graphs, the vertical axis indicates the fraction of the reaction that has taken place.)

The volume of blood pumped by the heart would therefore have to be enormously greater than it actually is to deliver the O_2 needed by the tissues.

Blood is thicker than water for a number of reasons, one being the presence of red cells, which give blood its characteristic color. The cells are red because they contain hemoglobin, an iron-containing protein that forms a red chemical complex with oxygen (when exposed to it in the lungs); the complex breaks up and releases the oxygen where it is needed (in the tissues). The presence of hemoglobin enables the blood to carry much more oxygen than can dissolve in an equivalent volume of water. In the lungs, oxygen gas (at a pressure of about 0.2 atm) first dissolves in the water of the blood. Once dissolved, it reacts with hemoglobin to form the oxygen-hemoglobin complex. Both processes are near equilibrium in the lungs (see Figure 11.3). In the tissues there is no gas phase: oxygen is present either as the oxygen-hemoglobin complex or as dissolved oxygen in the water (serum) phase. As the dissolved oxygen is used by the metabolizing cells of the tissues, the oxygen-hemoglobin complex breaks up to release more. The system, though a little more complicated than the pop bottle, lends itself to a similar analysis.

Using the symbols Hb for the hemoglobin molecule and $Hb{:}O_2$ for the complex it forms with oxygen, let us write the reaction in which the complex is formed as

$$Hb + O_2 \text{ (aq.)} \rightarrow Hb{:}O_2$$

O_2 (aq.) is oxygen dissolved in the water of the blood. In the oxygen-hemoglobin complex, the oxygen molecule is bound to an iron atom of the hemoglobin molecule, a process chemically like the rusting of iron and thus a kind of combustion. The binding of oxygen by hemoglobin leads to a release of heat and an entropy increase in the surrounding environment. In the vicinity of the metabolizing cells, the complex breaks up to release oxygen:

$$Hb{:}O_2 \rightarrow Hb + O_2 \text{ (aq.)}$$

This is the reverse of the first reaction. When it takes place the combustion of the iron atom in the hemoglobin is thus reversed, and heat must be supplied from the surrounding environment to make it happen, with a consequent decrease of entropy there.

A quantitative treatment of the various entropy changes in this process predicts the binding of oxygen by hemoglobin in fair agreement with experiment. Better agreement is obtained when it is recognized that several oxygen molecules are absorbed by a single hemoglobin molecule and that absorption of the first makes absorption of additional ones easier.

In more qualitative terms, this is how blood does its job:

1. Gaseous oxygen in the air is inhaled into the lungs. It is at a pressure of 0.2 atm.
2. It dissolves in the water (serum) of the blood. When blood is in equilibrium with air at the normal human body temperature of 37°C, the concentration of dissolved oxygen is only about 0.005 gram per liter.

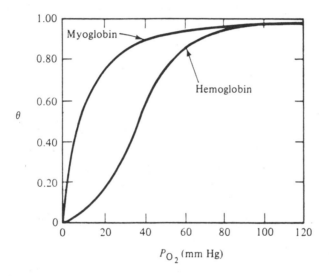

Figure 11.3 Binding of Oxygen Gas by Blood Proteins
Two experimentally observed curves for the binding of oxygen by a blood protein as the pressure of oxygen gas is changed: myoglobin is the oxygen-binding protein in the blood of whales, hemoglobin that in humans. The curve for myoglobin fits a simple theory for the binding, in which one oxygen molecule is assumed to bind randomly to each myoglobin molecule. The curve for hemoglobin does not fit this simple theory. It appears that in humans each hemoglobin molecule has several possible sites for binding oxygen; binding at one site makes binding of a second oxygen in a neighboring site more likely. The more complicated behavior of hemoglobin actually meets the biological needs of the human organism better than the simpler behavior would. (θ = fraction of protein reacted. O_2 pressure measured in millimeters of mercury; see pages 175–176.)

3. Once dissolved in the serum, oxygen is then bound to hemoglobin in the red cells. Typically human blood contains enough hemoglobin to bind up to 0.3 gram of oxygen per liter.

4. In the metabolizing cells, O_2 combines chemically with carbon from glucose to form carbon dioxide, in which form it diffuses out of the cells and, dissolved in the serum, is returned by the blood to the lungs to be exhaled.

The entropy per gram of gaseous oxygen is very high, of dissolved oxygen in water much less, and of oxygen bound to hemoglobin less still. If these entropy changes were the only ones to be considered, hardly any gaseous oxygen would dissolve in the blood at all. It is the heat given out to the environment, a small quantity when oxygen dissolves in water and a larger quantity when it binds to hemoglobin, that increases the entropy of the environment and leads to an increase of the total entropy, thus making the process possible.

Carbon Monoxide Poisoning

The danger of exposure even to low levels of carbon monoxide (CO) is well known. Some carbon monoxide is formed whenever carbon-containing fuels—wood, charcoal, coal, gasoline, oil—are burned, more when insufficient oxygen is available for complete combustion. An automobile engine running in a closed garage soon produces a lethal concentration of CO.

Some gases—carbon dioxide, helium, nitrogen—are not poisons but can asphyxiate us if they displace the oxygen we need to breathe. Air is approximately 80 percent nitrogen and 20 percent oxygen. A mixture of 80 percent helium with 20 percent oxygen will support life just as well. In contrast, carbon monoxide can be fatal in less than one hour at levels of 0.35 percent in air, even though oxygen remains at 20 percent. The reason is that hemoglobin binds carbon monoxide more strongly than it binds oxygen. Other things being equal, a hemoglobin molecule is 200 times more likely to bind CO than O_2. As a result, if we are breathing air with the usual 20 percent oxygen and 0.1 percent carbon monoxide, half the hemoglobin molecules that would normally have carried oxygen are carrying CO instead. Our blood in turn is bringing to our tissues only half the oxygen it usually carries.

The net entropy increase for the binding of CO to hemoglobin

$$Hb + CO \rightarrow Hb:CO$$

is greater than the increase for the binding of O_2. There is a greater heat output for the formation of Hb:CO than for Hb:O_2, and thus a larger increase of entropy in the surrounding environment. The other entropy changes are not much different from those for O_2 binding.

The Flexibility of Equilibrium

Unlike the equilibria reached in ice-water or graphite-diamond systems, but like soda pop, the oxygen-hemoglobin equilibrium responds only slightly to slight changes in temperature or pressure. A small rise in temperature does change the entropies of each of the substances present a little, but equally small changes in the concentrations of the dissolved oxygen and the oxygen bound to hemoglobin molecules produce compensating entropy changes so that a new equilibrium is quickly reached, with a little more Hb and a little less Hb:O_2 than before. A dramatic change—the disassociation of all the Hb:O_2 molecules—resulting from a small rise, say, in temperature would be a disaster: a slight fever would kill us. Similarly, a small decrease in air pressure—the difference between New York (at sea level) to Denver (a mile high), for example—reduces the solubility of oxygen gas in blood only slightly. A small number of Hb:O_2 complexes disassociate to free their oxygen molecules and a new state of equilibrium is reached, one in which the concentrations are only slightly changed from before. If that small decrease in air pressure produced a dramatic change in the solubility of oxygen gas in the serum, traveling from one altitude to another would be fatal.

A reaction involving gases or dissolved substances does not usually go to completion: both reactants and products will be present at equilibrium. We may describe the limited extent of the reaction by a percentage: the percentage of the total carbon dioxide that is in the gas phase, the percentage of hemoglobin molecules that have bound an oxygen molecule. Thermodynamics shows that in general, when there are at least two gases or dissolved substances taking part in a reaction, the extent is never 0 and never 100 percent. In the example of carbonated soda we gave, carbon dioxide plays a double role—as *gas* and as *dissolved substance*. For some reactions, the percentage may be so close to either 0 or 100 percent that we may think the reaction has gone to completion. When we burn a gaseous fuel like propane in sufficient oxygen, it looks as though all the propane burns. This is not strictly true; there is some unburned propane at equilibrium but the quantity is far too small to detect, and for many purposes we can disregard it.

For many other reactions, known as *limited* reactions, the incompleteness at equilibrium is striking, as, for example, in the extraction of iron from its ores. Iron was presumably discovered when rocks containing Fe_2O_3 (chemical name: ferric oxide) were exposed to the heat and combustion gases of a wood fire. Hematite, one of the most common ores of iron, is a blackish-red to brick-red mineral, essentially Fe_2O_3. The name comes from a Greek root meaning "blood": hemoglobin is the protein pigment of blood, and iron plays an essential role in its structure, its function, and its color.

In a modern blast furnace a mixture of iron ore, limestone (to react with silicates from the ore that would otherwise interfere with the desired reaction), and coke (made by heating coal in the absence of air) are fed in at the top and then a stream of very hot air is blown in at the bottom. First the coke reacts with carbon to form carbon monoxide and then the carbon monoxide reacts with the oxide of iron to remove the oxygen. The products are metallic iron and carbon dioxide.

The net reaction between carbon monoxide formed from coke and iron ore can be written:

$$Fe_2O_3 \text{ (solid)} + 3CO \text{ (gas)} \rightarrow 2Fe \text{ (liquid metal)} + 3CO_2 \text{ (gas)}$$

In the nineteenth century, it was observed that the exhaust gases from the furnace still contained large quantities of unreacted CO, more in fact than CO_2. This seemed wasteful: more coke was being used in the process than was strictly needed to convert the ore to metal. The response of the iron and steel industry in France and England to this problem was described by the French thermodynamicist Henri LeChatelier in 1887:

These investigations [of chemical equilibria using thermodynamics] of a rather theoretical sort are capable of much more immediate practical application than one would be inclined to believe. Indeed the phenomena of chemical equilibrium play a capital role in all operations of industrial chemistry . . . It is known that in the blast furnace the reduction of iron oxide is produced by carbon monoxide, according to the reaction

$$Fe_2O_3 + 3CO = 2Fe + 3CO_2$$

but the gas leaving the chimney contains a considerable proportion of carbon monoxide, which thus carries away an important quantity of unutilized heat. Because this incomplete reaction was thought to be due to an insufficiently prolonged contact between carbon monoxide and the iron ore, the dimensions of the furnaces have been increased. In England they have been made as high as thirty meters. But the proportion of carbon monoxide escaping has not diminished, thus demonstrating, by an experiment costing several hundred thousand francs, that the reduction of iron oxide by carbon monoxide is a limited reaction. Acquaintance with the laws of chemical equilibrium would have permitted the same conclusion to be reached more rapidly and far more economically.

What was at the frontier of scientific research in 1887, "investigations of a rather theoretical sort" of which the captains of industry were completely unaware, are now covered in full in first-year chemistry courses. Freshmen are routinely asked, as an exercise in the application of the entropy concept, to show that at equilibrium, at the temperature of the blast furnace, there will be about three times as much carbon monoxide in the gaseous phase as carbon dioxide. The reaction is a limited one, and bigger furnaces can't change the concentrations at equilibrium.

Chemistry goes on all around us, and within us as well. Substances are reacting to form new ones, in our bodies, in the atmosphere and in the oceans, in factories and in automobile engines. All reactions have a natural direction in which they must proceed: that of increasing net entropy. Thermodynamics enables us to predict that natural direction, and to be sure that the reactions cannot go in the reverse direction: ashes, smoke, and carbon dioxide cannot spontaneously re-form into wood and oxygen.

Thermodynamics does more: it enables us also to predict what substances are present and in what quantities if chemical equilibrium is reached. But the "if" is a big "if"—it raises questions we cannot answer with thermodynamics alone.

12
Biology:
Muscles, Kidneys,
Evolution

~~~~~~

Does the second law of thermodynamics apply to living things as it does to inanimate matter? To answer this question we will first examine the inputs and outputs of energy for an organism using muscles to do work—and show that net entropy increases, just as it does when an automobile engine runs or a battery is used to start it. Next we will go beyond the macroscopic bookkeeping approach and discuss what happens on the molecular level, when living organisms convert the energy of the foods they eat into the work they must do to stay alive. While organisms "burn" food in oxygen to get the energy they need, they are not heat engines but engines that convert chemical energy directly into the energy of motion. There are three main kinds of biological work: the work done by the muscles to produce motion; osmotic work, such as is done by the kidneys when they transfer urea from the bloodstream to the urine; and synthetic work, done to build large molecules from smaller ones. For all these kinds of work, a particular molecule, adenosine triphosphate, carries chemical energy, obtained by the reaction of the sugar glucose with oxygen, to the molecules that do the work.

Finally, we will examine the claim, made by some who believe in the literal truth of the biblical account of creation, that the evolution of species and an origin of life from inanimate matter contradict the second law. We will show that a proper application of that law—one that takes into account all entropy changes, including those in the surrounding environment—leaves us with no reason to believe that either evolution or current scientific ideas on the origin of life need violate it.

Muscular Work

The first biological process to which we will apply the laws of thermodynamics is the work done by the muscles of an animal. As in our example of the heat engine, we must begin with certain measurements:

How much work is done?
How much fuel is consumed to do the work and what are the
 energy and entropy changes due to the consumption of the fuel?
How much heat is given out to the environment?

The *way* we measure work and energy for a living organism will differ from the way we study a heat engine, of course, but nevertheless we can calculate the net entropy change when an animal does a certain amount of work and verify that it represents an increase. In addition, the second law permits us to calculate the efficiency of fuel consumption, the ratio of the actual work done to the maximum possible.

The Work Done

How do we determine the amount of work done by the muscles of a living organism? One way is to equip an exercise bicycle with an "ergometer," a device that calculates the work done, usually in Calories for the health-conscious exerciser, by multiplying the force exerted on the pedals by the distance the pedals travel in their circular path. A person climbing a staircase or a mountain certainly does at least an amount of work given by the product of his weight and the height climbed. The same calculation applies to a person lifting weights. As we will see, it is not difficult to measure fuel consumption simultaneously with exercise, if the subject does the exercise while remaining in the same place. This gives exercise bicycles and treadmills pride of place in laboratories where work output and fuel input are to be compared. The quantitative measurements of work done that we describe were mostly done with such devices.

Fuel Consumption

The fuels the body uses include carbohydrates (for example, the sugar glucose, $C_6H_{12}O_6$, to which the body converts other carbohydrates), fats, and under conditions of starvation or prolonged exercise, proteins. Fats and glucose do not provide equal quantities of energy per gram: fats

Is Work Always Measurable?

Not all ways of doing work lend themselves to quantitative measurements as easily as do exercise bicycles. Some are more difficult, and for some we have no good measuring procedure at all.

Let us look more closely, for example, at the process of putting one foot in front of the other on level ground. The average result over many steps is no height change, but if we examine the process instant by instant we realize that work is being done after all. First of all, at regular intervals a leg is lifted and its potential energy increased. Then the body, originally stationary, is set into motion, gaining kinetic energy. When there is motion, there will also be friction, in this case friction within the joints and muscles, with the air, against the ground.

Both kinetic and potential energies can be measured directly by cinematographic techniques: how much the leg is raised, how fast the forward motion of the body is at each instant. Additional information about gains and losses of both forms of energy are obtained when the walker or runner steps on "force platforms," instruments that measure both downward and lateral forces exerted by the foot.

As a person walks, the kinetic and potential energies rise and fall with each complete step but in opposition. The physicist's term is *out of phase:* one is at a maximum when the other is a minimum. A large share of the potential energy is converted to kinetic as the foot is lowered; walking has been described as a series of falls arrested just in time. Muscular work is done both to increase the potential energy of the leg when it is raised and to resupply the kinetic energy lost to the various forms of friction.

Bipedal walking, with its alternation of kinetic and potential energy, has been compared to the rolling of an egg about an axis perpendicular to the long axis. The center of gravity of the egg rises and falls as it rolls. If the egg is given an initial push, and thus an initial kinetic energy, its kinetic energy diminishes as its center of gravity (and therefore its potential energy) rises, so that it is rolling more slowly at the moment it is standing on its end. As the center of gravity descends, potential energy is converted to kinetic and the speed of rolling increases again. The alternation of kinetic and potential energy is similar to what happens when a pendulum swings.

During other kinds of activity, the muscles are certainly used but it is not obvious whether work is done at all. Clearly we cannot attempt to apply thermodynamics when work done can neither be directly measured nor accurately calculated, but the questions are worth thinking about anyway. We might think walking down a mountain should not require work at all, but we do get tired walking down, which is only a little more surprising than the fact that we also get tired simply holding a weight in the air. In these activities the product of the weight and the height it is raised is either zero or, in the case of coming down a mountain, negative (as shown by the fact that we can go downhill on a bicycle or a taboggan and arrive at the bottom not only not tired but with kinetic energy to spare). Has work been done nevertheless?

Consider as an example an individual downhill step. The leg is moved out horizontally first, so that it is above the descending ground, and then lowered. As the leg is lowered it must resist the pull of gravity to remain under the control of the climber; no one wants to go downhill in free fall. Fuel energy is spent to control the leg, as can be demonstrated from oxygen consumption, but unlike the energy cost of climbing, it is hard to calculate from the laws of macroscopic physics. The muscular work of controlling the leg, or of holding an object off the ground, is done at the molecular level, in the tensed muscles. This subject is too complex to deal with here, but we can measure the work done quantitatively from the oxygen consumption.

give more, but they require more oxygen per gram to burn. In strenuous exercise it is the rate at which we can breathe in oxygen that limits our performance, so we tend to burn glucose, though for slower, steadier, more prolonged exercise we burn fat also. Fortunately, how much of each we are using during exercise can be determined by measuring oxygen intake and carbon dioxide output.

We will simplify matters somewhat by considering only one chemical reaction during exercise, the combustion of glucose:

$$\text{glucose} + \text{oxygen} \rightarrow \text{carbon dioxide} + \text{water}$$

For each glucose molecule burned, six oxygen molecules are used up, and six carbon dioxide and six water molecules are produced:

$$C_6H_{12}O_6 + 6O_2 \rightarrow 6CO_2 + 6H_2O$$

By laboratory procedures described in Chapters 5 and 11 we can determine the energies and entropies of each of the substances in the equations above, and hence calculate the changes in energy and entropy from the chemical reaction alone. From this information we can also calculate the resulting changes in the energy and entropy of the environment.

While it would be difficult to measure directly how much of the glucose or other carbohydrates in the subject's body were actually used during exercise, we know that six carbon dioxide molecules are produced for each glucose molecule metabolized. So we can infer glucose consumption by monitoring the carbon dioxide output of the subject. This can be done either by performing the experiment in a closed room equipped with appropriate monitoring devices—a so-called human calorimeter, to be described below—or by having the subject wear a face mask with input and output tubes. Fats and carbohydrates give different numbers of liters of carbon dioxide for each liter of oxygen used to burn them: one liter of carbon dioxide for carbohydrates and 0.7 liter for fats. Thus the mix of the two fuels actually being used can be determined if oxygen consumption is measured simultaneously with carbon dioxide output.

Heat Output

In Chapter 3 we described Lavoisier and Laplace's calorimeter. In one of their experiments, the heat given out by a resting guinea pig was measured and compared to its CO_2 output. Lavoisier's conclusion was that the heat output was approximately the same as would be given out by burning an amount of carbon sufficient to produce the same quantity of CO_2. The result is approximate because animals don't metabolize carbon but rather carbohydrates, and the heat output from burning carbon happens not to be very different from that from burning carbohydrates, per liter of CO_2 produced. The heat output of a *resting* human being can also be measured in a "human calorimeter," a similar but larger and more comfortable device than Lavoisier's, and compared with the heat output of food that is directly burned in oxygen. The two are found to agree within the limits of error of the measurements, confirming that the first law applies here. The reason the human calorimeter is used only for a resting subject is that errors are introduced when a subject exercises; for example, the subject sweats, putting out more water than is produced by the combustion of the food and causing heat loss by evaporation.

In the human calorimeter, heat output is measured not by a quantity of ice melted but by the small rise in temperature of water circulated through pipes passing through the insulated chamber in which the subject sits. The results using the human calorimeter are, as mentioned above, only for resting, not for exercising subjects, but they are needed as a baseline: after all, even a resting animal metabolizes food, breathes in oxygen, and breathes out carbon dioxide and water, so these must be subtracted from the inputs and outputs during exercise to determine those due to exercise alone.

For the exercising subject, a procedure called "indirect calorimetry" is used. No chamber is needed, only a breathing mask worn by the subject. First we establish the relations among fuel consumption, heat output, O_2 intake, and CO_2 output in the human calorimeter. Then by measuring O_2 intake and CO_2 output through the breathing mask, we can infer how much glucose and fat are metabolized and how much water is produced. The calculation of the heat output of the exercising subject is trickier, and it requires an application of the first law. The resting subject in the human calorimeter, doing no work, releases the same quantity of heat per gram of glucose metabolized as is given out by that same gram of glucose burning in oxygen in a laboratory flask. For each *additional* gram of glucose metabolized by the exercising subject, the heat output is *less* because some of the energy is used to do work. Since the work done is measured as described earlier, the correction to the heat output can be calculated.

The Thermodynamics of Exercise

The obvious way to test whether the laws of thermodynamics apply to an exercising individual is to calculate from the various measured quantities described above the net energy change, which should be zero, and the net entropy change, which should be positive. There is, however, an alternative and equivalent test of the laws. This alternative is more useful here because it focuses on work, an easily visualized concept, rather than the more elusive entropy, but it is applicable only to processes taking place in a system, living or inanimate, in contact with an environment whose temperature and pressure remain constant. We may assume that living organisms interacting with their ambient environment, at least for short periods of time, meet this requirement.

Common experience and common sense suggest that an infinite amount of work cannot be done on one pound of food or fuel. There must be an upper limit, no matter how carefully we do the conversion, no matter how well we avoid friction and other kinds of waste. Let us call this upper limit the *maximum work*, W_{max}. How can we tell, for any given quantity of food, what that maximum work is?

It is plausible that the maximum work any process can perform is done when the process is carried out "reversibly"—in a state of near-perfect balance between opposing forces, so that the direction of the process can be reversed with negligible effort and all motion takes place extremely slowly, minimizing the effect of friction (See Chapter 5). It would be easy to determine the maximum work if it were easy to carry out the chemical reaction reversibly. This is possible for some chemical reactions, but not for the majority, including the combustion of glucose in oxygen.

Fortunately, the two laws show (under the conditions of constant environmental temperature and pressure) that the reversible work can be calculated *without actually having to run the reaction reversibly* from the energies and entropies of the chemical substances involved, as determined in the laboratory. The two laws not only enable the reversible work to be calculated, they also predict that the actual work done in any process under constant temperature and constant pressure cannot exceed the reversible work calculated in this way. This provides a proof of what we guessed to be plausible: the reversible work *is* the maximum work. If we define the *efficiency* of this constant-temperature process as the ratio of the actual work to the calculated reversible work, it follows from the two laws that the efficiency cannot exceed 100 percent. This may seem a self-evident statement, but it isn't. Suppose we found by experiment that the efficiency was 150 percent—what would we conclude? That either the first law, or the second, or both have failed an experimental test.

Let us consider the work done, and the efficiency with which it is done, by an organism using glucose as the primary fuel.

The chemist's unit of quantity of material is the *mole*. It represents a fixed number of molecules (6.02×10^{23}; see Chapter 4) rather than a fixed weight. For glucose, a mole is 180 grams, a little less than half a pound. The maximum work this much glucose can do on combustion in oxygen to form carbon dioxide and water is found from the net energy and entropy changes to be 2,870 kilojoules, about 0.8 kilowatt-hour.

In various experiments comparing the actual work done with the amount of glucose consumed to do it, we find that the efficiency of muscular work is a fairly consistent 25–30 percent. That the efficiency is less than rather than greater than 100 percent shows that the two laws of thermodynamics are not violated when an animal works. It does, however, raise another question. Of the 0.8 kilowatt-hour of work a mole of glucose burning in oxygen could ideally do, most is wasted. Why?

Why the Waste?

The reason is that to do the maximum work possible, the body would have to consume the glucose reversibly, which implies that it would have to operate infinitely slowly. The goal of an organism is to maximize the amount of work done in a definite time period, such as a day, and the question is, what is the optimal efficiency of fuel utilization to reach this goal? We can see immediately that if the efficiency is 100 percent we get no work done in a day because we are too slow; nor would we get anything done if efficiency were 0 percent. The optimal condition is somewhere in between.

This is not to say that we could operate at 100 percent efficiency if only we worked very, very slowly. Over the course of evolution our cells have settled on one particular sequence of chemical reactions and one particular set of enzymes for oxidizing glucose. The work we actually do per molecule of glucose depends primarily on this sequence and only to a lesser extent on how fast we move. Conceivably evolution under other conditions could have produced a different sequence of chemical reactions, that would have given us, say, an efficiency of 35–40 percent rather than 25–30 percent. The chances are, however, that the efficiency we have is the best compromise between the availability of fuel, the time

"Free Energy"

Let us consider processes taking place spontaneously in contact with a surrounding environment whose temperature and pressure do not change. We can imagine a chemical or physical process taking place in one of two extreme ways: perfectly reversibly, doing the maximum possible work (W_{max}), or completely irreversibly, doing no useful work whatever. For the first process, the net entropy change is zero; for the second it must be positive, otherwise it would not take place spontaneously. An example of the second kind of process is gasoline burning in the open air rather than inside an automobile engine, or a storage battery running down while standing on a shelf. Let us call the entropy change when the process takes place in this least efficient fashion ΔS_{max}. Then it can be shown from the first and second laws that

$$W_{max} = T\Delta S_{max}$$

ΔS_{max} includes, as usual, both the entropy changes of the chemical substances taking part in the reaction and any changes in the entropy of the environment that result from heat given out to it or absorbed from it. It can be calculated accurately from the entropies and energies of the starting substances and the products. For the reaction to be spontaneous ΔS_{max} must represent an increase: that is, it must be positive in sign. Thus the maximum work is positive also, which tells us that the reaction can do work for us.

If on the other hand ΔS_{max} were negative, the maximum work would also be negative. This means that the reaction will not occur spontaneously but can be made to occur *if work is done on the system to bring it about.*

An example: the reaction between lead, lead oxide, and sulfuric acid to form lead sulfate and water, which powers an automobile battery, has a positive net entropy change when it takes place spontaneously (when the chemicals are mixed). When the reagents are arranged properly in a battery (the metallic lead must be used for the electrodes), the reaction can do work for us (turn the engine over). The reverse reaction, the conversion of lead sulfate and water to metallic lead, lead oxide, and sulfuric acid involves a net entropy decrease. It does not take place *unless we do work on it to recharge the battery*. In the recharging process, the net entropy change—including the change in the environment, which would have been *negative* if the battery had spontaneously recharged itself—will be *positive* (unless the recharging process is carried out reversibly and therefore with infinite slowness, in which case it would be zero).

The equation for W_{max} given above shows that the criterion of possibility given us by the second law—nothing is possible that leads to an entropy decrease—is clearly equivalent (when the temperature and pressure of the surrounding environment are held constant) to the statement that nothing is possible unless it can do work for us. That is why a discharged auto battery left on the shelf is never found to have spontaneously recharged, whereas a charged battery left to itself long enough is often found to have spontaneously discharged, doing none of the maximum possible work.

Chemical substances that can undergo a reaction may be thought of as storing a potentiality to do a certain amount of work, calculated as above from the net entropy change when the reaction occurs spontaneously. As the substances react, the potentiality to do work decreases, whether or not the work is actually done. The "potentiality to do work" is just the negative of the maximum work the reaction could do. It is called the *Gibbs free energy*, usually symbolized by the capital letter G (in honor of its discoverer, the American scientist J. W. Gibbs).

In an equation,

$$\Delta G = -T\Delta S_{net}$$

For any reaction taking place spontaneously in an environment at constant temperature and pressure the following three statements are essentially equivalent:

1. The net entropy must increase.
2. The reaction must be capable of performing work for us.
3. The total free energy of the substances taking part in the reaction must decrease.

needed to process it and transport it within the body, and the need to get things done expeditiously.

The Chemical Engine

There are different ways energy stored in one form or another can be made to do work. We have discussed at some length the operation of heat engines, which do not work unless they receive heat at a high temperature and discharge some of it to a colder environment. Other kinds of engines—those that use the potential energy of water behind a dam, electric energy supplied by a power plant, the elastic energy of a wound spring, or compressed air—do work at a single temperature. Since these engines use forms of energy other than heat, it might be thought thermodynamics has nothing to say about them, but we shall see it says a lot.

Chemical energy, though it can be converted to thermal energy to power a heat engine, can also be converted to other forms of energy directly. The energy supplied by a battery is produced by a chemical reaction and is first converted to electrical and then to mechanical energy. But it is also possible for chemical energy to be converted *directly* into mechanical energy. No battery does this, however—it is almost exclusively the mechanism used by living organisms.

The work done by and in living organisms falls into three broad categories.

1. *Muscular.* The muscles of animals do work in both the scientific and common senses of the term, as when they raise legs against the force of gravity to propel the body forward against the friction of the ground or circulate the blood against the resistance of the blood vessels.

2. *Osmotic.* Another kind of work is exemplified by the job the kidneys do: transferring urea and water from the blood to the urine and preventing the loss of salts and sugar from the blood. For both urea and salt, the largest possible entropy would occur when the concentrations of each in both blood and urine are equal. Concentrating either substance in one of the two liquids at the expense of the other is something like compressing a gas originally distributed uniformly throughout a container into one-half of it, thus decreasing its entropy by doing work on it. The work of extracting a dissolved substance from one liquid and concentrating it in another is called *osmotic work,* and thermodynamics enables us to calculate the minimum amount of work required and the entropy price to be paid.

3. *Synthetic.* A third kind of work living bodies perform is synthesis, also called *chemical work.* Large, biologically important molecules are made from smaller ones, and the entropy of the large molecules is smaller than the entropy of the small ones.

Proteins, for example, are made by joining amino acids together. When we eat meat, or cheese, or beans, the animal or vegetable proteins are broken down by digestive enzymes into their constituent amino acids. This is an entropy-increasing and therefore spontaneous process; the digestive enzymes—catalysts—merely make it happen faster than it would in their absence. The entropy increase on breaking a large molecule down into many smaller ones can be understood as arising from the greater freedom of spatial arrangements and of motion the small, now-independent molecules acquire. The human body then re-constructs protein using both the amino acids from food and amino acids synthesized in the body, combining them in the right quantities and the right order to make human protein. This is an entropy-decreasing process, and it would not occur unless there is a simultaneous compensating entropy-increasing combustion of a fuel.

We could take a "black box" approach to the study of living organisms—by disregarding all molecular and biological details and concentrating on macroscopically observable quantities of heat, work, and chemical change—but we want to know more—to see the molecular processes going on inside the "black box." How does the combustion of glucose lift a weight or pump water and urea out of the blood stream? Many processes are now understood, but much of what goes on in a body remains unexplained. In what follows we examine those biological processes that run the "chemical engine" of a living organism and the types of work they can do.

ATP, the Intermediate Fuel

The energy made available by the metabolism of foods such as sugars is not used directly to perform work. Instead, it is used to produce a kind of intermediate fuel, a carrier of the energy from the site where food is metabolized to the site where it is used; this intermediate fuel may be compared with the electrical energy generated at a power plant, which is a carrier to homes and factories of the energy made available by the combustion of coal or oil.

The chemical name of this carrier substance is adenosine triphosphate, abbreviated ATP (Figure 12.1). What is remarkable about ATP is not only that it is the primary agent through which all kinds of biological work are done, but also that it is common to all forms of life: bacteria and plants as well as humans. It must have been "invented" by living organisms very early in the history of life.

When ATP is used by the body to do work, it reacts with one molecule of water and loses one of its phosphate groups. The products of this reaction then return to that part of the cell where fuel, usually sugars,

(a)

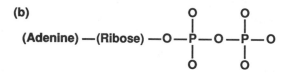

(Adenine) —(Ribose) —O—P—O—P—O—P—O

(b)

(Adenine) —(Ribose) —O—P—O—P—O

(c)

O—P—O

Figure 12.1   The ATP Molecule

A schematic representation of the ATP (adenosine triphosphate) and ADP (adenosine diphosphate) molecules. Adenine, a compound of carbon, oxygen, hydrogen, and nitrogen having basic (acid-neutralizing) properties, is the "organic base." Ribose is a sugar (a simple carbohydrate molecule). Adenine is combined with ribose to form a molecule called *adenosine.* Phosphate groups (a phosphorus atom with associated oxygen atoms) may be attached to the ribose portion of an adenosine molecule. For ATP, three phosphate groups are attached *(a),* and for ADP *(b)* one of the phosphate groups has been detached and is found in the aqueous solution as an independent molecule combined with two hydrogen atoms taken from a water molecule. Phosphate groups *(c)* may bear negative electric charges not shown in the figure.

is being metabolized and ATP is reconstituted, to be used again. The first reaction, a net entropy-increasing process, we will write as

$$ATP + H_2O \rightarrow ADP + P_i$$

Regeneration, a net entropy-decreasing process, is written as

$$ADP + P_i \rightarrow ATP + H_2O$$

ADP is adenosine diphosphate, the ATP molecule with one less phosphate group. The phosphate group consists of an atom of phosphorus surrounded by four oxygen atoms (Figure 12.1). This group usually has a negative electric charge as well. When the phosphate group is split off from ATP, it combines with hydrogen atoms from the water to form the free molecule H_2PO_4, which we symbolize by P_i and which bears a single negative charge.

The first reaction is a spontaneous one, in which the net entropy increases. It follows that the reverse action is not spontaneous: it can take place only when some other spontaneous reaction takes place with a compensating entropy increase to "drive" it. The regenerating reaction takes place in an organ of the cell called the *mitochondrion*, where glucose or other fuels are reacted with oxygen to power the process.

Synthetic Work

Animals store their carbohydrate supplies in the form of a large molecule, glycogen, which is essentially a long chain of glucose molecules linked together. Using the abbreviation G for a glucose molecule, we can represent glycogen as G–G–G–G– . . . or equally well, G_n, where n is the number of glucose units in the glycogen chain.

Let us assume that in living organisms glycogen is made stepwise from glucose molecules:

$$G + G \rightarrow G_2$$
$$G_2 + G \rightarrow G_3$$
$$G_3 + G \rightarrow G_4$$

and so on. But each of these steps involves a small net entropy decrease, so they should not occur spontaneously. Chemical studies show, however, that glucose molecules do react in the body to form glycogen in spite of this entropy decrease and, moreover, that one ATP molecule is simultaneously converted to ADP for each glucose added on to the glycogen. The ATP reaction has a large enough net entropy increase to compensate for the entropy decrease when glucose molecules form glycogen, so the overall net effect of both reactions is an entropy increase. But how do the glucose molecules "know" that?

Consider the following two reactions:

$$(1)\ G + G \rightarrow G_2$$
$$(2)\ ATP + H_2O \rightarrow ADP + P_i$$

Their net effect is that two glucose molecules form a double molecule and an ATP molecule breaks down into ADP and phosphate. The latter reaction involves an entropy increase and will occur spontaneously, but the former involves an entropy decrease and will not. What actually happens in the cell is a sequence of two spontaneous reactions: a reaction between ATP and glucose that transfers a phosphate group to one glucose

$$(3)\ ATP + G \rightarrow ADP + G\text{–}P$$

followed by a second reaction between the phosphated glucose and a second glucose

$$(4)\ G\text{–}P + G + H_2O \rightarrow G_2 + P_i$$

These reactions are catalyzed by enzymes. *Both* of them involve small entropy increases, so both are spontaneous. Yet the net chemical change of the pair of spontaneous reactions—(3) and (4)—is exactly the same as the net chemical change of the original pair—(1) and (2)—only one of which could occur spontaneously. In both pairs we start with ATP and two glucoses and end up with ADP, phosphate, and G_2.

	Reaction	Net entropy change (joules/kelvin)
(1)	$G + G \rightarrow G_2$	-52
(2)	$ATP + H_2O \rightarrow ADP + P_i$	$+91$
Net:	$G + G + ATP + H_2O \rightarrow G_2 + ADP + P_i$	$+39$
(3)	$ATP + G \rightarrow ADP + G\text{–}P$	$+26$
(4)	$G\text{–}P + G + H_2O \rightarrow G_2 + P_i$	$+13$
Net:	$G + G + ATP + H_2O \rightarrow G_2 + ADP + P_i$	$+39$

Since the *net* chemical change of (1) and (2) is the same as (3) and (4), the net entropy increase is also the same, but reaction (1) is not spontaneous because of the entropy decrease while (3) and (4) are both spontaneous.

This is how ATP does the "chemical work" of making large carbohydrate molecules out of small ones: it substitutes one set of chemical reactions, not all of which can occur spontaneously, with a different set, each of which can occur spontaneously and which accomplish exactly the same net effect. The synthesis of proteins and fats is carried out in an analogous way.

Weights and Pulleys

To visualize more clearly how a spontaneous reaction can make a nonspontaneous one occur, consider a mechanical analogy.

A weight will not rise upward spontaneously, but it will fall spontaneously. A weight can, however, be made to rise if we have a heavier one connected to it by a pulley and rope. If the two weights are exactly equal, neither will move, but if one is a little heavier—enough to overcome friction in the rope and pulley—it will descend and lift the lighter one. How much heavier the second weight must be to raise the first one depends on how fast we want to do it. The heavier the second weight, the faster the first rises, but the more energy is wasted in friction. The changes of potential energy of the two weights, though opposite in sign to the entropy changes of the two chemical reactions, are analogous to changes in entropy. If a heavy weight is used to raise a lighter one, some potential energy is wasted. To make a reaction with an entropy decrease occur rapidly enough, we need a driving reaction whose entropy increase is *greater*, not just *equal to*, the decrease of the reaction to be driven.

Of course, to make one weight raise another we need a rope and pulley. The "rope and pulley" in glycogen synthesis is the enzyme-catalyzed reaction of ATP and glucose to form glucose-phosphate, followed by its reaction with another molecule of glucose to join the two glucoses together and produce a free phosphate molecule.

Osmotic Work

Osmotic work—the work of extracting dissolved substances from one liquid and transferring them to another, more concentrated one—is less familiar than muscular work, but it consumes a large share of the energy provided by food intake. We have used the operation of the kidney as an example; let us give another.

Whatever is dissolved in the fluid of a living cell tends to leak out through the cell wall into the surrounding liquid; whatever is dissolved in the surrounding liquid tends to leak into the cell interior. In a state of equilibrium, the concentrations of all dissolved substances would be the same inside and out: so decrees the second law.

In spite of the second law, the red cells floating in the serum of the blood normally have much higher potassium concentrations and much lower sodium concentrations than the serum. These concentration differences are required for the cells to perform their biological functions. The cells can be brought nearly to an equilibrium state, in which potassium and sodium concentrations are the same inside the cell and outside in the serum, by storing blood at low temperatures. But warm the blood up and allow glucose and oxygen to enter, and the cells quickly pump out sodium and pump in potassium to reestablish the original, unequal concentrations. In this nonequilibrium state, the entropy of the sodium and potassium is less than it would be if concentrations inside and out were equal, but careful measurements show that more than enough glucose was oxidized to pay for it. Again, we want to know more than that the books balance. How do the molecules do it?

The Motion of a Large Molecule

Muscle tissue is composed of protein fibers that are aligned parallel to the direction of stretch or contraction. There are two chemically different types of protein fiber, which lie next to one another in the tissue. In contraction the fibers of one type move relative to the other.

The wall of a cell is mainly composed of fatty substances insoluble in water. Protein molecules are embedded in this rather thin wall, with each molecule extending out on both sides of the wall.

In both muscle work and osmotic work there is motion. Muscles are attached at their ends to the joints, so that when they contract the limbs are made to move. In osmotic work small molecules are transferred from one side of a cell wall to the other by the motion of a protein molecule. ATP is also the carrier of the energy for these processes. There are no heat engines, pistons, or flywheels involved, only molecules undergoing chemical reactions. How does the chemical reaction of an ATP molecule cause motion?

Proteins are large molecules built up by connecting smaller ones to form a chain. The individual units are called amino acids, and there are some twenty different kinds found in proteins. In addition, some proteins may have other kinds of molecular units than amino acids: the blood protein hemoglobin contains such units, called *hemes*, which contain iron atoms that bind oxygen. A protein molecule, as well as other large molecules like polystyrene, is long enough and flexible enough to bend around and make contact with itself at one or more places. Sometimes additional chemical bonds, called *cross-links*, hold the points of contact of these chains together (see Figure 12.2).

Chain is a useful metaphor for these molecules. A chain of only two links would have some freedom of movement but one would hardly

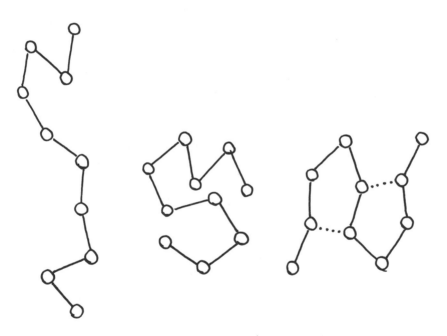

Figure 12.2 Polymers
Polymers are large molecules made of a large number of smaller molecules connected together. Both synthetic products, such as plastics, and naturally occurring molecules, such as proteins, carbohydrates, and nucleic acids (DNA), are examples. Such molecules, when they are connected together in a chain, possess considerable flexibility. Here we show a protein molecule made up of individual amino acids, symbolized by circles. The chains can take many different spatial configurations because of their flexibility: the figure shows two, an extended and a compact one. When additional cross-links are introduced (as shown at right), much of the flexibility is lost.

consider it "flexible." When we string hundreds of links together, however, the limited freedom between adjacent links makes the long chain very flexible indeed. Flexibility implies that many different conformations (spatial arrangements) of the links are possible: this is what gives rubber its stretchability. Large molecules share with simpler molecules the property of having kinetic energy by being in constant motion. They move from one place to another, they rotate and vibrate, but the large flexible molecules also undergo internal motion at each link, so that they are constantly undergoing changes from one spatial conformation of the chain to another. They have been compared to strands of spaghetti in rapidly boiling water.

The protein molecules in both muscle tissue and cell membranes have so many cross-links that much of their flexibility is lost. However, those that can perform muscular or osmotic work can take at least two different conformations, and as time goes on their constant erratic motion causes them to hop from one conformation to the other and back again. The Maxwell-Boltzmann formula for molecular probabilities shows that if the two conformations had the same energy, each molecule would spend about half its time in each. If, however, the two conformations—call them A and B—have different energies, with B higher, the molecules would average more time in A. The relative energies of the two conformations can be altered by a reaction between the protein molecule and ATP. The basic ATP reaction is

$$(1) \ \text{ATP} + \text{H}_2\text{O} \rightarrow \text{ADP} + \text{P}_i$$

and results in a net entropy *increase*. However, ATP can also react with the protein in a cell wall to add a phosphate group to the protein:

$$(2) \ \text{ATP} + \text{Protein} \rightarrow \text{ADP} + \text{Protein–P}$$

This can be followed by a reaction of the protein-phosphate with water:

$$(3) \ \text{Protein–P} + \text{H}_2\text{O} \rightarrow \text{Protein} + \text{P}_i$$

The net chemical effect of reactions (2) and (3) is identical with (1) alone; the protein is restored to its original state, and ATP has been converted to ADP and P_i. It follows that the net entropy change of (2) and (3) together is the same as that of (1). Let us now suppose that the two spatial configurations A and B of the original protein are also present in the protein-phosphate molecule. Let us also suppose that when the phosphate group is attached to the protein, it changes the relative energies of the A and B configurations so that now B is energetically favored. Of course when the protein loses its phosphate group, A is again energetically favored (Figure 12.3).

We can write the reactions above as

(4) ATP + Protein (A form) → ADP + Protein–P (B form)

(5) Protein–P (B form) + H_2O → Protein (A form) + P_i

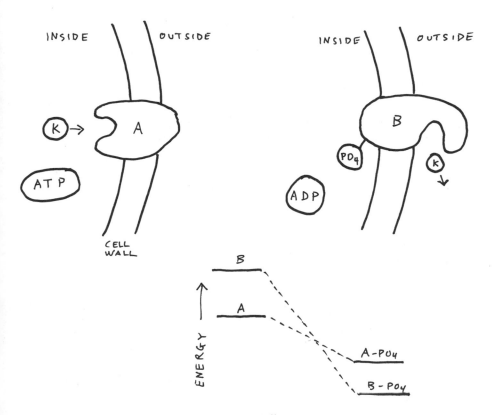

Figure 12.3 Cell Pumps

A highly schematic representation of a process that is not too well understood: The protein molecule embedded in the cell wall can exist in two different spatial arrangements, which we designate A and B. In the normal protein one arrangement (A) is favored, but when a phosphate group is attached to the protein, the other (B) is favored. The addition to the protein of a phosphate from an ATP molecule, followed by the detachment of the phosphate group to form a free phosphate molecule in solution, causes back-and-forth transitions of the protein molecule between its A and B configurations. The transition enables the protein to pump a potassium ion (K) from one side of the cell membrane to the other.

The net effect is the same as for reactions (2) and (3), but now we can see the protein undergoing transitions back and forth between the two configurations as the ATP reacts, like a piston moving back and forth in a cylinder. This is the basic step in the process by which the work of the muscle or the kidney is done. In the muscle, reaction (4) is the step that leads to a contraction of the muscle. In the kidney the conformational change of reaction (4) transfers a small molecule or a sodium or potassium atom across the cell membrane.

Thermodynamics and the Creationists

Is the evolution of species consistent with the second law of thermodynamics? When Darwin proposed his theory of evolution by natural selection in the nineteenth century, it was perceived as inconsistent with religious belief. Certainly it was inconsistent with a literal reading of the account of creation in the Hebrew Bible. Though in time many religious individuals and institutions have reconciled themselves to the apparent inconsistency, some have not. "Creationists" are those who believe in the literal truth of the Bible and therefore believe also that life was created in the relatively recent past by a deity.

Creationists have not only relied on biblical evidence but have also attempted to find weaknesses in the scientific evidence for evolution. They have formed organizations to foster their efforts, such as The Creation Research Society and The Institute for Creation Research. One of their first claims, made in the 1940s and repeated frequently since, is that evolution would be a violation of the second law.

We will deal narrowly with the creation-evolution controversy: we focus only on possible violations of the second law and ignore all the nonthermodynamic issues involved, such as the mechanism of natural selection, the adequacy of the fossil record, the evidence for the age of the earth and the universe, biochemical similarities between humans and apes, and so on.

The creationist criticism of evolution has been stated forcefully by Robert E. Kofahl, Ph.D., in a book, *Handy-Dandy Evolution Refuter* (San Diego, 1977), published under the auspices of an organization called Creation-Science Research. Dr. Kofahl's field of science is not identified. Early in the book he gives a brief but clear and accurate exposition of the second law for the nonspecialist reader (pp. 39–40). Following this he states (pp. 40–41) that evolution "is supposed to be a natural process which occurred according to the laws of physics, yet transformed completely disordered matter into highly ordered, energy-rich living organisms. And supposedly these organisms continually increased in order and complexity by evolution. Surely in this there is an enigma: scientists who accept the second law of thermodynamics as valid believe nevertheless that a natural process called evolution, which violates that law of physics by causing entropy to decrease, has continued for billions of years."

Kofahl is raising two distinct points: one concerns the possible origin of life from inanimate matter; the second is the evolution of complex, many-celled, many-organed creatures—monkeys, trees—from one-celled organisms. He is right that if either could be shown to violate the second law, the case for special creation by a deity would be stronger.

The Entropy of a Mouse

To apply thermodynamics to the problem of how life got started, we must ask what the net energy and entropy changes would have been if simple chemical substances, present when the earth was young, were converted into living matter.

For each of these processes, we may identify an initial state and a final state, a "before" and an "after." Our test of thermodynamics asks: In the process as it actually occurred, has there been any change at all in the total energy of everything involved? Is the total energy "after" the same as it was "before"? Has there been a net increase in the entropy? Is the total entropy after greater than it was before?

To answer the questions we must be able to determine the energies and entropies of everything present initially and of everything present in the final state. Energy determinations are—at least in principle, and usually in practice—quite straightforward: the energy change of a system undergoing any process is given by the sum of the heat and work inputs and outputs; how the process is carried out doesn't matter. Entropy determinations are more difficult. We can make them by laboratory measurements only when the processes meet two conditions:

1. The values of such macroscopic properties as temperature, chemical composition, and pressure in both the initial and final states must be uniform throughout the system and must not be changing with time. In living organisms, where the chemical reactions necessary to sustain life must always be going on, this criterion is not usually met.

2. Entropy changes taking place during a process that a "system"— here a functioning organism—undergoes are the sum of changes in the

system itself and in the environment. To determine the entropy changes in the system, we must find an alternative and reversible process for going from the initial to the final state—one in which frictional and other dissipative losses are eliminated and the work done is the maximum possible—and measure heat inputs or outputs and temperatures during this process. The real process we are interested in need not be reversible, and it may occur wastefully, doing less than the maximum work possible, or even none at all. Once we have determined the entropy change *of the system* along the idealized reversible path, however, then *that* entropy change during the real process must be the same, although the entropy changes in *the environment* need not be, and usually will not be, the same.

There is as yet no known reversible process for converting living organisms to those simple chemical substances they revert to on death—carbon dioxide, water, salts—nor a process for reversibly converting the simple substances back to living matter. How, then, can we apply the second law to evolution, or to the origin of life?

Approximating the Entropy

There are several assumptions or approximations we may make that permit an estimate of the entropy changes on assembling a living cell from its chemical constituents.

The chemical composition of even single cells are extraordinarily complicated, and the more they are studied the more new substances are discovered. However, for the purpose of an entropy estimate for the formation of a living cell we do not need to know all of them. The number of simple substances that make up 95 percent of the mass of a cell is limited to about 20 amino acids, a few sugars, fatty acids, organic nitrogen-containing bases, salts, and a lot of water. In the cell most of these simple substances are combined together to form proteins, starch or other carbohydrates, and nucleic acids, dissolved in the watery environment of the cell, and fatty, insoluble membranes. The energies and entropies of the simple substances are known from laboratory measurements, as are the changes when they are dissolved in water and then joined together to form large molecules. What is not known accurately and can as yet be estimated only roughly are the entropy changes in the last steps: the assembly of these large molecules into the functioning organelles of the living cell and the assembly of living cells into a whole organism, the steps that distinguish a chicken from concentrated chicken soup. This rules out an accurate calculation, but the uncertainty in the estimated entropy is not thought to be very large.

The results of the analysis show that the process of making living organisms out of simple molecules requires a net *decrease* of entropy. Equivalently, the process of converting living matter to carbon dioxide, water, nitrogen, and a few salts leads to a net entropy increase and can take place spontaneously, a thermodynamic prediction sadly in accord with common experience. This implies that life could not have arisen spontaneously under conditions of equilibrium, a point on which both evolutionists and creationists would agree.

How Did Life Begin?

The earth, however, was never in a state of thermodynamic equilibrium. Four-and-a-half billion years ago, when it was formed, and before life existed on it, it was, as it is today, hotter inside than out and bathed in radiant energy from the sun. There were then, as there are now, winds, lightning, volcanoes, and earthquakes. The radiant energy of the sun corresponds to black-body radiation at the sun's surface temperature of 6,000°C, and its absorption at the earth's surface represents a flow of heat energy from 6,000°C to about 20°C (the mean temperature of the earth's surface). There is thus a great increase in entropy every day the sun shines; it may be estimated as a twenty-fold increase of the entropy of the incoming radiation. No net decrease when life began need have taken place: a plausible scenario which does not contradict the second law is that there would be merely a smaller rate of increase of entropy than if life had never come into being.

Our argument is weakened, though, by the fact that we don't as yet have a plausible hypothesis as to just how the sun's radiation, a lightning bolt, or a volcanic eruption could have got life started. All the processes we regard as characteristic of life—self-replication being the most dramatic—require the cooperative interaction of large and complicated molecules. These in turn seem to require living organisms to make them. Laboratory experiments in which simple chemical substances of the type that may have been present in the atmosphere of the earth in its early years were exposed to radiation or electric discharges have yielded a few substances found in living matter, but not the variety and complexity needed to make a plausible case for a mechanism for converting inorganic to organic matter. In particular, only two amino acids—glycine and alanine—of the fifteen or so found in proteins were produced in appreciable amounts, and no sugars or fats. Other hypotheses have been proposed, one of the more interesting being that the surface layers of clay particles, which have surprisingly complicated atomic arrangements, may have served as templates for the synthesis of large biological molecules, and one of the more bizarre being that seed molecules for life were formed in outer space and carried to earth on comets. None of these hypotheses has so far convinced an appreciable number of scientists. Creationists consider this as evidence that no plausible hypothesis will ever be found; most scientists are much more optimistic.

Entropy Changes during Evolution

Complex creatures, whose bodies contain many different organs with specialized functions, are in some sense more "ordered" than one-celled organisms. It is not clear, however, whether the greater order they possess implies a significantly lower entropy. We are up against a problem referred to earlier: we cannot tell if the entropy is really lower or by how much because we do not as yet know of a reversible process for "assembling" complex organisms from single cells. The rough estimate we have described above suggests that the entropy difference is not large, but there is no experimental confirmation as yet. But the complexity of living organisms is only one source of the entropy changes occurring when species evolve. A more significant part is the entropy increases living organisms constantly produce, either as they utilize the sun's radiant energy by photosynthesis or as they ingest food (whose ultimate source is the sun's radiant energy) and oxygen and discard waste products. These entropy increases go on in the daily life of the individual, and they have gone on throughout the whole history of life on earth.

Kofahl recognizes (pp. 46–47), as do other creationists, that in the ordinary processes of the life of an animal from birth to death there is no violation of the second law: the entropy increases when food is metabolized balance the entropy decreases of certain biological process. Even giving birth to young doesn't change the balance: creationists have not claimed that the birth of a fully formed, multicellular human baby only nine months from the fertilization of a one-celled ovum by a one-celled spermatozoon is a violation of the second law. And if giving birth to young does not violate the law, it is hard to argue that giving birth to an individual whose genetic makeup gives it a survival advantage would violate it either.

To summarize: it is plausible on the face of it that living organisms produce enough entropy increases to account both for their continued existence and their evolution from single-celled species. By now the reader may be tired of hearing that to test the second law we must consider all entropy changes in the process: decreases of entropy in one part of a system we are studying are possible if compensating increases occur elsewhere. While believers in the theory of evolution are not in a position to prove by exact calculation or show by experiment that the evolution of complex organisms does not violate the second law, creationists can neither prove that it does or even show that a violation is plausible.

Evolution and Maxwell's Demon

Now let us end this chapter with a change in focus. Instead of using the second law to test the theory of evolution, let us assume evolution is a fact, and use this fact to provide one more experimental test of the second law.

We have discussed in Chapter 8 the challenge to the second law posed by Maxwell, in his invocation of a "demon": an intelligent being with very fine perceptions who could separate individual molecules in such a way that ultimately work could be performed by them at the expense of the thermal energy of the environment. We argued there that anything an intelligent being can do by perception and the exercise of choice can be done equally well by a device the intelligent being can invent. This argument turns the attention to devices rather than beings, and the question becomes: can we imagine a device of fiendishly clever design that can do in reality what Maxwell's demon could do in fantasy? Is it possible to make a perpetual-motion machine of the second kind? Ever since the second law was discovered, attempts to invent such a device have been made. Most often, the inventions have been imagined but not built: tiny doors with springs just strong enough to distinguish fast molecules from slow ones; a metal sheet with conical holes whose dimensions are comparable to the distance a molecule typically travels between two collisions, and so on. All devices so far proposed have been shown sooner or later to involve fallacies of one sort or another; most often the mistake is to ignore the inevitable Brownian motion of the device itself.

When faced with some new proposal for such a device, most scientists who are familiar with the second law will have the confidence to predict the device will fail, even if they cannot immediately point out the particular fallacy that it embodies, and they would prefer to wait for a working model before concluding that the second law is in trouble.

One of the things that makes living things fascinating is their extraordinary adaptability. Over the billions of years that they have evolved, they have solved, by means of natural selection, such problems as how to move from one place to another, how to metabolize sugar to get the energy to do so, how to see both their prey and their predators as well as the path ahead of them, how to make copies of themselves, how to convert radiation from the sun into sugar, and many more. Given this adaptability, it is significant that with all that thermal energy available in the environment, no known life form has ever evolved in such a way as to put it to use for doing work. Think what a competitive advantage it would be not to need food to stay alive! So much of the effort of living things is spent to get food that any organism that could dispense with eating would have more time to spend on reproductive effort and would dominate the earth. *If* evolution can produce a hemoglobin molecule, an eye, or an ant colony, *and if* any escape from the second law—a molecular process or cell doing the work of a Maxwell's demon—were possible, *then* evolution should have produced it by now also. It didn't, so our confidence in the second law is further strengthened.

13
Geology:
How Old Is the Earth?

A literal reading of the Hebrew Bible gives the age of the earth as some six thousand years. A careful analysis of the biblical account by James Ussher, a seventeenth-century Protestant bishop of Meath, in Ireland, and Archbishop of Armagh, put the year of the creation as 4004 B.C.E. Evidence of the biblical flood was found in the fossil shells of marine animals occurring in rock strata at great elevations. Some of the first practitioners of the science of geology, to explain the mode of formation of the various rocks and minerals then being distinguished and studied, assumed that a tremendous flood had in fact covered the whole surface of the earth at one time. The first to suggest that the earth might be older than implied by the biblical account was the French naturalist the Comte de Buffon, who in the eighteenth century put forward some speculative ideas about the development of the earth and the evolution of species, which did not have much immediate impact on geology or biology. Buffon assumed that the earth had been formed originally as a molten ball of rock and then cooled down to its present state, and he gave a figure for its age as about 80,000 years, based on his estimate of how long it should have taken to cool.

In 1788, a Scottish amateur student of geology, James Hutton, published a paper with the unwieldy title "Theory of the Earth; or an Investigation of the Laws Observable in the Composition, Dissolution, and Restoration of Land upon the Globe." Hutton had studied chemistry and medicine at Edinburgh and the Sorbonne and received a doctorate in medicine at Leiden, but he never practiced medicine. First he farmed in England, and then he became a partner in a successful chemical manufacturing firm, which gave him financial independence. He returned to Edinburgh in 1768 to spend his time on scholarly pursuits for the rest of his life. His 1788 paper and a subsequent book, *Theory of the Earth, with Proofs and Illustrations* (1795), initiated a revolution not only in geology but in man's picture of the universe and his place in it. Just as the Copernican revolution enlarged space enormously and shrank the earth to a microscopic region of it, so Hutton's geology enormously lengthened time and shrank to very little the share of it belonging to human history. Prior to Hutton, geology had been catastrophe-minded. The surface of the earth was covered with dramatic features: mountain peaks, chasms, rushing rivers, unfathomable oceans. These were believed to be the relics of a past era during which cataclysmic events such as vast floods, earthquakes, and volcanic eruptions were common. Hutton saw clearly that these features most likely arose from the simple, slow action of commonplace forces; such as flowing water, gradually but inexorably sculpting mountains out of solid rock and depositing the detritus at the mouths of rivers to form sedimentary rocks, which in turn were uplifted and underwent once again the cycle of carving and deposition.

Hutton's concepts have been given the name *uniformitarian*. They describe an earth in which the slow and unexciting changes we see taking place today are the same as those that have gone on in the indefinite past: there is no prior era of catastrophes that shaped the earth to what it is; simple, steady processes are quite sufficient, given time.

But the time required was almost beyond the powers of human beings to conceive. As John Playfair (see page 306) put it:

> The mind seemed to grow giddy, by looking so far into the abyss of time; and while we listened with earnestness and admiration to the philosopher [Hutton] who was unfolding to us the order and series of these wonderful events, we became sensible of how much further reason may sometimes go than imagination may venture to follow.

Hutton concluded his 1788 paper as follows: "the results, therefore, of our present enquiry is, that we find no vestige of a beginning, no prospect of an end."

When Hutton's views were first presented, they were attacked not only on religious grounds, as would be expected, but also by his scientific colleagues as well. But the more geological evidence accumulated, the more plausible they became. By the time Charles Lyell published his enormously influential *Principles of Geology* in 1830–1833 (in three volumes), supporting and extending Hutton's work, the opposition, both on religious and scientific grounds, had weakened, and controversies among geologists on this issue were carried on in a spirit of mutual respect and friendliness. Within a short time the uniformitarian views had basically prevailed over the catastrophic ones, although certain details were still argued about. The geologist's earth was one that had existed an indefinite but enormously long time already and would continue to do so in the future.

Charles Darwin was greatly impressed with Lyell's book and carried a copy with him on the expedition of the *Beagle*. He made a number of geological observations, along with his biological ones, on this journey; most notably, he proposed an explanation of the formation of coral reefs, involving an interaction of both biological and geological factors, still considered correct today. When Darwin came to develop his theory of evolution, he was able to conclude from his study of Lyell that the earth had existed for the vast periods of time required for evolution to have occurred by natural selection, a process taking place, in Darwin's view, by a gradual accumulation of very small changes. The first (1859) and every subsequent edition of *On the Origin of Species* contains the sentence: "He who can read Sir Charles Lyell's grand work on the Principles of Geology, and yet does not admit how incomprehensibly vast have been the past periods of time, may at once close this volume."

In the *Origin*, Darwin made one estimate of an age for the earth. There is a large eroded valley on the south coast of England called the Weald. The volume of matter that must have been carried off by marine erosion could easily be calculated from the dimensions of the valley. From a rough calculation of the rate at which denudation was taking place, Darwin concluded that 300 million years was needed for the formation of the valley. Unfortunately for him, a number of geologists showed that his calculation was too crude to be reliable and was probably a gross overestimate of the time required. Many of the scientific criticisms of *The Origin of Species* focused on Darwin's treatment of the denudation of the Weald, and he withdrew it for the third edition. In a letter to Lyell, he remarked that he had "burned my fingers so consumedly with the Wealden."

The Flow of Heat in the Earth

William Thomson, later Lord Kelvin and commonly referred to today as "Kelvin," was thirty-five years old when the *Origin* was published, and he was already an accomplished scientist. Educated, along with his older

The Effects of "Action Long Continued"

James Hutton published his *Theory of the Earth* in 1795, but his ideas received more attention when they were explained by his friend and associate, John Playfair, in a book that appeared after Hutton's death. Here is Playfair, in *Illustrations of the Huttonian Theory of the Earth* (1802), describing the difference in perspective between the theory of "catastrophic" or "cataclysmic" geological change and the uniformitarian view associated with Hutton.

> It is, however, where rivers issue through narrow defiles among mountains, that the identity of the strata on both sides is most easily recognised, and remarked at the same time with the greatest wonder . . . There is no man, however little addicted to geological speculations, who does not immediately acknowledge, that the mountain once was continued quite across the space in which the river now flows; and if he ventures to reason concerning the cause of so wonderful a change, he ascribes it to some great convulsion of nature, which has torn the mountain asunder, and opened a passage for the waters. It is only the philosopher [scientist], who has deeply mediated on the effects which action long continued is able to produce, and on the simplicity of the means which nature employs in all her operations, who sees in this nothing but the gradual work of a stream, that once flowed

brother James, by their self-educated father, a professor of mathematics at Glasgow University, Kelvin matriculated at the university at age ten. He began winning prizes that same year (one for a translation from Latin), and at sixteen won a university medal for an essay, "On the Figure of the Earth."

One of Kelvin's professors at the university introduced him to Fourier's mathematical theory of the flow of heat, and he studied it avidly: it was a major influence on his career as a scientist. His first scientific paper was an exposition of Fourier's theory, which he published under a pseudonym. He published several other important papers on topics in physics while attending Cambridge but did not use his own name on them until he had graduated, presumably to avoid embarrassing his professors.

One of the conclusions he derived from Fourier's theory was important in his thinking about the problem of the age of the earth: it is a conclusion resembling but not the same as the second law he was to as high as the top of the ridge which it now so deeply intersects, and has cut its course through the rock, in the same way, and almost with the same instrument, by which the lapidary divides a block of marble or granite.

Playfair cites the shapes of river valleys as evidence for Hutton's ideas:

> Every river appears to consist of a main trunk, fed from a variety of branches, each running in a valley proportional to its size, and all of them together forming a system of valleys, communicating with one another, and having such a nice adjustment of their declivities, that none of them join the principal valley, either on too high or too low a level, a circumstance which would be infinitely improbable if each of these valleys were not the work of the stream that flows in it.
>
> If, indeed, a river consisted of a single stream without branches, running in a straight valley, it might be supposed that some great concussion, or some powerful torrent, had opened at once the channel by which its waters are conducted to the ocean; but when the usual form of a river is considered, the trunk divided into many branches, which rise at a great distance from one another, and these again subdivided into an infinity of smaller ramifications, it becomes strongly impressed upon the mind that all of these channels have been cut by the waters themselves; that they have been slowly dug out by the washing and erosion of the land; and that it is by the repeated touches of the same instrument that this curious assemblage of lines has been engraved so deeply on the surface of the globe.

discover independently of Clausius some years later. Fourier's mathematical description of the flow of heat enables us to predict, for a body in which a flow of heat is taking place, how the temperature in various parts of the body will change with time until the final equilibrium state of uniform temperature is reached. As an example, consider a sphere in which the temperature is initially high and uniform being exposed suddenly to a cold environment. First the surface cools, and then the interior: the final state is reached when the center has cooled down to the temperature of the environment. The mathematics enables us to look backward in time as well as forward. Suppose we consider the cooling sphere at some intermediate time: the temperature in the sphere is not uniform: the surface is cold and the inside still warm. Suppose further we do not know the prior history of the sphere but only its state at this particular moment; we know only how the temperature varies from the center to the surface and the temperature of the environment. Fourier's mathematics enables us not only to predict the *future* course of the temperature distribution to its final uniform state, but also to infer the *past* course of the temperature distribution back to the starting moment when the sphere was hot and the temperature within it was uniform. In brief, Fourier's theory implies a starting time for any situation involving heat flow. An infinite past is impossible.

This reasoning led Kelvin to the idea that the earth's age must be finite and that Fourier's theory might provide a clue to how old it could be. In 1846 the twenty-two-year-old Kelvin was elected to a professorship of Natural Philosophy at Glasgow University, a position he held for the rest of his life. It was customary for a newly elected chair-holder to give an inaugural lecture on a topic selected by the faculty. The topic chosen (suggested, in the opinion of his biographer, to the faculty either by him or his father) was "Age of the Earth and Its Limitations as Determined from the Distribution and Movement of Heat Within It." Unfortunately, no copy has survived.

The Second Law: "A Finite Period of Past Time"

For the next few years Kelvin was occupied with the work on which his current scientific reputation rests: the independent discovery of the second law. One of the papers he wrote at this time, applying thermodynamics to a variety of physical systems, was called "The Universal Tendency in Nature to the Dissipation of Mechanical Energy." He notes in it the tendency of mechanical energy to be dissipated by friction or similar processes, and he concludes that once dissipation has occurred, the original state of mechanical energy of some body—in which there was a potential to do work, some or all of which was wasted—cannot be restored without work from an outside source of energy being done on the system. He wrote:

> Any restoration of mechanical energy, without more than an equivalent of dissipation, is impossible in inanimate material processes, and is probably never effected by means of organized matter, either endowed with vegetable life or subjected to the will of an animated creature . . . Within a finite period of past time the earth must have been, and within a finite period of time to come the earth must again be unfit for the habitation of man as at present constituted, unless operations have been or are to be performed which are impossible under the laws to which the known operations going on at present in the material world are subject.

In brief, a decrease of total entropy is impossible.

The term *arrow of time* came later, but the idea it embodies, discovered by Kelvin (and by Clausius), formed the foundation of all his later thinking about the age of the earth. An indefinite past for the earth is impossible. Whatever state it may have been in the past, the processes that take place on it and that constitute its geological history must involve dissipation—the flow of heat from hot to cold regions, the flow of air and water on the surface, the ebb and flow of the tides, the grinding of rocks by glaciers, and the eruptions of volcanoes. As these processes take place, the total entropy increases.

The earth does not have a fixed quantity of energy: it receives energy from the sun and radiates energy in turn into space. If the geological processes taking place on earth were powered by the energy received from the sun, the problem would only be shifted from here to there: what is the source of the sun's energy? Again, the two laws tell us even the sun's energy is not unlimited. As the sun radiates enormous amounts of energy into space, either it must be receiving energy from some other source, or it is using up its own supply and must eventually burn out. Whatever the case, neither the past nor the future of the solar system can be identical with its present state.

Kelvin's Estimates of the Earth's Age

But Kelvin went further than to assert that the past of the earth cannot be infinite. He applied the laws of physics to set quantitative limits on the possible age of the earth. He made three different estimates; the one in which he had the greatest confidence was based on the rate of cooling of the earth. From available data on the current temperature distribution of the earth, he was able to use Fourier's theory to infer how long in the past it would have been hot enough to be in a molten state. The second estimate was of the age of the sun, since the earth was not likely to be older. It was obvious that the energy radiated by the sun could not be accounted for by ordinary chemical combustion: no fuel known to Kelvin could supply so much energy for more than a few thousand years. The only hypothesis he could propose was that the sun was surrounded by a large cloud of meteors falling in upon it and generating thermal energy in the process. There were problems with this hypothesis, as the amount of meteoric material available could only be guessed at and there was no independent way to prove it was there at all. Kelvin later supplemented this calculation by allowing for some additional thermal energy produced by gravitational shrinkage of the sun.

The third estimate started with the assumption that the earth and the moon were once a single body and that some time in the past the moon split off from the earth. Now the moon revolves about the earth, just as the earth revolves about the sun, because there is a balance between its kinetic energy that tends to make it escape and the gravitational potential energy that pulls it back. In the absence of friction of any sort this state of affairs would continue indefinitely, but friction there is. The gravitational attraction of the moon causes the tides, and in their ebb and flow there is a dissipation of the kinetic energy of the motion of the water by friction. A mathematical analysis of the effect of this friction on the motion of the moon leads to the conclusion that the moon will slowly recede from the earth. Turning the analysis backward in time, Kelvin, by making some assumptions about the rate of tidal dissipation of the energy of the earth-moon pair, could calculate how far back in the past the moon must have split off from the earth.

The three methods did not give exactly the same age, and the age given by each depended on particular assumptions made in the absence of definite knowledge. This uncertainty is not surprising, nor is it necessarily a poor reflection on Kelvin's endeavor. Very often in science we are in the position of having to apply a theory we have every reason to trust to data of uncertain accuracy. The predictions of the theory will have a great range of uncertainty because of the shaky data. A good theory cannot compensate for that.

The first method, based on the cooling of the earth, was believed by both Kelvin and his contemporaries to be the most reliable, but there were problems encountered in applying it.

The interior of the earth is warmer than the surface. We are made dramatically aware of this when volcanoes erupt, but it was also known from measurements of temperatures in deep mines, data with which Kelvin had familiarized himself. The temperature in the earth rises with depth below the earth's surface at a rate of 1°F per 50 feet. Now if the inside of the earth is warm and the surface cold, it follows that heat must be flowing from the inside to the surface. The surface is cool because outer space is cold, and the heat lost by radiation to outer space is not sufficiently compensated for by the heat received from the sun to keep the earth's temperature constant.

So the earth is a cooling ball of matter. From Fourier's theory and the kind of mathematics that was Kelvin's forte, he could infer both how hot the earth was at its center and how much hotter it must have been in the past. And at some calculable time in the past the earth must have been molten. The history of the earth inferred from geology begins with a solid earth whose surface temperature is below the boiling point of water, so the time elapsed since the earth was molten gives an upper limit to the geologic age.

Now there is no reason to doubt Fourier's theory: it is too well confirmed in the laboratory. But to apply it one needs to know the conductivities for heat of all substances of which the earth is composed, their melting points, and their latent heats of melting. The heat conductivity of those rocks found at the surface are easily measured, as are their melting points, and it was reasonable to use these values in the calculations. But one could not claim with confidence that the interior of the earth is made from rocks like those on the surface, nor could one claim that at the extremely high pressures and high temperatures at the core the properties of rocks need be the same as they are in the moderate conditions in the laboratory.

Kelvin, who was aware of these problems and was constantly revising his age estimates, always allowed considerable leeway in his conclusions. In one of his first estimates he bracketed the earth's age as lying between 20 million and 400 million years, with 100 million as the most probable value. As time went on, he kept returning to his calculations, using new data on the properties of matter or modifying the assumptions, and his estimates tended to shrink below 100 million. He recognized the sensitivity of the estimates to the assumptions made, but felt that they did not compromise the main point he wanted to make: that the earth's age was finite, and with proper application of the laws of physics it could be determined; in other words, the unlimited time span invoked by Hutton and Lyell was impossible. Kelvin was not only candid about the extent to which his results depended on the starting assumptions, but he was also willing to consider as a possibility that those laws of physics he had helped establish and in which he had such confidence might themselves not be the last word. In one of his papers on the age of the sun he wrote:

It seems, therefore, on the whole most probable that the sun has not illuminated the earth for 100,000,000 years, and almost certain that he [sic] has not done so for 500,000,000 years. As for the future, we may say, with equal certainty, that inhabitants of the earth cannot continue to enjoy the light and heat essential to their life, for many million years longer, unless sources now unknown to us are prepared in the great storehouse of creation.

The Geologists' Response

Kelvin's first papers setting forth these ideas, published in the 1850s and 1860s, initially did not have much influence on geologists. But he persisted in his attempts to make them aware of the implications of the laws of physics for their field. In 1868 he gave a lecture to the Geological Society of Glasgow with the title "On Geological Time," summarizing his views and his estimates, and finally his ideas began to receive attention. He continued to work on this problem and publish his results for the rest of his long professional life. As his scientific prestige, already high when he began this work, continued to rise throughout his career, so also did his influence. His first major convert in the geological world was a personal friend and fellow Scot, Archibald Geikie, Director of the Geological Survey for Scotland. Soon others followed. Their initial reaction was to accept Kelvin's 100 million years as an authoritative answer, based as it seemed to be on a fundamental law of physics applied to this problem by one of the world's premier physicists. It was a figure not really contradicted by any geological data then available, and they felt they could live with it.

But Kelvin's ideas drew the interest of geologists toward the question of whether purely geologic evidence might not be used to make an independent estimate of the earth's age. A number of methods were developed, based on the stratification of rocks, the rate of deposition of water-borne sediments to form new rocks, the rate of denudation (erosion: as noted earlier, Darwin had tried this method, but not to the satisfaction of geologists), and the rate the rivers bring salt to the oceans together with the known salinity of the oceans. It is interesting that the results all tended to agree rather too well with Kelvin's figure of 100 million years. It suggests, unhappily, that since each of these methods of estimation, like Kelvin's, required the making of assumptions about the magnitudes of many quantities that could not be directly measured, the authors tended to choose those values that agreed with what was believed to be the "right" answer.

But this acquiescence of geologists in what physics handed down did not continue. Kelvin, as mentioned earlier, continued to revise his analyses on the basis of new data and new assumptions. As he did so, he shortened the age of the earth. In 1862 he had allowed an upper limit of 400 million years. By 1868 he cut it to 100 million, by 1876 to 50 million, and in 1897 he concluded that 40 million was too high and that about 20 million years was the most likely age of the earth. The geologists' confidence in the more quantitative methods they had developed was growing, and they felt anything much less than 100 million would contradict their results. They did not doubt the second law or Fourier's theory, but had begun to realize Kelvin's method depended, as much as their own, on unsupported assumptions that might reasonably be doubted. They came to the conclusion that while the history of the earth could not violate the laws of physics, it could not violate the evidence of geology either. If there was a contradiction, it need not necessarily have to be resolved in favor of the physicist's viewpoint.

"An Odious Spectre"

Although the geologists could live comfortably with the apparent conflict between their discipline and Kelvin's physics, Darwin couldn't. In 1869 he wrote to Alfred Russel Wallace, the co-discoverer of the theory of evolution, that Kelvin's "views on the recent age of the world have been for some time one of my sorest troubles." Wallace had proposed hypotheses of an accelerated rate of evolution due to the stressful conditions during the ice ages, but Darwin was not persuaded. In the sixth edition of the *Origin* he did express similar views, but in the same edition he asserted very strongly the unpalatability (to him) of the shortened age of the earth:

> With respect to the lapse of time not having been sufficient since our planet was consolidated for the assumed amount of organic change, and this objection, as urged by Sir William Thomson, is probably one of the gravest as yet advanced, I can only say, firstly that we do not know at what rate species change as measured in years, and secondly that many philosophers are not as yet willing to admit that we know enough of the constitution of the universe and of the interior of our globe to speculate with safety on its past duration.

Darwin's friend Thomas Henry Huxley, who did so much to publicize the theory of evolution and to answer its critics, gave a lecture to the Geological Society of London in 1869, defending geology, and, implicitly, evolution, from Kelvin's criticisms. His case, however, was a weak one: the strongest point he made was to stress how the results of Kelvin's application of indubitable laws of physics depended on quite doubtable starting assumptions:

> Mathematics may be compared to a mill of exquisite workmanship, which grinds you stuff of any degree of fineness; but, nevertheless, what you get out depends upon what you put in; and as the grand-

est mill in the world will not extract wheat-flour from peascod, so pages of formulae will not get a definite result out of loose data.

As noted by the historian A. Hallam, the modern, less delicate phrasing of this principle is "garbage in, garbage out."

Darwin died in 1882 with this cloud still hanging over his theory. How he felt about it is revealed neatly in an 1871 letter to Wallace, in which he describes his efforts to reconcile the geological and biological time scales: "I can say nothing more about missing links than I have said. I should rely much on pre-Silurian times; but then comes Sir W. Thomson like an odious spectre."

Radioactivity and the Age of the Earth

In 1895 Wilhelm Roentgen discovered X rays. A year later Henri Becquerel, at the Ecole Polytechnique in Paris, was investigating the possibility that salts of uranium, which were phosphorescent (capable of glowing in the dark after exposure to light), could emit X rays after exposure to sunlight. His procedure was to wrap an unexposed photographic plate in black paper, place a uranium salt next to the wrapped film, and expose the salt to sunlight. On developing the film, he found it had darkened where the uranium salt had been placed. Since the black paper was opaque to all then-known radiations except for X rays, it appeared to follow that the sunlight had caused the uranium to emit X rays. But by chance Becquerel, while repeating the experiment, ran into some cloudy days (to be precise, the twenty-sixth and twenty-seventh of February, 1896, according to his own account). He developed the film anyway, expecting to find much less darkening, but to his surprise the intensity of darkening was every bit as great. He concluded that uranium emits rays capable of penetrating black paper without the stimulation of sunlight. Radioactivity had been discovered.

Pierre and Marie Curie, at the suggestion of Becquerel, began a search for other substances that emit "Becquerel rays." They found, first, thorium and then, on the clue that certain minerals containing uranium or thorium emit more intense Becquerel rays than can be accounted for by their uranium or thorium contents, discovered the more active elements polonium and radium. In 1903, Pierre Curie and an associate, Albert Laborde, announced the discovery that radium salts steadily emit heat: they are always warmer than their surroundings. As soon as this was announced, physicists and geologists both recognized it as a possible solution to the conflict between Kelvin's later, shortened estimate of the age of the earth and that provided by the geological evidence. In brief, if the earth contains radioactive minerals that are constantly providing heat, Kelvin's assumption that the earth is merely a passively cooling body, producing no heat within itself but only losing it at the surface, must be wrong. The same reasoning applies to the sun.

Ernest Rutherford, then a professor of physics at McGill University in Montreal, was one of the leaders in the scientific exploitation of radioactivity. He describes a lecture he gave in 1904 at the Royal Institution in London, with Kelvin in the audience:

> I came into the room, which was half dark, and presently spotted Lord Kelvin in the audience and realized that I was in for trouble at the last part of the speech dealing with the age of the earth, where my views conflicted with his. To my relief, Kelvin fell fast asleep, but as I came to the important point, I saw the old bird sit up, open an eye and cock a baleful glance at me! Then a sudden inspiration came, and I said Lord Kelvin had limited the age of the earth, provided no new source of heat was discovered. That prophetic utterance refers to what we are now considering tonight, radium! Behold! the old boy beamed upon me.

Kelvin, by that time in his eighties, recognized the relevance of the discovery of radioactivity to his assumptions but was not convinced that it invalidated his approach. It was not yet clear that the radioactive process was a *decay;* that the tremendous energies being given out were the result of a process in which one atom having a high energy changed into another having much less. Kelvin, who could not reconcile the energy output with his concepts of the energies of atoms, speculated that radioactive atoms are really re-emitting energies they have previously received by thermal radiation from surrounding matter. This explanation would have implied that radioactive materials are not themselves primary sources of heat but only concentrate and re-emit thermal energy already available in the surroundings. If this were so, the radioactive minerals in the earth would not be producing thermal energy, only recycling it, and his estimate of the age of the earth would not have been invalidated. It is ironic that this explanation of radioactivity, offered by him to save the first law of thermodynamics, would almost certainly violate the second. In private conversation with the physicist J. J. Thomson, Kelvin showed a recognition that his position on the age of the earth had been undermined, but he never published a retraction.

It was a sad end to an extraordinarily distinguished career. His own view, expressed in his conversation with J. J. Thomson, was that his work on the age of the earth was the most important piece of scientific research he had done, until the discovery of radioactivity had compromised it.

In the eyes of his contemporaries Kelvin was one of the towering scientists of all time. He had made significant contributions to the theories of mechanics, fluid dynamics, electricity and magnetism, and elasticity, in addition to his work in thermodynamics. Although he was primarily a theorist, applying mathematical analysis to derive new results from the established laws of physics, he had a hand in several important practical applications of scientific laws also. He was involved in the design of the first Atlantic Telegraph cable and sailed as technical director of the project on the ships that laid the cable. After a number of failures, the laying of the cable was successfully completed by the famous steamship the *Great Eastern,* in 1866, and he was knighted soon after. He patented a large number of commercially successful inventions, mostly of electrical apparatus, and was a wealthy man at his death. His scientific stature, in the view of at least one eminent physicist of his time, could be compared only to that of Newton.

Kelvin's reputation now is not so high. Rather than a discoverer of new insights and new laws, he was a thorough exploiter of the known. He was late in drawing the implications of Joule's work, and he never fully accepted the grand synthesis of electromagnetic theory achieved by Maxwell, although some of his own work had influenced Maxwell's thinking. He was not a careful reader of the work of other scientists or sensitive to views different from his own. He was not wrong in pointing out to geologists the implications of the second law and Fourier's theory, nor was he unreasonable in his choice of starting assumptions for his estimates of the age of the earth and the sun. But he was unduly closed-minded in his refusal even to examine carefully the geologists' evidence and to recognize that the conflict could not be resolved by insisting on the primacy of physics over geology. In the view of some (but not all) historians of science who have examined his influence on the growth of geology, the net effect of his activities was harmful, but in our view this position is not tenable unless one believes Kelvin should have anticipated radioactivity. In short, a great, but flawed, figure.

The Age from Radioactive Decay

The discovery of radioactivity not only wrecked Kelvin's hypothesis of a passively cooling earth, it also provided a quantitative means of determining its age. In brief, once it is established that uranium, for example, decays (according to a negative exponential curve) through a long sequence of transmutations into lead at such a rate that it takes 5 billion years for half the original uranium to be converted, the age of a rock can be determined by measuring the ratio of its uranium content to its lead content. The chemical processes by which minerals are formed tend to separate uranium from lead, so that when a uranium-containing mineral is freshly formed, none of the isotope of lead (lead of atomic mass 206; the common form is mostly of atomic mass 207) formed by uranium decay is present, and the clock starts. The oldest rocks found on the earth's surface (in Greenland) are 3.7 billion years old. Estimates from meteorites suggest the earth itself is about 1 billion years older: no rocks from this earlier period have survived. The best estimate for the age of the earth is 4.6 billion years (plus or minus 0.5 billion).

Darwin would have been pleased.

14

Quantum Mechanics and the Third Law

Newton discovered his laws of motion in 1666 and published them in his *Principia Mathematica* twenty years later. During the next two hundred years these laws were applied with success to so great a variety of problems of motion that by the middle of the nineteenth century most scientists could not imagine that they might fail. Newton's theory of light—that it consisted of small particles or "corpuscles" of some kind of matter traveling at high speeds—had not fared so well. Particles of light, according to Newton's mechanics, should travel in straight lines, consistent with his observation that opaque bodies cast sharp shadows, whereas waves tend to bend around obstacles. In the beginning of the nineteenth century more careful experiments than Newton could perform showed that in fact shadows are not that sharp, and it was concluded that light must be a wave. A deeper insight into the nature of light was provided by Maxwell around 1860, with his theory of electricity and magnetism, which showed that a changing electric or magnetic field produces waves whose speed of travel, calculated from certain electric and magnetic measurements, turned out to be exactly equal to the known speed of light. The conclusion: light itself must be an electromagnetic wave.

Maxwell's theory is universally regarded as one of the greatest of all scientific discoveries, but within thirty years it seemed to be in trouble. Some new experimental observations on light—accurate measurements of its speed of travel, spectroscopic studies of how it is emitted and absorbed by gases and solids—had raised questions that neither Maxwell's electromagnetic theory nor Newton's corpuscular theory could answer.

In the early years of the twentieth century a number of scientists, Einstein prominent among them, replaced Newton's laws of motion with new theories: special relativity, for things moving very fast; general relativity, a theory of gravity; and quantum mechanics, for very small things. Maxwell's electromagnetic theory required changes to absorb the concepts of quantum mechanics, but it fit in well with the special theory of relativity. Newton's laws were reduced to a useful approximation, valid only for bodies that are intermediate in size between molecules and stars, do not have very high velocities, and exist in weak gravitational fields. Nevertheless, these restrictions allow the laws a vast area of applicability: in most problems of motion we still use them. It is only when speeds approach the speed of light, or when strong gravitational fields are present, or when the objects in motion are molecules, atoms, and their component protons, electrons and other particles, that Newton's laws do not hold.

So far quantum mechanics and the general theory of relativity have not been united into a single comprehensive theory: their unification is a goal of current research. One of the problems is that the two theories seem incompatible: they can't both be right, though both may be wrong. Both theories have been confirmed in the experimental tests so far performed, though the experimental evidence for quantum mechanics is much more far-ranging and impressive.

Earlier, we made a distinction between thermodynamics, a macroscopic science concerned with measurable quantities of heat and work, and its microscopic justification from a kinetic-molecular theory. The new theories of the twentieth century have had an impact on both. Quantum mechanics left the two laws of thermodynamics unchanged but established the kinetic-molecular theory on a sounder foundation. The general theory of relativity has led to fundamental revisions of the very concept of energy and the law of its conservation. These twentieth-century discoveries provide a more accurate description of the physical world than the theories they replaced, as well as predict and explain previously unknown and unsuspected phenomena, but they do so at a price: they deliver a painful jolt to our common sense.

Important scientific discoveries are always surprising, but they do not necessarily conflict with common sense. Once we understand how energy is defined and measured, we are not shocked to learn that it is conserved, nor are we startled out of our wits to learn that an inverse square law of gravitational force leads to elliptical orbits for the planets. When discoveries *do* challenge common sense, though, it is common sense that usually must yield.

Ideas that once were revolutionary—the world is round; the sun does not move around the earth once a day, it is just the rotation of the earth that makes it seem to—are so commonplace now that we do not usually think of how hard it must have been to accept them when they were new. Why don't people on the bottom of the earth fall off? Why doesn't the rotation of the earth make a wind blow from the east at a thousand miles an hour?

Both relativity and quantum mechanics contradict our ordinary common-sensical understanding of what space, time, and matter are like. There is not much that we can do about it, except realize that common sense is based on common experience, and when we expose ourselves to extraordinary experiences we must expect shocks.

Quantum Mechanics: A New Molecular Theory

The laws of thermodynamics are macroscopic laws that describe how bodies subjected to heat and work inputs behave, without making explicit reference to the molecules that make up the bodies. Even so, their discoverers were inspired by a molecular picture of matter, and attempts to show why the laws of thermodynamics follow from a molecular hypothesis began immediately.

Although the nineteenth-century kinetic-molecular theory, which applied the theory of probability to molecules moving according to Newton's laws, made a number of outstandingly successful predictions, it made some outstandingly wrong ones as well. The most striking error was the failure of the equipartition theorem to give the correct average energies for vibrating molecules and its absurd prediction that hot bodies should emit radiation at an infinite rate. These failures, together with a then-fashionable philosophic distaste for speculative theories, led many scientists to an increasing skepticism about the existence of molecules. This did not affect their confidence in macroscopic thermodynamics, which seemed then to be comprehensive, infallible, and solidly grounded in an observable reality (see Chapter 4).

Fortunately, there were others who persisted in believing that molecules were real. From their attempts to understand the failures of the molecular theory began the development of a new molecular theory, quantum mechanics, which replaced Newton's laws in the atomic domain.

Quantum mechanics (QM for the rest of this chapter) is one of the most successful of scientific theories. Its quantitative predictions have been repeatedly confirmed to extraordinary accuracy. In addition, it has shed light on some fascinating philosophical issues: free will vs. determinism, the impossibility of neutral observation, the nature of reality. A recent journal article exploring some of these issues was entitled "Is the Moon There When Nobody Looks?"—and the author's answer appears to be, "Probably not."

Here we must confine ourselves narrowly to those conclusions of QM that have a fundamental bearing on the molecular justification of thermodynamics. We will first discuss the equipartition theorem and its failures and show how attempts to account for those failures led directly to the discovery of two basic hypotheses that form the foundation of quantum mechanics:

1. Light energy comes in "discrete" bundles, or "quanta," of definite energies. The energy of each bundle depends only on the frequency of the light.
2. When the energies of atoms and molecules are measured, they are usually found to take on only discrete values; in other words, energies do not usually vary continuously as they do in Newton's mechanics.

We will explain the meaning of "discrete" values and describe some of the experimental evidence for the discrete variation of the energies of both matter and light. Then we will show that recognition of the discreteness of energy gives correct results for average molecular energies, which Newton's laws and the equipartition theorem failed to do.

Finally we will show how QM provides correct answers to several other questions on which nineteenth-century science gave wrong ones, specifically: Why do experimentally measured entropies of pure substances at equilibrium tend to zero as the absolute temperature goes to zero, a universal behavior now known as the third law of thermodynamics? And does all atomic motion cease at absolute zero?

The Equipartition Theorem and Its Failures: Matter

The nineteenth-century kinetic-molecular theory, from which the equipartition theorem is derived, was based on three assumptions:

1. Molecules move according to Newton's laws.
2. The theory of probability can be applied to large collections of molecules.

3. For such large collections, all distributions of molecular positions and energies having the same *total* energy are equally probable.

Any sample of matter large enough to be visible or weighable is necessarily a large collection of molecules. Even at equilibrium, identical molecules differ in energy (see Chapter 4): oxygen molecules in air are not all traveling at the same speed; some are moving fast and therefore with large kinetic energies, others more slowly, with smaller energies. It is useful, therefore, to be able to calculate the average energy of the molecules, as a means of characterizing the collection and predicting the macroscopic properties.

To calculate an average molecular energy in a collection of molecules, we need to know the energy of each molecule, then add all the energies up and divide by the number of molecules. Since the molecules differ in energies, we need to know how many have each energy, from zero energy on up. The answer is given theoretically by the Maxwell-Boltzmann "negative exponential" formula (see Chapter 4). This formula is derived using the theory of probability and the assumption that all ways of distributing a fixed quantity of total energy among the molecules are equally likely, but it does *not* assume Newton's laws. We will find, therefore, that it is also valid if the molecules obey the laws of quantum mechanics.

The Maxwell-Boltzmann formula is one of the most basic in molecular theory (and is discussed in more detail in the appendix "Math Tools"). Here we will describe what it predicts in qualitative terms. It gives the numbers of molecules in a collection having any possible energy *relative to* the number having zero (or the lowest possible) energy. It shows, first, that the numbers of molecules having higher energies are always smaller than the numbers having lower energies and, second, that when temperature is raised, the numbers of molecules having higher energies increase (Figure 14.1, top). If to this formula we add the assumption that molecules obey Newton's laws—and Maxwell and Boltzmann had no reason to assume otherwise—the average energies can be calculated. We list in Table 14.1 the results of energy calculations for several kinds of molecular motion and also for electromagnetic waves in a furnace (black-body radiation; see Chapter 10). The predictions for rotational and vibrational motion depend on the number of atoms in the molecule and their arrangement in space. For simplicity, we give the predictions for rotational and vibrational energies only for two-atom molecules such as O_2.

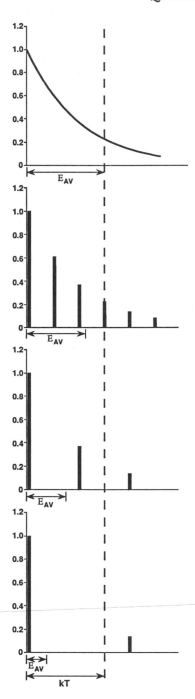

Figure 14.1 Why and When Equipartition Fails
The Maxwell-Boltzmann distribution formula, giving us the relative numbers of molecules having any permitted energy in a collection of molecules at equilibrium, enables us to calculate the average energy of the molecules. This calculation, like any other calculation of an average, is a matter of adding up all the energies and dividing by the number of molecules, although to do the adding some advanced mathematics is needed. We consider the distribution among possible energies for an idealized vibrating molecule. The top figure shows what the energy distribution would be if the molecules obeyed Newton's laws: any energy is possible, so the distribution is continuous, and the average energy E_{AV} is kT. The other figures show the distribution for vibrating molecules as described by quantum mechanics. The possible energies are discrete and evenly spaced: the spacing in the second figure was assumed to be about $\frac{1}{3}kT$, and the average energy is found to be a little less than kT. In the third figure the spacing is larger, about $\frac{2}{3}kT$, and the average energy is considerably less than kT. In the bottom figure the spacing is now larger than kT, and the average energy very much less. We can conclude that when the ratio of the spacing between permitted energies is very small, equipartition is obeyed and the average energy is kT. When the ratio is very large, the average energy falls far below kT, and it can even be nearly zero. This result implies two things: (1) if two different kinds of molecules, say H_2 and N_2 molecules, are compared at the same temperature, the average energy of the molecule with the *greater* energy spacing between its permitted energies will fall further *below* the equipartition value of kT; (2) if one kind of molecule is studied at different temperatures, its average energy will fall further below kT the lower the temperature. (See pages 340–341 for further discussion.)

These results are specific examples of the *equipartition theorem*. The predictions of equipartition for the average energies of molecules or crystals are tested by measurements of specific heat (the energy needed to raise the temperature of 1 gram of the substance by 1°C). The comparison between predicted and experimental values also requires knowledge of the relative masses of the molecules (see Chapter 4).

The experimental results on gas molecules were found to disagree with the theory, and it quickly became apparent that the vibrational motion was the source of the discrepancy. An analysis of crystals according to Newton's laws shows that the specific heat of a crystal is solely due to the vibrations of the constituent atoms. The equipartition theorem (see Table 14.1) predicts an average energy per atom in the crystal of $3kT$. Measurements made in the early nineteenth century on the specific heats of crystals at ordinary ambient temperatures were compared with this theoretical prediction and had given fairly good agreement, with a few exceptions, such as diamond (one form of crystalline carbon) having lower specific heat than predicted. When experimental methods for cooling to very low temperatures were developed later, however, it was found that the more one cooled any crystal, the more its specific heat fell below the equipartition value (see Figure 4.11).

Equipartition and Its Failures: Radiation

Equipartition failed not only for the vibrational motion of atoms and molecules; it failed even more seriously for calculating the energy of radiation in temperature equilibrium with matter.

Radiant energy (light) was regarded in the nineteenth century as composed of electromagnetic waves having different wavelengths and

Table 14.1

Predictions of the Equipartition Theorem

Kind of motion	Average energies
Translation	$\frac{3}{2}kT$
Rotation (diatomic molecules only)	kT
Vibration (diatomic molecules only)	kT
Vibration in a crystal, per atom	$3kT$
Radiation, for each permitted wavelength	kT

correspondingly different frequencies. A furnace at temperature equilibrium is filled with radiation, whose energy spectrum (how much energy is present at each frequency) depends on the temperature but not on the material of which the walls of the furnace are made. This equilibrium radiation is called "black-body" radiation (Chapter 10). Its energy spectrum was known experimentally—the challenge was to calculate it theoretically. Two problems had to be solved: first, to find from Maxwell's electromagnetic theory what wavelengths of radiation are present in the furnace and, second, to calculate how the electromagnetic energy in each wavelength depended on temperature. The answer to the first problem can be visualized more easily if we consider a simpler problem first.

Violin Strings

The strings of musical instruments vibrate only at frequencies which are multiples of a base frequency. The A string of a violin has its base frequency at 440 Hz (when properly tuned), and this is the predominant note we hear when we pluck it. But analysis of the sound it makes shows that it is also vibrating (though less strongly) at 880 Hz (one octave higher), 1,760 (two octaves higher), and so on, without any upper limit (Figure 14.2); it does not vibrate at intermediate frequencies such as 479, 1,013, or 1,591 Hz. The base frequency of a string depends on the mass per unit length of the string (compare the different strings on a violin or guitar), the tension in the string (which is how we tune it), and the length (what we change when we place a finger on the string).

Finding the possible wavelengths or frequencies for electromagnetic waves in a furnace is somewhat more complicated, but the results are qualitatively similar to the violin string:

1. The possible frequencies depend on the dimensions of the furnace; not all frequencies are permitted.
2. There is no upper limit: very high frequencies, corresponding to infrared, visible, and ultraviolet light, X rays, and even gamma rays, are permitted.

This kind of a distribution of frequencies, in which only certain values are possible, is the first example we will encounter of a "discrete" variation. For the violin string and the furnace, discreteness can be deduced from, respectively, Newton's laws and Maxwell's theory. The

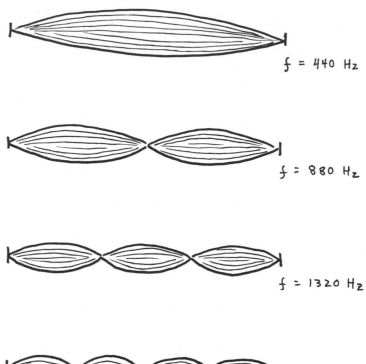

$f = 440 \text{ Hz}$

$f = 880 \text{ Hz}$

$f = 1320 \text{ Hz}$

$f = 1760 \text{ Hz}$

Figure 14.2   The Discrete Vibration Frequencies of a Piano String
A string can be made to undergo wave motions with waves of different wavelengths, but not of *any* wavelength. On a piano, for example, the string for the
A above middle C is tuned to vibrate with a basic frequency of 440 Hz. But when
it is struck with the hammer operated by the key, it vibrates simultaneously at
higher frequencies also, but only at frequencies which are multiples of 440 Hz.
There are ways to make the string vibrate at one of these frequencies only, and
the appearance of the string for each vibration frequency is indicated schematically in the figure. The wavelength for 880 Hz is one-half that for 440 Hz, for
1,320 it is one-third, and for 1,760 one-fourth. Only whole-number multiples of
the base frequency are possible for a string. The successive frequencies are one
octave higher than the preceding ones, so the second permitted frequency of
the A string corresponds to the base frequency of the A one octave higher. The
tone of a piano, distinguishing its sound from that of a violin or a harp, reflects
the amplitudes of the various permitted frequencies produced when the strings
are bowed or plucked.

common denominator in the two situations is that both involve wave motion in systems with definite boundaries: the two fixed ends of the string, the walls of the furnace.

It should be noted that although only discrete frequencies are permitted in the furnace, the spectrum of black-body radiation *appears* continuous; no gaps are visible between different frequencies (Figure 10.8), but that is because the many permitted frequencies occur so close together that the spectrometers used to measure the energies cannot separate individual frequencies and lump large numbers of them together.

We turn now to the second problem: how much energy should each frequency have at a particular temperature? An electromagnetic wave oscillating at a particular frequency is a kind of vibration, and in Maxwell's theory the energy of such a wave could vary continuously, just as the energy of a vibrating molecule does, so it was reasonably assumed that each frequency should have the same energy as a vibrating diatomic molecule, kT. This implies that in black-body radiation, all permitted frequencies should have the same energy. Since frequencies corresponding to ultraviolet light, X rays, and so on are permitted, the radiation should have as much energy in these frequencies as it does in, say, each infrared frequency. In fact, however, no detectable X rays or gamma rays are emitted by red-hot bodies (temperatures around 600–1,000°C). Even the sun, with a surface temperature around 6,000 K, emits a much smaller quantity of ultraviolet light than the equipartition theorem predicts and not enough X rays to be easily detectable.

The drastic discrepancy between theory and experiment can be stated even more strikingly. Since there is no upper limit to the possible frequencies, there is an infinite number of them. If each has the same kT of energy, then the total energy in the furnace would be infinite. This applies to any furnace at any temperature.

To summarize: experimental measurements of the energy spectrum of black-body radiation (Figure 14.3) showed that the lower frequencies had the energies predicted by the equipartition theorem, but the higher ones had less energy than expected; the higher the frequency, the worse the discrepancy. There was almost no high-frequency radiation at equilibrium.

To summarize more graphically: the equipartition theorem tells us our own warm bodies should visibly glow in the dark. The discrepancy between the equipartition calculation and the observed reality was so dramatic it was given a dramatic name: "The Ultraviolet Catastrophe."

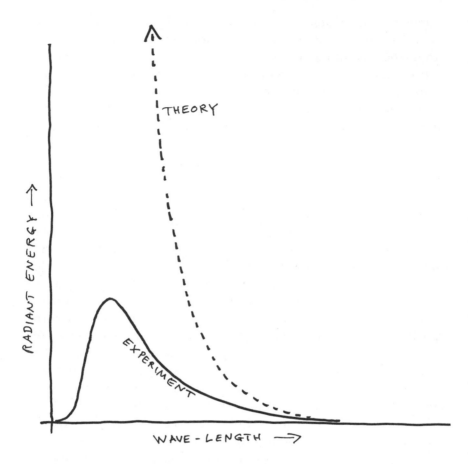

Figure 14.3 The Ultraviolet Catastrophe

The graph compares the experimentally observed energy spectrum of radiation and the theoretical curve calculated from the equipartition theorem. Experiment and theory agree at long wavelengths, but the experimental curve has a maximum and then decreases at shorter wavelengths, while the theory predicts more and more energy at decreasing wavelengths. In fact, the theory predicts an infinite quantity of radiation radiated per second from a black body. This egregious discrepancy was first observed when the shortest known type of electromagnetic radiation was ultraviolet radiation, whence the name "ultraviolet catastrophe."

Planck's Conjecture

To resolve this discrepancy, Max Planck, a German physicist, made what seemed to everyone, including himself, a wild conjecture that had no physical justification. It had only one merit: it gave the correct answer. The conjecture was that when heated matter emits radiant energy, it does so in discrete bundles rather than continuously. In particular, the energy of the light emitted at each frequency had to be a multiple of a basic unit that was proportional to the frequency. In an equation, the quantity of energy in the basic unit (which we symbolize by e) is given by:

$$e = h \times f$$

with h being one more constant of nature. This constant has the same value for all frequencies and for all radiating substances: 6.63×10^{-34} joules per hertz. Multiplying the frequency of the radiation measured in hertz (cycles per second) by the constant h gives the energy in joules of each bundle. Planck determined the value of h from the observed black-body energy spectrum. It is called, naturally, Planck's constant.

We recognize the possibility of confusion here. The permitted frequencies inside the furnace are discrete, but this much is Maxwell's theory, not QM. Planck's hypothesis, the founding hypothesis of QM, was that the *energy* radiated in each discrete frequency can itself vary only *discretely*, in multiples of a base energy given by $h \times f$. This kind of discreteness is not consistent with Maxwell's theory, in which the energy in each frequency could have any value from zero up. Once this bizarre (in those days) assumption was made, everything else fell into place. The experimental black-body radiation curve corresponded within experimental error with the curve given by the new theory.

Planck himself did not think the discreteness of energy was a property of light but only a specific property of the process of matter emitting radiation. He spent a considerable effort trying to reconcile it with nineteenth-century physics. Einstein, on the other hand, saw more deeply: he guessed that discreteness is a general property both of the way matter stores energy and of the energy carried by electromagnetic radiation. He realized that if this concept was applied to the calculation of the average energies of vibrating solids and molecules, it followed at once, for reasons we will explain later, that their energies should fall below the equipartition value of $3kT$ per atom more and more the colder they got, again bringing theory and experiment into accord (see

Figure 4.11). Einstein's analysis represents the beginning of the conceptual revolution that led to the complete formulation of QM in the 1920s.

Basic Concepts of Quantum Mechanics

Discrete vs. Continuous

What do we mean when we say that the energies of atoms and molecules vary *discretely* rather than *continuously*?

We have given one example of a familiar discrete variation, the value of the permitted frequencies of vibration of a violin string. The distinction between continuous and discrete variation can be visualized by comparing a digital clock with the old-fashioned "analog" kind, having two hands that sweep around the dial. Many digital clocks read time to the nearest minute, so one may read either 1:17 or 1:18, but never anything in between. Time, according to such a clock, varies discretely. In contrast, the hands of an analog clock move continuously. At one particular moment the minute hand will point to 1:17, and at a later moment to 1:18, but in the intervening period it points to each and every time on the dial between these two times.

Another example of a discretely varying quantity is the size of a particular family, which may have begun with two members in 1954, increased abruptly to three in 1956, to four in 1959, and to five in 1961, followed by equally abrupt decreases as members left to start families of their own.

Continuous variation, which is more familiar to us, seems somehow more "natural." Bodies in motion move continuously rather than discretely—or at least appear to do so—not only clock hands but billiard balls and planets: Newton's laws assume continuity for both motion and energy. Once the concept of discrete variation is grasped, many examples will come to mind, but there is an element of artificiality about it.

The two kinds of variation may seem to form a dichotomy: if something is continuous it can't be discrete, and vice versa, but as usual the boundaries are not that sharp. Matter appears continuous in space unless we examine it very closely: we do not usually see that it is made up of discrete atoms. When we pour water out of a bucket, we assume that we can pour any quantity whatever, without thinking that we can only pour out whole numbers of H_2O molecules. On the other hand, there are continuously varying quantities that we may choose to treat as discrete. The height of a person varies continuously with time, not only in the course of a lifetime, but even, within a small range, in the course of a day: we are slightly taller when we get up out of bed in the morning than when we fall into it, tired, at night. Yet we usually measure height to the nearest centimeter or to the nearest inch, which serves well enough for most purposes.

There are mathematical methods for dealing with continuous variation and mathematical methods for dealing with discrete variation. Some situations demand to be treated as continuous, others as discrete, and still others can be either, and the way they are handled is merely a question of the convenience of the investigator.

To portray more vividly the difference between our intuitive expectations and quantum-mechanical reality, let us contrast the dying away of the vibration of a tuning fork and that of a molecule set vibrating with a large amplitude. Both amplitudes diminish in time, the fork by the loss of energy to sound waves and by macroscopic friction, the vibrating molecule by collisions with other molecules or by the emission of light. In Figure 14.4 we show how the amplitude of vibration of the fork changes continuously with time and how its molecular equivalent, the vibrational energy of the molecule, changes in discrete jumps. One may

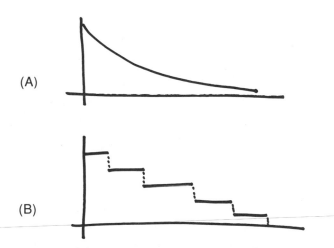

Figure 14.4 Discrete vs. Continuous Loss of Energy
The energy of a vibrating tuning fork gradually diminishes *(A)* as the energy is carried away by the sound waves and by various frictional losses, so we hear the sound gradually die away. In contrast, the vibrational energy of a molecule set vibrating with a large amplitude and losing its energy by collisions with other molecules decreases in discrete jumps *(B)*.

reasonably ask, what about the vibrational energy of the molecule *during* the collision? Doesn't it change continuously, even if very rapidly, from the earlier value to the new one? We will answer this question when we discuss the uncertainty principle, later in this chapter.

Quanta of Energy

The experimental evidence for the discrete variation of energy is provided by studies of the energy changes when atoms absorb or emit light, a technique called spectroscopy (see also Chapter 10). Figure 10.6 shows portions of the energy spectrum of the light emitted by hydrogen atoms excited by an electric discharge; the figure clearly indicates that the hydrogen atom emits radiation *only at definite, discrete frequencies.* QM enables us to calculate the discrete energies permitted to the hydrogen atom, and the frequency of the light it emits, from an equation formulated by Erwin Schrödinger, which takes the place of Newton's laws. To solve this equation, we need to know, just as we needed for Newton's laws, the relevant masses—here the masses of the proton and the electron that make up the hydrogen atom—and the strength of the forces between them—here forces of electrical attraction. The calculated frequencies for the hydrogen atom agree with the experimental ones of Figure 10.6 to within 0.05 percent. The slight discrepancies between theory and experiment can be accounted for by the special theory of relativity.

Another illustration of the quantization of energy is provided by the vibrational, rotational, and translational energies of an oxygen molecule, O_2. As we discussed in Chapter 4, an oxygen molecule in the gaseous state is moving simultaneously in three ways: translation (movement from one location to another), rotation, and vibration (see Figure 4.5). How is it possible to separate the different kinds of motion so cleanly by spectroscopy that we can study the rotational motion all by itself? The answer is that the spacing of the permitted energies for rotation are very different from those of vibration. In consequence, the frequencies of light emitted or absorbed when the molecular rotational energy changes, frequencies determined by the spacing between the permitted rotational energies, are much smaller than those frequencies determined by changes in vibrational energy. The former show up in the microwave region of the spectrum, the latter in the "near" infrared (frequencies lower than but close to those of visible light). The spacing of translational energies is much smaller even than the rotational energy spacing. These differences in spacing will be seen to be the reason why the translational energy of molecules in a gas at ambient temperature always agrees with the equipartition theorem, rotational energy almost always, and vibrational energy not very often.

Schrödinger's equation is, like Newton's, an equation for material bodies. It might therefore be expected to look something like Newton's equations, even though it is an improvement on them. Oddly enough, it resembles Maxwell's equation for electromagnetic waves more closely, although it is not identical with it either. The result is that the behavior of the atoms or molecules we describe by Schrödinger's equation has something in common with the behavior of waves, a fact strikingly confirmed in various laboratory experiments (Figure 14.5). QM is sometimes called wave mechanics because it predicts the wavelike properties of matter. The uncertainty principle, which we will be discussing shortly, is a consequence of those properties.

Bundles of Light

Evidence that the energy of light comes in little bundles is provided by the simplest of experiments. The image formed on a black-and-white photographic film after exposure and development is composed of tiny black grains of metallic silver. Each grain has grown from a single atom

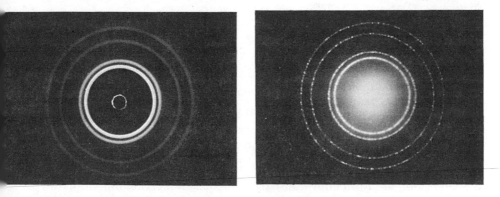

Figure 14.5 The Wave Nature of Matter: Electron Diffraction
A beam of electrons striking a solid object is bent by it not as Newton's laws would imply but rather exactly as though it were a beam of light; in other words, the beam displays both diffraction and interference. These photographs were produced by directing beams onto an object (a piece of aluminum foil), one a beam of X rays *(right)* and the other a beam of electrons *(left)*.

Permitted Energies for Certain Kinds of Motion

When the energies of atoms and molecules are measured, they are found to vary discretely rather than continuously; they are said to be "quantized." This figure shows, for illustration, the energies possible for an electron attracted to the proton in a hydrogen atom by electrical forces and the possible vibrational, rotational, and translational energies of an oxygen molecule.

The first thing to note is the great difference between the energy spacings for the different kinds of motion. The spacing between the lowest permitted electron energy and the next higher energy is about 10^{-18} joules, that between the first two vibrational energies is about 10^{-20} joules, that between the first two rotational energies about 10^{-22} joules, and that between the first two translational energies about 10^{-42} joules. Thus the electronic energy spacings are about 100 times the vibrational spacings, which in turn are about 100 times the rotational spacings, which finally are an unimaginable 10^{20} times translational spacings. Because of these enormous differences in the energy spacings, each kind of motion requires a different energy scale. The scale of energy on the left must therefore be multiplied by different powers of 10 to give the actual energies. For example, the energy corresponding to 10 joules on the scale is really 10×10^{-42} joules for the energy of translation, 10×10^{-22} joules for rotation, 10×10^{-20} joules for vibration, and 10×10^{-19} joules for the electronic energy in the hydrogen atom. To illustrate the enormous differences in scale, we have shown by dashed lines how the spacings would look if they were plotted on the same scale as the type of motion with the next higher spacing. The energy of the first vibrational spacing is just visible on the scale for the electronic energies, and the energy of the *ninth* permitted rotational state about as visible on the scale for the vibrational energies. The energy of the thirteenth translational state, however, is not visible on the rotational scale; this energy would require a magnification of 10^{18} times to be visible.

Also note that the spacings for the different kinds of motion form quite different patterns. The spacing between the first and second electronic energies is very much larger than the spacing between the second and third, and in fact the spacings diminish so rapidly that the maximum possible energy of an electron in a hydrogen atom does not exceed a definite limit. If the electron is given more energy than this, it flies out of the atom, away from the attractive force of the proton. The vibrational spacings, on the other hand, are the same between each pair of adjacent energies (the pattern shown here is actually an idealization of vibrational motion, but it works well for the lower energies). The rotational spacings, in contrast to both electronic and vibrational spacings, get larger and larger at higher energies. Still a fourth type of behavior is shown by the translational energies, which seem to vary erratically. All of these, surprisingly, are direct consequences of Schrödinger's equation solved for the system in question, and all are in agreement with experimental observation.

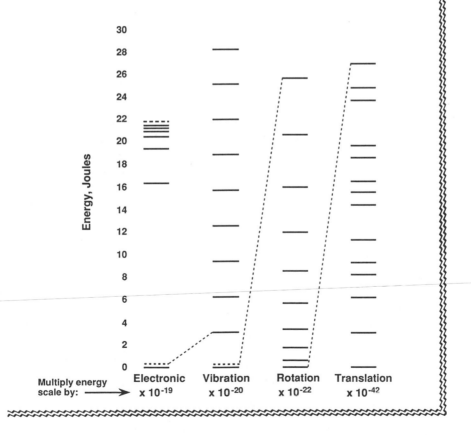

Atomic & Molecular Permitted Energies

of metallic silver formed from a compound of silver in the emulsion when the light struck the film. If the film is uniformly illuminated, and light were assumed to act much like an ocean wave, the wave energy reaching each point on the film would be the same as that reaching any other point, just as the ocean waves striking a beach bring about the same energy to each point on the beach. We would expect, therefore, that if the film is exposed for a short time there should be a faint but uniform darkening of the film.

Examination of the film with a microscope shows us a different picture, however. We find a small number of those tiny black grains distributed randomly on the film. The only difference between the film image of a bright light and a faint one, or between a short exposure and a long one, is the number of grains, but not the size or blackness of the grains. Each grain is like every other.

The occurrence of grains shows that the energy of the light is highly localized in space, more like a bullet than an ocean wave. Newton's "corpuscular" theory of light was not totally wrong. The modern term, coined by Einstein, for a quantum of energy carried by light is *photon*.

Where does this leave Maxwell's wave theory? It is still correct, if we interpret it differently from the way that Maxwell did originally. *First* we use the theory to calculate what the *intensity* of the wave energy at any point on the film should be, and *then* we interpret this theoretically calculated intensity as the *probability* of finding a grain of silver at that point after the film is developed, the grain being the indication that one quantum of light energy hit that particular spot on the film. In any small area of the film where Maxwell's calculated intensity is low, the number of grains will be small, but they will be randomly distributed in that area. In any area where the calculated intensity is great, the number of grains will be large, and they too will be distributed at random. This interpretation doesn't square with the ordinary perception of light as grading continuously from faint to bright, and we discover it only by a microscopic examination of the film. A similar procedure is followed to predict where an electron or a proton moving according to Schrödinger's equation is to be found.

The Edison Effect

Einstein provided the first direct evidence that the energy of light is quantized by his analysis of a previously mysterious experimental observation by Thomas Edison. Light can cause an electron to be ejected from the surface of a metal (the photoelectric effect). Edison had found that the color of the light used determined whether electrons are ejected or not. Red light, even of very high intensity (a large quantity of energy falling per second on each square centimeter of metal surface), ejected no electrons. Violet light, even of very low intensity, ejected many.

Einstein's reasoning, based on Planck's work and applying the principle of conservation of energy to the interaction between light and electrons, followed from three assumptions:

1. Each electron is ejected by acquiring energy from a single photon.
2. The energy acquired by an electron when it is ejected by a photon should be proportional to the frequency of the photon: $e = h \times f$.
3. It takes a certain minimum energy to eject an electron from a particular metal.

From these assumptions it follows that any photon whose energy is less than this minimum energy will not eject any electrons, no matter how much the total energy in the beam of light used, the total energy being the product of the energy of one photon and the number of photons in the beam.

From Planck's equation one can show that the energy of a photon of red light of wavelength 7×10^{-7} meters and frequency 4×10^{14} Hz is 2.8×10^{-19} joules; a photon of violet light of wavelength 4×10^{-7} meters and frequency 7.5×10^{14} Hz is 5×10^{-19} joules, almost twice as great. The ejection energy for electrons in barium metal is 4.0×10^{-19} joules, which means that the energetic violet photons have more than enough energy to expel an electron from the metal. The red photons have less energy, and so no electrons are ejected no matter how many red photons strike the surface of the metal.

Note the contradictory character of this reasoning: a "bundle" of energy is a corpuscular concept, "frequency" a wave concept. What must we do about the contradiction? The answer is that we must learn to live with it. The idea that "corpuscle" and "wave" are mutually exclusive concepts, that if anything is one it can't be the other, went out when QM came in.

It is interesting to note that Einstein was awarded his Nobel Prize primarily for this work, published in 1905, rather than for his special theory of relativity, then lacking strong experimental support.

Light Frequencies and Molecular Energies

We said earlier that the discrete frequencies of light emitted by atoms and molecules give us two kinds of information:

1. They *show* that the possible energies of atoms and molecules are discrete.
2. They *permit us to determine* these discrete energies from the light frequencies absorbed or emitted.

When an atom or a molecule emits light, the law of the conservation of energy is obeyed: the energy lost by the atom or molecule is exactly equal to the energy of the light emitted. If the atom or molecule is only permitted discrete energies, then when it loses energy, the lost energy must correspond to the *difference* between the two permitted energies. It follows that the possible energies an atom may lose must also be discrete.

$$\text{Loss of energy} = \Delta e = e_{final} - e_{initial}$$

The light emitted by an atom or a molecule, which carries the energy away, carries it away as a single photon. The energy of this photon (e_{photon}) is related to its frequency according to Planck's equation:

$$e_{photon} = h \times f$$

But this energy is exactly equal to the energy loss of the atom:

$$e_{photon} = \Delta e = e_{final} - e_{initial} = h \times f$$

The energies of the photons emitted, therefore, give us directly the *differences* between permitted atomic or molecular energies. Once we know these differences, it is straightforward to calculate the energies themselves.

Newton's Laws Still Work—Sometimes

We began this section by pointing out that great errors can result from using Newton's laws. Why, then, do they seem to work so often and so well? Even large rotating wheels obey QM, not Newton's laws, but Newton's laws are good approximations when the mass of a body is large or when its energy is high. The tuning fork is actually permitted only discrete energies, but because of its large mass the quantized energies are so close together that we cannot detect the tiny jumps so obvious with the vibrating molecule of Figure 14.4.

In general, a typical macroscopic system, with 10^{23} molecules, has quantized energy levels, just as its individual molecules have, but the experimental error with which the energy of such a large system can be measured—even if it is as small as 0.01 percent or even 0.0001 percent— is so much larger than the differences between the discrete permitted energies of the system that we perceive the total energy as varying continuously. We have mentioned earlier the apparent continuity of the spectrum of black-body radiation; the reasons are similar.

QM and Thermodynamics: What Changes, What Doesn't

The discovery of QM has changed many things, but not everything.

1. Classical mechanics is still the simplest way to describe and predict the motions of macroscopic objects in the absence of very strong gravitational fields. Newton's laws even work fairly well for many molecular problems, especially when the molecules are massive and the temperature and thus the average energies are high.
2. The first and second laws of thermodynamics remain unchanged.
3. The principle of the conservation of energy applies in QM as it did in Newton's mechanics.

Quantum mechanics does, however, give us a different and experimentally more accurate description of molecular behavior, and so we expect, and find, that when we calculate macroscopic thermodynamic properties from the properties of molecules, we will get different answers and better ones, in fact, than we did using Newton's laws.

Consider, for example, what QM tells us about the equipartition theorem. This theorem calculates average molecular energies by assuming that molecules can take up energy continuously. Quantum mechanics tells us that they can't. How can we explain why an erroneous approach, one that assumes continuity, sometimes works and sometimes doesn't?

Equipartition Still Works—Sometimes

We have said earlier that discrete problems can sometimes be treated to an adequate approximation as though they were continuous, and vice versa. While molecular energies are *always* discrete, there are conditions under which pretending they can take up energy continuously will give nearly the correct answer. Vibrational motion happens to be a particularly easy case to describe, because the permitted discrete energies of a (somewhat idealized) vibrating molecule have a simple regularity. The possible energies increase stepwise: if we call the lowest energy 0 and the first permitted energy above it e, then the next permitted energy is $2e$, the next $3e$, and so on, successive energies all differing by the same amount e. Now the molecules in a very large collection will not all have the same energy except at absolute zero, when they all have their lowest permitted energy. At any temperature other than zero, some will have 0, some will have e, some $2e$, and so on.

To calculate the *average* molecular energy in a collection we need, as before, to know how many molecules have *each* energy, and, as before, the Maxwell-Boltzmann formula gives the answer. As we noted earlier, the formula is not specifically based on Newton's laws but applies equally well to molecules obeying QM. However, we no longer get the equipartition result for the average energies unless the temperature is relatively high. Figure 14.1 shows what the formula predicts for vibrational motion in several situations: first, if vibrational energy could vary continuously, and then for several examples of the real, *discrete* variation. In our idealized version of the O_2 molecule (see pages 334–335), the permitted energies are equally spaced. The idealization assumes two things about an oxygen molecule that are not strictly correct: first, that the force holding the two oxygen atoms together changes with the distance of separation between the atoms in a particularly simple way and, second, that the rotational and vibrational motions take place independently.

Let us imagine varying temperature from a very low value to a very high one. As temperature is raised, the molecules, initially crowded into the state of lowest possible energy, spread out among the states of higher energy more and more, and the average energy increases from a very small fraction of the equipartition value of kT to a value nearly equal to it.

The spacing of vibrational energy levels is typically smaller for crystals than for two-atom molecules, which is why crystals often have almost the equipartition energy around room temperature and deviate from it only on cooling to low temperatures.

Translational energy spacings in gases are smaller than $\frac{3}{2}kT$ at all temperatures high enough for anything to be a gas, so equipartition always works for translational energy in gases. Rotational energy spacings are larger the lighter the molecule doing the rotating. Only the spacing in the very light hydrogen molecule (H_2) is large enough for average rotational energy to be noticeably less than kT, and that only at lower temperatures than ordinary ambient ones.

The black-body energy spectrum is never approximated correctly by the equipartition theorem, even at very high temperatures. Figure 14.1 enables us to see why. At any temperature T, no matter how high, the radiation will always have frequencies high enough for the energy of photons of that frequency to greatly exceed kT. For them, equipartition *always* fails badly, and their contribution to the energy of the radiation is negligible.

Like the vibration of the idealized oxygen molecule we considered above, each possible vibration frequency for electromagnetic waves in the furnace has only discrete, equally spaced possible energies; a photon escaping from the furnace must have an energy corresponding to such a spacing. Figure 10.4 shows the correspondence between the temperature of a black body and the frequency at which it emits the most energy (given by Wien's law). Photons with this maximum emission frequency can be shown to correspond to energies ($e = h \times f$) of about kT. Photons of higher energies do not obey equipartition and are nearly absent from the radiation.

Uncertainty

In the world of familiar objects, to which Newton's laws apply, we are able to measure without difficulty where anything is, how fast it is moving, and in what direction. We measure these properties to a limited accuracy, of course—whatever we need or can manage for the purpose at hand—but we assume that there is no limit to the accuracy possible: the more accuracy we need, the more we could get, as long as we are willing to pay the price for it. Given the results of the measurements, we can then use Newton's laws to predict what will happen in the future: where the objects of our study will be and how fast they will be moving at each moment.

In QM things are different: we cannot measure all the properties we thought we could, because measuring one affects our ability to measure others. In particular, an accurate measurement of where an electron is at a certain moment precludes an accurate measurement of the direction and speed it has, and vice versa. There are thus questions to which experiment cannot provide an answer, and therefore there is no point asking them. We have already given an example: what is the energy of a vibrating molecule while it is in transition from one energy state to another?

The principle that certain properties cannot be simultaneously determined with high accuracy, because measurement disturbs the system measured, is called the *uncertainty principle*. It has its origin in a basic feature of QM: the wavelike character associated with matter as embodied in Schrödinger's equation. But why uncertainty emerges from this "waviness" is more than we can answer here.

As an example of how the uncertainty principle works, let us consider again the permitted energies of the idealized oxygen molecule. The molecule is composed of two oxygen atoms known to be, on the average, 1.207×10^{-10} meter apart. During vibration, the distance varies about this average. When the molecule is in its state of lowest possible energy of vibration, we might expect intuitively that it has *no* energy and is therefore not vibrating at all. This would imply that we know two things about the molecule:

1. The two atoms, not in motion, have therefore exactly zero speed.
2. The two atoms are exactly 1.207×10^{-10} meter apart all the time.

It doesn't seem surprising that we might know this much, but if we did we would have violated the uncertainty principle, which forbids knowing both speed and position. The principle is, however, satisfied if the lowest permitted energy is not zero energy. In fact, Schrödinger's equation shows that the molecule is still vibrating at the absolute zero of temperature, having a quantity of energy equal to *half* the energy difference e between successive vibrational energies. This in turn ensures that the two atoms are not fixed in location exactly 1.207×10^{-10} meter apart but continue a small back-and-forth motion about this distance.

The excess energy that vibrating molecules in gases cannot get rid of even at absolute zero is called the *zero-point energy*. Crystals, in which the atoms or molecules vibrate about fixed positions, also have this kind of energy. In both, its existence can be confirmed in the laboratory. Gas molecules confined in containers have zero-point energies for their translational motion also, which are larger the smaller the container (the more localized in space the molecule is). In containers of the usual (1 cubic meter, or even 1 cubic centimeter) size, however, this zero-point energy is too small to detect.

Calculating W

When we calculated W for the ideal gas, we slurred over some difficulties with the counting (see Chapter 7). For W_V, the number of different spatial arrangements of the gas molecules, we divided the container up into imaginary, discrete little cubical boxes, each just big enough to hold one molecule (see Figure 7.4). There was an element of arbitrariness in this arrangement. Molecules of a gas are not really confined in little boxes: they could as well be on the imaginary wall between two imaginary boxes as in the center of either. When Maxwell and Boltzmann calculated W_E, the number of ways the total kinetic energy could be divided up, they made a similar arbitrary division of the energies a molecule could have: instead of a continuously varying energy, the possible energies were divided up into discrete little boxes also, but the size of these little boxes was assigned even more arbitrarily. It is ironic that they had found it convenient to treat a property that they believed to be continuous by approximating it as discrete while in fact it really is discrete. They can't be blamed, though, for not knowing that. The calculation of W_T was found to be insensitive to these subdivisions of space and of energy, provided the boxes were made small enough, and the results often agreed with experiment, but the arbitrariness is troubling.

QM does away with all these problems. First, the separation of the calculation into a spatial part and an energy part is neither necessary nor realistic. The molecules, once we know which of their possible states of definite and discrete energy they are in, do not need to have their positions independently specified. The uncertainty principle implies that knowledge of the energy of a molecule *includes* adequate knowledge of its position for calculating W. Further, the discreteness of the possible energies means that we don't think of the molecule as somewhere within a vaguely defined small "energy box": it has only one of a number of quite definite energies. We can thus calculate W without any of the earlier arbitrariness.

The Third Law

Calculating W *near Absolute Zero*

At low enough temperatures, the spacing between the permitted energies of molecules is always too large for anything but QM to describe the behavior. At extremely low temperatures, the Maxwell-Boltzmann formula shows that all systems tend to have close to their lowest possible energies, and at absolute zero, they would have just that lowest energy. This state of lowest energy is called the *ground state*. When the system is in its ground state, what is the number W, the number of different possible arrangements of the system? We want to know this in order to calculate the entropy at absolute zero from the formula $S = k \log W$.

When a system is not in its ground state of energy but has extra energy corresponding to a few quanta, we can easily visualize that there are many arrangements of even these few quanta among the molecules. Intuitively we would expect that in the ground state itself, where there are no quanta to distribute, there is only one arrangement of the system. For once our intuition is nearly correct. An analysis of the physics of various systems in the light of QM shows that most of them for which W can be calculated at all give $W = 1$. At absolute zero, then, all substances should have zero entropy (because $\log 1 = 0$). There are some special cases for which W is greater than 1 but still small, 2 or 4 or even 1,000. Even here the calculated entropy $k \log W$ is so small that it cannot be experimentally distinguished from zero (because k is very small: 1.38×10^{-23} joules per kelvin). We will give an example shortly.

This reasoning leads to one of the decisive experimental tests of the kinetic-molecular theory as modified by QM. The entropy S of a system is one of its macroscopic properties. Changes in entropy when a system undergoes any changes in temperature, pressure, or any other variables are determined by measurements of heat inputs and temperatures during the process of change. No molecular theory is involved. On the other hand, the kinetic-molecular theory gives us a relation between S and W, the number of microstates of the system when it is in a particular macroscopic state, which we can calculate from detailed knowledge of how the molecules of the system can take up energy. Does the formula $S = k \log W$ agree with the result we determine in the laboratory?

The number W can be calculated most easily for matter in the "ideal gas" state: gases under low pressures and at high temperatures (compared with the boiling point of the substance studied). Information needed for the calculation includes the number of molecules, the volume of the container, the mass of the molecule (all needed to calculate the permitted translational energies); the permitted vibrational and rotational energies, which can be found from spectroscopic studies; and the permitted energies for any other types of motion, if any are relevant.

At absolute zero, the system is no longer an ideal gas but a crystalline solid. We do not need to know how the crystalline solid takes up energy for this discussion, because as noted above QM permits us to conclude even without such knowledge that the entropy S should be zero. The theoretically calculated change in entropy (ΔS) when the temperature of the solid is increased from absolute zero until it becomes an "ideal" gas is thus simply equal to the entropy S calculated for this high-temperature state. On the other hand, the experimental entropy change ΔS is determined in the usual way from knowledge of how the specific heats for the solid, liquid, and gas depend on temperature and of the latent heats and temperatures of melting and boiling.

Theory vs. Experiment

Does the change in entropy calculated from the kinetic-molecular theory as corrected by QM agree with the measured entropy? Not always. For many substances it does, but for many others it does not. The entropy change directly determined in the laboratory is less than the calculated entropy change, not usually by much, but not negligibly either. Table 14.2 gives some examples of both. What is wrong?

Careful analysis of the discrepant systems shows something often takes place on cooling that thermodynamics cannot explain: the rates at which certain kinds of molecular rearrangements occur become

Table 14.2
Comparison of Theoretical and Experimentally Measured Entropies

Substance	Entropy		
	Theory	Experiment	Discrepancy
Nitrogen (N_2)	191.5	192.0	0.5*
Oxygen (O_2)	205.1	205.4	0.3*
Hydrogen chloride (HCl)	186.8	186.2	0.6*
Carbon monoxide (CO)	197.9	193.	5.
Water (H_2O)	188.7	185.3	3.4
Nitrous oxide (N_2O)	220.0	215.2	4.8

Note: Theoretical entropies were calculated from $S = k \log W$; experimental entropies were measured with calorimeters. Entropies are given in joules per kelvin at 25°C and 1 atm pressure for 1 mole of substance (see Chapter 4).

*These discrepancies are within experimental error.

slower and slower at lower and lower temperatures. Some molecular systems require such rearrangements to reach equilibrium, and therefore they often remain trapped in some state of higher energy when they are cooled; in other words, they can no longer reach an equilibrium state. Since systems with higher energies invariably can have the energy distributed in many different ways, they will have an entropy different from zero as a result, even at absolute zero.

Consider carbon monoxide, CO. The liquid form of carbon monoxide boils at about $-190°C$ and solidifies at about $-210°C$. In the solid crystalline form at $-210°C$, the CO molecules lie in a periodic arrangement in space, as illustrated in Figure 14.6A, but they can point at random in either of two directions. A crystal in this arrangement is said to be "disordered." If there are N molecules, and each has two choices for its orientation in space, the number of arrangements $W = 2^N$, contributing $S = k\log W = kN\log 2$ to the entropy. For one mole of carbon monoxide, this corresponds to 5.8 joules per kelvin. The disordered state of the crystal, however, has a higher energy than an ordered state, in which each of the molecules must point in a single direction, as illustrated schematically in Figure 14.6B. If the crystalline form always remained in equilibrium as it is cooled, it would necessarily reach this lowest energy state at absolute zero. The ability of the molecules to turn around in their places is lost on cooling, however, and a considerable degree of "frozen-in" disorder remains at absolute zero, so that the entropy does not become zero.

The carbon monoxide crystal is an example of a system that even at equilibrium and in its lowest energy state does not have $W = 1$. Every molecule at equilibrium should point in the same direction, but this direction can be either to the left or to the right, so W is 2 rather than 1. The entropy associated with the two possible arrangements of the whole crystal cannot be experimentally distinguished from zero.

Ice, a crystalline solid, does not reach an equilibrium zero-entropy state at absolute zero because the hydrogen atoms do not find unique positions of the lowest possible energy on cooling. Glasses are liquids trapped by the slowness of molecular rearrangements in nonequilibrium states of high energy, and their entropies also never reach zero.

To summarize, molecular systems that show the discrepancies between theoretical and experimental entropies have always been found to show specific molecular or atomic rearrangements taking place too slowly at low temperatures for equilibrium to be reached at absolute zero. In these cases, the occurrence of the discrepancy can be predicted from knowledge of the structures and properties of the molecules, and its magnitude can be predicted quantitatively. The discrepancies, instead of raising doubt about the correctness of the molecular approach, serve instead to confirm it more strongly.

The knowledge that the entropy of all substances is zero at absolute zero if they are at equilibrium at that temperature is useful because it permits us to calculate the entropy changes in chemical reactions from measurements on the substances reacting and the substances formed, as we did in our discussion of the graphite-diamond reaction in Chapter 11.

One might wonder if the disappearance of entropy on cooling is a

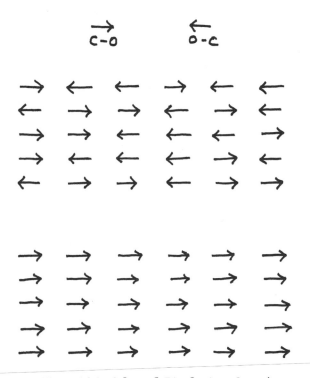

Figure 14.6 Order and Disorder in a Crystal
In crystalline carbon monoxide, the CO molecules (schematically represented by arrows) occur in a periodic arrangement *(bottom)*. At higher temperatures the molecules point randomly in one of two possible directions *(top)*. At lower temperatures, if the system were at equilibrium the molecules should all point in the same direction, but as the crystal is cooled they are frozen into nearly random orientations and do not reach equilibrium.

result for which QM was necessary. Even if Newton's laws applied, we would expect that at absolute zero there is no more thermal energy in a collection of molecules, and therefore no energy to distribute among them. Should not W be 1 here also? It may seem that it should, but if Newton's laws applied, the equipartition theorem would be valid all the way down to absolute zero. If it were, the average molecular energy, being proportional to kT, would go to zero as we would expect, but we would find that the entropy at absolute zero could not be calculated at all! This is a purely mathematical consequence of the equipartition theorem, but it makes no physical sense. We are lucky that QM has spared us this difficulty.

The tendency of the entropy at equilibrium to go to zero when the temperature is lowered is called the *third law of thermodynamics*. It was inferred from experimental measurements before it was shown to be a consequence of quantum mechanics.

15
Relativity and the Fate
of the Universe

~~~~~~

We now turn to the application of thermodynamics to the expanding universe—an expansion that was discovered by direct experimental observation but that can be interpreted only with the general theory of relativity. In this expanding universe the first law, the law of the conservation of energy, breaks down. The second law, the law of increasing entropy, may still hold, but there are difficulties in describing the course of entropy in such a universe. We will begin with a discussion of the special theory of relativity, which changed the concept of energy but left the principle of its conservation unchanged. Then we describe the general theory, how it applies to the expanding (and possibly someday contracting) universe, the conditions under which conservation fails, and the problems of applying the second law to such a universe.

Special Relativity

The Speed of Light

Light travels very fast. As far as the first scientists to study it were concerned, it seemed to travel with an infinite speed. That its speed is finite was demonstrated in the seventeenth century by astronomical observations. The first successful terrestrial measurement was carried out by the French physicist Armand Fizeau in 1849. More and more accurate measurements have been made since then and are still being made.

The speed of light in a vacuum is about 186,000 miles per second, very close to 3.0×10^8 meters per second. When light travels in matter—in air, water, or glass, for example—its speed is less, in agreement with Maxwell's electromagnetic wave theory (Newton's corpuscular theory required that it be greater).

Common sense suggests that the speed of light an observer measures should depend on the observer's motion. Let us imagine, for example, two observers both measuring the speed of light coming from a distant source (the sun). One (male) is standing still relative to the sun, and the other (female) is moving at a high speed away from the sun—let's say at two-thirds the speed of light, or 2×10^8 meters per second.

The stationary observer finds the light from the sun is passing by him at 3×10^8 meters per second and also notes that the moving observer is heading away from the sun at 2×10^8 meters per second. He concludes that the light must be passing by her at a relative speed of only 1×10^8 meters per second. Then he contacts the moving observer and asks her what she found for the speed of the passing light. The answer comes back: 3×10^8 meters per second, the same result as his. This fits neither the wave theory of Maxwell nor the corpuscular theory, to say nothing of common sense.

The stationary observer can make an inference that might save his sanity. To measure the speed of light one needs two basic instruments, one to measure a time interval (a clock) and one to measure a distance (a meter stick). He may account for the bizarre result of the moving observer by inferring that either her clock, or her meter stick, or both, are reading incorrectly. But why should they? In particular, why should their errors always just compensate for each other to give the same answer found by him?

What we have described is only a thought experiment; space ships traveling at two-thirds the speed of light have not yet been constructed. Real experiments conducted at much slower speeds give the same result, however: the speed of light an observer measures does not depend either on the speed of the observer or on the speed of the source of the light relative to the observer.

The first experiments of this type were done by an American physicist, A. A. Michelson, who hoped to determine the speed of the earth's motion through space from differences in the speed of light measured in different directions relative to the earth's presumed motion. As the earth's speed in its orbit is one ten-thousandth of the speed of light, an extraordinarily accurate determination of the speed of light was needed. Michelson had invented an instrument capable of such accuracy in 1880. He found no differences whatever, whether he measured the speed in the direction of the earth's motion or at right angles to it. Later, he and Edward Morley, an American chemist, improved the accuracy of the instrument. Again, no effect of the earth's motion on the speed of light was observed.

The Special Theory of Relativity

Copernicus would not have put the sun at the center of things if he had not realized that motion can be an illusion, that the rotation of the earth can fool us into thinking that the sun rises and sets. Newton's laws embody this insight in that they work equally well in laboratories moving at a constant speed in a straight line as in stationary laboratories: a pendulum swings with the same period, balls bounce, liquids flow in the same way. Although Newton believed that there is a difference between a laboratory at rest in some sort of absolute space and one in uniform motion, no experiment using his laws can be devised to distinguish them. But light, an electromagnetic wave, obeys Maxwell's equations rather than Newton's, and according to these equations it *should* have been possible to detect absolute motion—motion relative to space itself—by the Michelson-Morley experiment.

Einstein's solution to this paradox was to elevate it into a law of nature. His special theory of relativity (SR) consists of two assumptions. The first, and more plausible, was as follows:

1. *All* the laws of nature must appear the same to all observers moving relative to one another in straight-line paths at uniform speeds.

This assumption already applied to Newton's laws specifically; Einstein expanded it to include not only the phenomena of light but all other phenomena of nature as well. It ensures that there is no way to detect absolute motion—only motion relative to other bodies has meaning. Even if only *relative* motion has meaning, however, the first assumption does not rule out the possibility that the speed of light could appear to depend on a motion of the source of light *relative* to the observer. Since it doesn't, Einstein added the following:

2. Any measurement of the speed of light (in a vacuum) must give the same result, regardless of any uniform motion of either source or observer.

These two assumptions constitute SR. Once they are accepted, logic and mathematics do the rest, with dramatic and counterintuitive consequences for our notions of space and time. The clocks of observers moving relative to us slow down, their meter sticks shrink in the direction of their motion, events that are simultaneous to them need not be to us. When we measure the mass of a body moving with them, we find a higher mass than they do. Further, the effects of motion operate symmetrically. Since, to them, we are moving, to them our clocks slow down, our meter sticks shrink, and so on. Space and time, completely separate kinds of things to most people today as they were to Isaac Newton, become intermingled in curious ways.

The experimental fact that the measured speed of light in a vacuum is the same regardless of motion of source or observer cannot be regarded as experimental evidence for SR. If the theory could explain nothing else, one would be justified in feeling that it is only circular reasoning: the speed of light is constant for all observers because it is a law of nature that the speed of light is constant for all observers. Einstein's achievement was to see perfectly clearly all the strange things such a law implied. The experimental evidence for the theory is just that all these strange things, one after another, have been confirmed in the laboratory. Let us look at some of them.

Though we are avoiding most of the mathematics of the theory, there are some formulas we will need. They involve a quantity symbolized by the Greek letter ß, which tells us, for a body moving relative to us, how time on that body slows down, length shrinks, and mass increases. The quantity ß depends on the ratio of the *relative* (to us) speed v of that moving body to the speed of light c. How it varies is shown in Figure 15.1. When v is much smaller than c, ß is nearly 1, and when ß is nearly 1, the special theory of relativity is not needed. As v increases, ß decreases and eventually becomes 0 when $v = c$, but it does not begin to differ noticeably from 1 until v is quite large (by human standards). It does not decrease to 0.99 until a speed of 1.2×10^7 meters per second (about 7,500 miles per second) is reached. Rockets escaping the gravity of the earth need a speed of about 7 miles per second to make it. This is about the best we have done with large objects, but we can bring electrons and protons to speeds very close to c in particle accelerators and confirm that they behave as SR says they should.

If we observe a moving clock, the time interval it registers, symbolized by t, is shorter than the time interval t_0 our own clock shows, according to:

$$t = ßt_0$$

For a body moving at half the speed of light, ß = 0.87, resulting in a noticeable slowing, according to us, of a clock it carries.

Imagine that a measuring rod whose length measured in our own laboratory is L_0 is placed in a laboratory moving relative to ours and pointing in the direction of relative motion of that laboratory. Now if we measure its length as it passes by, we find it appears to be shorter, with a length L given by

$$L = ß L_0$$

Mass and Energy

One of the striking consequences of SR is that no material body can move faster than the speed of light. Newton's laws imply no such cosmic speed limit: the acceleration—the increase in speed in one second—produced by a given force is the same whether the body is moving slowly or fast. It is smaller the more massive the body, but since the mass (in Newton's view) doesn't change, a steady force should produce a steady acceleration, and in sufficient time any speed whatever can be reached. In SR, the mass of a moving body no longer seems fixed and unchanging, but apparently increases as the speed increases, so that a steady force no longer produces a steady acceleration. It becomes harder and

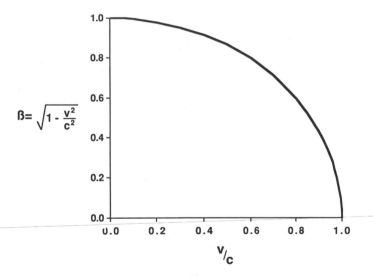

Figure 15.1 How Much Length Shrinks and Clocks Slow Down
The factor ß, giving the shrinkage of lengths and the slowing of clocks for bodies moving relative to us at a speed v, depends on v. At low speeds ß $= 1.00$, but it falls rapidly as the speed approaches that of light and becomes 0 when $v = c$.

harder to increase the speed by, say, 1 mile per second the faster the body is moving. The formula for the *apparent* mass m of a body whose mass is m_0 when at rest relative to us is:

$$m = \frac{m_0}{\text{ß}}$$

Note that ß divides m_0 whereas it multiplied t_0 and L_0. When ß becomes very small (as v approaches c), m becomes very large, and it would reach an infinitely large value when $v = c$. No further acceleration is possible no matter what the force, so the speed of light cannot be exceeded. The quantity m_0 is called the rest-mass of the body. Figure 15.2 contrasts the speed reached by a body accelerated by an unchanging force according to Newton and to Einstein.

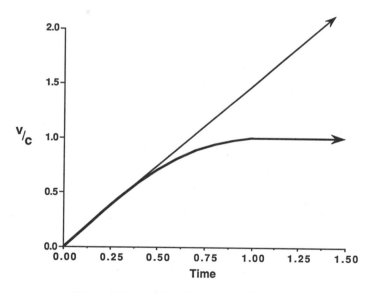

Figure 15.2 Acceleration in Special Relativity

According to Newton's laws, a constant force applied to a body produces a constant acceleration no matter what speed has been reached, so the speed v increases steadily as long as the force is applied. This predicted behavior is shown by the straight line inclined at a 45-degree angle to the two axes. In special relativity, the acceleration produced by a force depends on the speed already achieved. The behavior is as though the mass increases with the speed v according to an equation given in the text. The result is that the speed levels off at the speed of light c (that is, $v/c = 1$), even if the force continues to act on the body.

In Newton's mechanics the work done to accelerate a body from rest to a speed v is exactly equal to its kinetic energy (in the absence of friction). SR also enables us to calculate the work done to reach the speed v, but the formula does not seem to resemble $\frac{1}{2}mv^2$ at all. Let us call this work to accelerate the body the relativistic kinetic energy (RKE). The calculation gives:

$$RKE = mc^2 - m_0c^2 = \frac{m_0c^2}{\beta} - m_0c^2$$

with m, as above, given by m_0/β. Note, however, that mc^2 increases as the speed v increases (because the quantity β depends on v). As we have learned to expect, when v/c is very small, we should get the same result as Newton's mechanics gives us, and indeed it can be shown mathematically that as an excellent approximation

$$mc^2 - m_0c^2 = \tfrac{1}{2}m_0v^2 \quad \text{when } v \ll c$$

When v gets larger, the relativistic and Newtonian results no longer agree (Figure 15.3). Energy, in SR, is not given by any of the nineteenth-century formulas, but it can be shown that it is still conserved.

Why $E = mc^2$

Once we accept the idea that speed increases the mass of a body, we should find it plausible that a rise in temperature increases the mass of a body also. Even in a body at rest the molecules are moving, and they are moving faster the higher the temperature. We already know that raising a body's temperature increases its energy, now we see its mass must increase also. This implies that hot bodies weigh more, even though this is contrary to common experience, but we have not yet said how much of a change SR actually predicts.

The above reasoning shows that the mass of a body must change to reflect at least a part of any change in its energy. Einstein was able to carry this argument to a dramatic conclusion: the mass increases with *any* increase of energy, no matter what form it is in—kinetic energy of the body's motion, kinetic and potential energy of its molecules, energy stored in the nucleus. When a body loses any kind of energy at all there is a corresponding decrease in the mass according to

$$\text{Change in mass } (m) = \frac{\text{Change in energy}(E)}{c^2}$$

Or:

$$\Delta m = \frac{\Delta E}{c^{2}}$$

Einstein chose to express this formula in terms of the total mass:

$$E = mc^2$$

This does not necessarily imply that *all* the mass of any body *can* be converted to energy in the surrounding environment, only that *if* it could, the energy so converted is mc^2. It is easy to calculate from the equation $E = mc^2$ how much the increase in mass should be when a body is warmed: for ordinary temperature changes it is too small for any conceivable weighing instrument to detect. The temperatures produced in nuclear explosions, of course, are another matter.

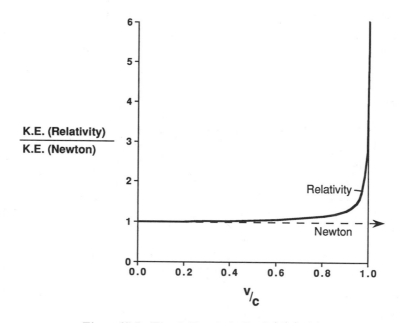

Figure 15.3 Kinetic Energy in Special Relativity
In this comparison of the kinetic energy of a body moving relative to us at a speed v according to Newton's mechanics and to the special theory of relativity, the ratio of the relativistic energy to the classical value is plotted against the ratio v/c. If the two theories gave the same answer, the ratio would be 1.00 at all speeds. Instead, the relativistic kinetic energy increases faster than the classical energy and becomes infinite when v equals the speed of light.

The Indestructibility of Mass

That mass is conserved—can neither be destroyed nor created—was assumed from the time of Newton. Lavoisier founded modern chemistry on the observation that in chemical reactions, even those with large outputs of energy, there was no change in mass that he could detect. A half-century later the principle of the conservation of energy was discovered. The two conservation principles were regarded as independent laws of nature. Since Einstein, we know that the two are really one: since the mass of a body increases in proportion to its energy, conservation of energy implies conservation of mass and vice versa. The changes in mass in Lavoisier's chemical reactions are too small for either him or us to detect. We test the equation $E = mc^2$ in the laboratory on processes that put out much more energy than any ordinary chemical reaction.

When an atom of uranium (U-235) absorbs a neutron, it decomposes to form fission products (such as atoms of the uncommon metals rubidium and caesium) and several additional neutrons. If other atoms of U-235 are in the vicinity, they can be decomposed by the neutrons released by the first, which means that there is a multiplication of neutrons and the possibility of a chain reaction. This is how the first man-made nuclear explosion was powered. Let us look at the energy and mass changes that occur.

First, the uranium is decomposed to form highly energetic, fast-moving fission atoms and neutrons. The original nuclear potential energy has been converted to kinetic energy of the fission products. There has been no loss of energy, and no change in mass yet.

Next, the fission-product atoms begin to lose their kinetic energy through collisions with other atoms nearby: atoms of oxygen and nitrogen in the air or of silicon or iron or aluminum in the earth. The fission products, slowed down, decrease both in energy and mass, and the atoms from earth and air gain both.

Eventually, the tremendous kinetic energy of the reaction products is dissipated in the environment, warming it and increasing its energy and its mass. If we collect the now-cold products of the original fission reaction and measure their total mass, we find a small but readily measurable decrease in mass compared to the original uranium: about 0.1 percent. If, however, we include in the measurement of mass the whole environment affected by the explosion as well as the original bomb ingredients, neither mass nor energy has been lost.

The Curve of Binding Energy

Atoms are made up of three kinds of particles: electrons, protons, and neutrons. Their masses and electric charges are listed in Table 15.1. The protons and neutrons are packed together tightly in the nucleus, which therefore contains most of the mass of the atom, while the electrons form a kind of cloud around the nucleus, taking up most of the space of the atom. The number of electrons is equal to the number of protons in all atoms under normal conditions, so the negative and positive charges just cancel. In all atoms of any one element—copper, for example, or oxygen—the number of protons in the nucleus is always the same, but the number of neutrons need not be. Carbon atoms always have six protons, but the number of neutrons may be 6 (most commonly), 7 (in about 1 percent of carbon atoms), or 8 (very rare: this form or "isotope" of carbon is radioactive).

One might expect that the mass of any nucleus is the sum of the masses of the protons and neutrons it contains, but it almost never is; it is almost always less. As an example, let us consider the common form of iron, which has 26 protons and 30 neutrons in the nucleus. From Table 15.1 we can calculate the mass of 26 protons and 30 neutrons to be 56.4492 atomic mass units, but the iron nucleus has a mass of only 55.9207 atomic mass units. The discrepancy is 0.5285 atomic mass units, a hefty 1 percent of the total. The discrepancy is called the *mass deficit* of this particular nucleus.

The deficit tells us that if 26 free protons and 30 free neutrons could be combined together to form a single iron nucleus, a tremendous quantity of energy, which we could calculate from the equation $\Delta E =$

Table 15.1
Mass and Charge of Subatomic Particles

Particle	Charge	Mass (amu)
Electron	$-e$	0.000549
Proton	$+e$	1.007277
Neutron	0	1.008665

Note: The magnitude e of the charge on the electron is 1.60×10^{-19} coulombs, a coulomb being the quantity of electric charge flowing per second past a point in a wire carrying one ampere of current. The amu, or "atomic mass unit," is 1.66×10^{-27} kilograms.

Δmc^2, would be given out to the environment. To convey the awesomeness of the amount, we can measure it not in joules but in the energy output of a ton of exploding TNT. Forming a kilogram of iron would give out the same energy as 200,000 tons of TNT. To visualize 200,000 tons, we note that the Woolworth Building in New York City, a 60-story skyscraper, weighs a little more than that. We are not claiming, incidentally, that it is easy or even possible to form iron by direct combination of the necessary protons and neutrons. The formation of iron in the universe has followed a stepwise route, with the nuclei growing by one or a few protons or neutrons at a time.

This large energy output tells us that the iron nucleus is extraordinarily stable: breaking it up into its constituent particles would require an input of the same quantity of energy given out when they are brought together. In effect, the mass deficit of a nucleus is a measure of its stability against *total* disintegration (into free protons and neutrons). However, in the nuclear reactions that supply us with energy—whether by fission (the breakup of very heavy nuclei into several lighter ones) or fusion (the combination of lighter nuclei to form a heavier one)—total disintegration does not take place. The nuclei of the elements we find in nature are *all* stable in the sense that energy would be given out if they could be formed from free protons and neutrons, but some are more stable than others, which makes energy-yielding fission or fusion reactions possible. The question to ask is simply: is the sum of the masses of the *products* of any possible nuclear reaction less than the sum of the masses of the *starting species?* If yes, energy will be given out when the reaction takes place.

As an aid in calculating the mass changes in any possible reaction it is useful to tabulate the mass deficit, calculated just as we did for iron, for each kind of atom. A particularly convenient way to express the mass deficit is to divide it by the total number of protons and neutrons in the nucleus, so that what is tabulated is the average mass deficit per particle (the generic term is *nucleon*) in the nucleus. For the iron nucleus previously considered, we divide 0.5285 by 56 (26 protons and 30 neutrons) to obtain 0.00944 atomic mass units per nucleon. The results of this procedure are shown in Figure 15.4.

From $E = mc^2$ we may easily calculate and make a graph of the stored nuclear energy per nucleon in the various kinds of atoms, arranged in order of the masses of the atoms. This graph is referred to as the *curve of binding energy*. From it we can see at a glance the large energies that are made available when hydrogen atoms fuse together to form helium, the process that powers the sun, and the lesser but still large energies made available when uranium atoms undergo fission to form such atoms as rubidium and cesium. We see also the greater stability of such atoms as iron and calcium, which represent, so to speak, the ashes of nuclear fires, from which no further fires may be kindled.

We must stress that in such processes mass is not "converted" into energy. Rather, the energy and mass of the starting species decrease as they are converted into the products, and the energy and mass of the surrounding environment increase by exactly the same amount.

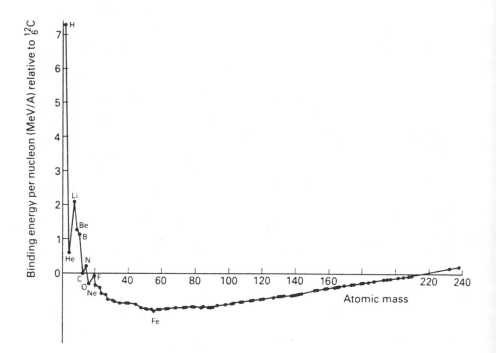

Figure 15.4 The Curve of Binding Energy
The binding energy per nuclear particle of the nuclei of the various elements. The binding energy of the nuclear particles in a carbon nucleus is used for reference in this figure, so helium and lithium have "positive" energies and iron and most other elements "negative" energies. Although this is the conventional way to represent binding energies, it can be misleading. All stable nuclei have "negative" binding energies in the sense that energy is given out when they are formed from protons and neutrons. The curve shows that energy is given out both when atoms like hydrogen or helium form heavier nuclei (fusion) and when heavy atoms like uranium break up to form iron and other elements of intermediate mass (fission).

Relative Time

While the direct evidence that $E = mc^2$ provided by nuclear reactions is the most familiar and certainly most practically important confirmation of SR, the slowing down of time that it predicts exerts a special power on the imagination. Do clocks run slower when they travel at high speeds? Since our bodies show the passage of time and are therefore clocks, do we age more slowly when we move fast? Experiments to test this possibility have compared extraordinarily accurate atomic clocks carried on jet aircraft on round-the-world trips with equally accurate clocks left at the starting point of the flight. The low speed of jetliners (relative to c) implies a time difference between the clocks of only about 10^{-15} seconds, but the accuracy of the clocks is good enough to confirm it.

More spectacular evidence comes from the behavior of *muons*, unstable particles that are often formed about 15 kilometers above the surface of the earth by collisions of cosmic rays with atoms. The muons formed by such processes travel at speeds almost equal to that of light. From laboratory experiments on muons traveling more slowly, it is known that their average lifetime is about two-millionths of a second. In this time interval, traveling at the speed of light c, they should only cover about a kilometer before they decay; almost none of them should reach the surface of the earth 15 kilometers below. In fact, however, large numbers of them do. Their internal clocks still tick off two-millionths of a second before they decay, but we see their clocks as running 20 times slower than our own, so that for us their decay takes forty-millionths of a second, and they make it all the way down.

The Conservation of Energy

Energy is still conserved in the special theory of relativity, but in a modified form that requires the introduction of a mathematical concept called a *vector* for its expression.

In talking about moving molecules until now, we have somewhat clumsily referred to their "speeds and directions of motion." In physical science it is useful to combine these two things into a single entity, called the "velocity vector," which includes both speed and direction. A 4 kilogram body moving north with a speed of 20 meters per second has a different velocity but the same speed as one moving southeast at 20 meters per second. Although the motion of a body has a direction, its kinetic energy does not. The two bodies above have the *same* kinetic energies, 800 joules ($\frac{1}{2}mv^2$). A vector may be visualized as an arrow, and in a number of figures (4.4, 4.7) we have represented the velocity vector in just this way. The *length* of the arrow represents the speed v, and the *direction* of the arrow the direction of actual motion.

Velocity is only one example of a vector. The property *momentum* is another. It is obtained by multiplying a body's velocity vector by its mass. In Newton's mechanics it is also a conserved quantity. In a collision of two hard bodies and in the absence of friction, both the energy (having no direction in space) and the momentum (having a direction) are conserved. Figure 15.5 shows why such a principle of conservation of momentum is needed.

SR has given us a new concept, space-time, which exists in a primitive form in Newton's mechanics. In Figure 15.6 we have represented the motion of a falling body by a graph showing how its position changes with time. The graph uses two perpendicular lines, one for the numerical value of time, the other for the numerical value of position. We are

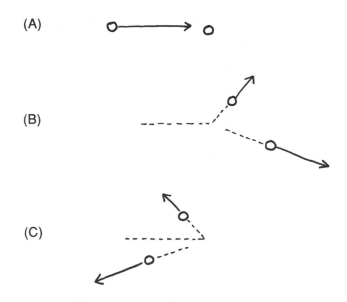

Figure 15.5 Conservation of Momentum

Even a schematic drawing shows the need for a principle of conservation of momentum. We show two examples of collisions between two balls, each collision satisfying the conservation of energy. The second collision, in which both the balls after the collision are moving to the left although the incoming ball was moving to the right, is clearly impossible.

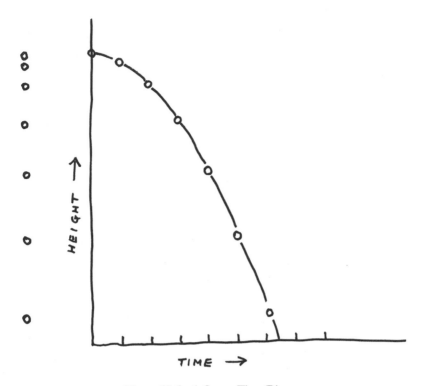

Figure 15.6   A Space-Time Diagram
A body dropped in the earth's gravitational field falls with an accelerating speed, which means that the distance it travels in equal time intervals, say one-second intervals, is greater the further it has fallen. Its path of fall is a straight vertical line, and on the left we show the falling body as though it were photographed stroboscopically, at equal time intervals. Another way to show its increasing speed is by plotting its height above ground on the vertical axis and the time elapsed on the horizontal axis. The resulting curve is called a *parabola* and the diagram a *space-time diagram*.

thus treating space and time as geometrically equivalent, representable by two straight lines, and in so doing it can be said that we have constructed a "space-time." In Newton's world doing so is only a convenience for graphing purposes; space and time are otherwise quite different kinds of things. In the world of SR the two cease to be so distinctly different, and the geometrical equivalence is not merely a convenience.

In the space-time of this theory, energy and momentum are now combined together into a single vector. As noted above, vectors have a length and a direction: the length of the velocity vector is just the speed. What takes the place of the conservation of energy is a conservation of the length of the energy-momentum vector. When bodies move slowly, Newton's mechanics and Einstein's reduce to the same thing, and the conservation of energy and the conservation of momentum appear to be separate principles. SR unites them, as it united the conservation of energy and the conservation of mass. All three are combined in a single general principle.

General Relativity

The special theory of relativity was founded on the impossibility of detecting absolute uniform motion. The general theory of relativity (GR) began with the question: if uniform motion has meaning only relative to another body, shouldn't nonuniform motion also? By "non-uniform" motion we include all kinds other than that with constant speed in a straight line, such as accelerated motion or motion in a circle. A possible paradox emerges in the statement that motion at a constant speed has meaning only relative to other bodies but that accelerated motion is not relative but absolute. There is, however, a powerful argument for making this distinction. One *feels* nonuniform motion with one's body, as any bumpy ride reminds us. We are all aware, when we ride an elevator, of an additional force added to gravity when the elevator begins accelerating upward; the feeling disappears when the elevator reaches constant speed, but then we notice a sensation of lightness when the ascending elevator slows down and stops. The forces we feel in nonuniform motion seem as real as the force of gravity.

In fact, in a completely enclosed elevator with no windows or other means of observing anything outside, there is no way to distinguish the force due to gravity from the forces due to acceleration or deceleration *except* from our knowledge that we are in an elevator in a building on the familiar earth and that the force of gravity acts all the time whereas forces arising from acceleration and deceleration are usually transitory. If we were unwittingly placed in an enclosed elevator in outer space, and the elevator were given an acceleration of 9.8 meters per second each second by some external agency—a rocket engine, perhaps—we would feel exactly as we do in the same elevator standing still on the earth's surface. In particular, if there happened to be a set of bathroom scales in the elevator we could step on, we would read our usual weight.

A more dramatic and yet feasible demonstration of the indistinguishability of gravitational and acceleration forces would be falling in an enclosed elevator from the top of a skyscraper. As the elevator accelerates downward, we would suddenly feel weightless. Objects held at arm's length and dropped would not fall but remain stationary in midair. If we step on the bathroom scales, we would find our weight to be zero.

None of this would have surprised Newton. He would have explained our weightlessness in the falling elevator by saying that although there is still a gravitational force acting on us, in free fall all bodies—including our own, objects held at arm's length, and the bathroom scales—are falling under its influence at the same accelerating speed. Under these circumstances we may have the illusion that gravity has been abolished (until we hit the ground).

As we all know, we can maintain a state of weightlessness for a longer time and with a less catastrophic outcome in a satellite circling the earth. The satellite is still in the earth's gravitational field—which is why it keeps circling the earth—but also in a state of free fall, so gravity is exactly neutralized.

Einstein's Conjecture

Let us return to the windowless elevator, within which a man standing on the bathroom scale observes his usual weight of 70 kilograms. If he knows Newton's laws, he knows that *either* he is, as usual, at rest on the earth *or* he is in outer space and the elevator is being accelerated so that the speed increases by 9.8 meters per second each second. Knowing Newton's laws, he knows also that no *mechanical* experiment he can do—drop tennis balls, time the swing of a pendulum, fire a projective horizontally and trace its downward curving path—can distinguish between the two alternatives.

It may occur to him, however, that light might offer a means of making the distinction. To him, light is a wave motion in empty space. If the elevator is standing still on the earth and he sends a beam of light in the horizontal direction, it will travel horizontally. After all, he would reason, light is not composed of small corpuscles of some material substance attracted by gravity: a wave in empty space has no mass and should not be deflected by gravity. On the other hand, if the elevator is accelerating upward, a beam of light traveling horizontally will appear to him, riding in the elevator, to be following a curved path (Figure 15.7). So he concludes that he should be able, without looking outside the elevator, to explain the gravity-like force he feels.

Einstein's conjecture, and it is a partial statement of the general theory of relativity, is that *no experiment whatever done inside the elevator should be able to distinguish between a state of rest in the presence of a gravitational field and an acceleration of the elevator in its absence.* This statement is called "the principle of equivalence."

Einstein's reason for making this conjecture was as much aesthetic as

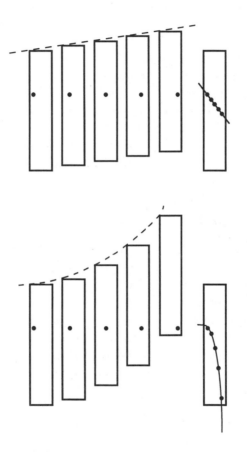

Figure 15.7 Why Light Is Deflected by a Gravitational Field
A bullet fired horizontally through an elevator rising at a constant speed *(top)* is believed by an observer inside the elevator to be traveling in a straight-line path, although the path according to the observer is not horizontal. If, however, the elevator is rising at an accelerating speed, the bullet appears to follow a curved path, which the observer attributes to a force acting on the bullet. He infers that there is such a force because in the accelerating elevator he feels a force on his body exactly as though he were subject to the force of gravity on the surface of the earth.

anything else. He proposed his general theory not because any experiments that he was aware of demanded it. The first successful experimental test of GR was his calculation of a deviation of the orbit of the planet Mercury from the path predicted by Newton's laws, but it was a small deviation and nobody had guessed that a major new theory would be needed to account for it. Einstein felt, rather, that there was something scientifically unsatisfactory about a situation where there was either a "real" force, gravity, or an "illusory" force (created by an accelerating laboratory) and a physicist inside the laboratory couldn't tell which it was by any mechanical experiment he could do. Surely the laws of physics should be expressed in a form that didn't depend on such contingencies.

If it had simply been a matter of finding a more satisfying language to talk about the same old things, we would not dignify the effort by the term *theory,* but there were startling experimental consequences to Einstein's conjecture. *If* no experiment whatever could distinguish the two possible explanations of the force observed, *and if* a horizontally directed beam of light follows a curved path in an accelerating laboratory, *then* such a beam must follow an identical curved path in a gravitational field (Figure 15.7). Light will therefore act as though it is attracted to massive objects like the sun, just as planets are.

An experiment to demonstrate the bending of a ray of light by the sun's strong gravitational field was performed during a solar eclipse in 1919, and it appeared to confirm Einstein's prediction. It was not the first experimental confirmation of the theory—the correction to the orbit of Mercury was—but it was the first to hit the headlines. The accuracy of the experiment left something to be desired, but more careful experiments done since then have confirmed the effect and its magnitude.

Another prediction of GR testable in the laboratory is that clocks appear to run slower in a strong gravitational field when observed from outside the field. This effect was confirmed with accurate atomic clocks, which tell time by the frequency of the light they emit. A "slowing down" of such a clock means that the frequency of its emitted light is lower than it would be in the absence of a gravitational field. From the relation between frequency and wavelength of light given in Chapter 10, it follows that visible light emitted by an atom at the surface of the sun is redder in color when observed from the earth than the light emitted by the same atom at the surface of the earth. The effect is called the *gravitational red shift.*

Curved Space-Time

As we indicated, to deduce all the consequences of the theory from the simple verbal statement of it we gave above, we would have to introduce some difficult concepts and difficult mathematics. The mathematics replaces the ordinary geometry of Euclid—the geometry taught in high school, which most of us have some intuitive feeling for—with a more general kind. In "ordinary" geometry, the geometry of flat, plane surfaces, the sum of the three angles of any triangle is equal to 180 degrees (two right angles). One knows what this means, even if one has forgotten how to prove it. It is easy to confirm experimentally with a protractor (a device for measuring angles) for any triangle drawn carefully with a good ruler on a sheet of paper. There is, however, a triangle for which it fails: a large triangle drawn on the surface of the earth, in particular a triangle with its base on the Equator and its vertex at the North Pole. The two angles at the base, both right angles, add up to 180 degrees, but the angle at the pole is not zero, so the sum is greater than two right angles. This isn't so surprising when one thinks of it: the surface of the earth is, after all, curved, and the geometry we characterize as "ordinary" applies only to flat surfaces. We need an alternative geometry for curved surfaces.

The experimental evidence for a round earth is quite overwhelming, and no one would go to the trouble of measuring the angles of our Equator–North Pole triangle in order to say "Aha!" We should not forget, however, that the roundness of the earth was surprising when it was first proposed, and whether either the surface of the earth or the space we live in may be curved rather than flat is a question for experiment rather than intuition to answer.

This example doesn't go very far toward making a curved space intuitively acceptable, and matters get worse when we include time together with the three dimensions of space as though it were a fourth spatial dimension.

GR requires us thus to deal with a four-dimensional space-time with a curved geometry. What causes the curvature?

Energy in the General Theory of Relativity

In the special theory of relativity, the energy of a moving body was combined with its momentum to form a mathematical entity called a vector. The concept of a vector as something having both a magnitude and a direction, like an arrow drawn on paper, is not beyond comprehension even for those without much mathematical background. In both SR and GR, we often need to deal with more complicated kinds of motion than that of a single body. For example, we wish to consider the flows in a large, continuous body of liquid or gaseous matter in which different regions are moving in different directions with different speeds, an ocean with distinct currents, and shorelines where waves break. Under such conditions, the vector that described the energy-momentum properties of a single moving body must be replaced by a more abstract mathematical construct called a *tensor*. This tensor can be regarded as giving, for each point in space, the local densities of energy, mass, momentum, and pressure. We will call it, somewhat inaccurately, the "energy density" tensor.

GR shows that the curvature of any local region of space-time can be calculated from this tensor, or in other words from the mass, energy, momentum, and pressure in its vicinity. In turn, the motions of bodies or the behavior of rays of light in the region can be predicted from this local curvature. The mathematical relations between the local energy concentrations and the curvature of space take the place of the "forces" of Newton's mechanics.

There is a major conceptual change involved here. In Newton's mechanics, and in SR, the flat geometry of space, or of space-time, was regarded as given *a priori*. It was not part of the laws of nature which we humans must discover. In GR the geometry is no longer a given but instead becomes, like the orbit of a planet in Newton's mechanics, an outcome of the operation of the laws of nature. The crucial question is not whether we humans, with intuitions formed by our limited experiences with space and time, can visualize a curved space-time, but whether the predicted behavior is confirmed by experiment.

First of all, GR predicts the same things as Newton's mechanics does when gravitational forces are not extremely large; it predicts the same elliptical orbits for all the planets except Mercury, but it also predicts the modified orbit for Mercury in the strong gravitational field close to the sun (where Newton's mechanics fails), the bending of light and the slowing of clocks by gravitational fields, and more, to be described below.

Now if we accept that space can be curved in the vicinity of matter and then recognize that matter is distributed throughout space—and there is every reason to think so—then space as a whole can be curved. We will return to this question when we consider the fate of the universe.

How General Relativity Changes the First and Second Laws

Energy retains an important role in GR. Combined with other quantities in the "energy density" tensor, it determines the local curvature of space and therefore what we call in Newtonian language "the force of gravity." However, the energy that is described by the tensor does not include an energy of the gravitational field, itself a consequence of the curvature of space. In fact, defining the "energy of the gravitational field" is so far an unsolved problem in the general theory of relativity—and there are reasons to doubt whether there is a solution at all. The consequence is that *the first law does not apply to situations in which changes in the curvature of space need to be taken into account.* An example of the breakdown of energy conservation is the change of the wavelength of the black-body radiation filling space as the universe expands, a point we will return to shortly. The first law still applies, however, in "local" regions, studied for "short" time periods. What is local, and what is short? Our galaxy, 100,000 light-years across, consisting of 10^{11} stars, is a local region, and the time until the sun burns out, 5×10^9 years, is short.

The second law does not "break down" when the general theory of relativity is taken into account, but it does require rethinking. Consider the laws as stated by Rudolf Clausius after he had discovered the second law:

1. The energy of the universe is constant.
2. The entropy of the universe tends to a maximum. (This has a more emphatic finality in the original language: *Die Entropie das Welt strebt einem maximum zu.*)

No one would quarrel with these statements for a small isolated system containing a finite amount of matter, but is the universe such a system?

As far as the scientists of the nineteenth century were concerned, the universe was infinite in extent and uniformly filled with stars. Many assumed also that it had been in existence an infinite length of time in much its present state, but with the discovery of the second law it was apparent that it could not have been. If entropy is increasing all the time, then we must envisage states of lower and lower entropy at earlier and earlier points of time and, somewhere far back in the past, a beginning. This was Kelvin's view, and he applied it not only to the history of the earth (Chapter 13) but to the history of the sun as well.

When Clausius formulated this statement of the second law, he must have had an infinite universe in mind, but he did not concern himself with the fact that if the universe is infinite and uniformly filled with stars, then the number of stars and therefore the entropy would necessarily be infinite also. If the entropy is infinite, it is not so clear what one means by saying it is "increasing." Perhaps Clausius thought of a uniform universe, with the *average densities* of energy and entropy—the quantity of each per unit volume—as having finite values (provided a large enough volume, containing large numbers of stars, was used for calculating the density).

If so, one might state the two laws as follows: the energy per unit volume remains the same, and the entropy per unit volume tends to a maximum. It is not obvious, however, even from a nineteenth-century perspective, that this is what the two laws really imply, and from a twentieth-century perspective it is even more doubtful.

We now know that the universe is expanding. This alone is enough to make energy and entropy densities decrease: the expansion thins matter out, and energy and entropy along with it. The only physically reasonable way to define a "density" is to allow the reference or unit volume used to calculate the density to expand in proportion as the universe expands, so that if nothing were happening other than the expansion, the matter inside the volume remains the same. Such a volume, with boundaries varying with time, is called a "co-moving volume," and in what follows, when we use the terms *energy density* or *entropy density* we will mean densities calculated this way.

There is, however, a more serious difficulty with the second law. Clausius's statement is an extension of the concept of an "isolated" system to the universe of his time. "Isolated" in this sense means a system that does no work on, and gains or loses no heat to, any other systems: the energy is therefore constant. A system not isolated may decrease both in energy and entropy, as happens every time a hot body in cold surroundings cools down. While the universe of GR is isolated in the sense that it is not losing energy to anything outside itself, energy is not conserved within it. Given this, a decrease in the entropy density of the system cannot be ruled out.

So far as we are able to estimate the net entropy changes that take place in various processes going on or that have gone on in the past, we find no evidence of such a decrease. We will give some examples later. In addition, for processes taking place in local regions and in short time periods (as we defined *local* and *short* earlier) we may safely expect entropy or entropy density to increase. Before we begin the history and fate of the universe according to GR, we would like to discuss a particular local and short process, first postulated theoretically as a consequence of GR and now believed really to occur: the formation of "black holes."

Black Holes

Newton's theory of light—that it is material in nature, composed of "corpuscles"—is certainly consistent with the possibility that light would be attracted by a gravitational field. John Michell, of Cambridge University, published a paper in 1783 suggesting that a star could be so massive that its gravitational field could prevent light from leaving its surface. Such a star would therefore be invisible, but it could be detected by its gravitational force on other bodies. When the wave theory of light replaced the corpuscular theory, this idea was forgotten.

Once general relativity and quantum mechanics became the prevailing theories of physics, it was discovered that objects with such strong gravitational fields, given the graphic name black holes, were not only theoretical possibilities but also, given the known histories and properties of stars, quite likely to exist. A number of objects in space have been identified as probable black holes by their intense gravitational fields and by the detection of high-energy radiation coming from their vicinity, plausibly attributable to the capture by the black hole of matter from other stars in its vicinity.

The gravitational fields at the surface of the largest known stars are not great enough to prevent light from escaping. The strength of such a gravitational field depends on two things, the star's mass and its radius. The gravitational force near any spherical object falls off as the square of the distance from the center. To produce a strong enough force to make the object a black hole, we need not only a very massive star, but also one compacted to a very small size. This in turn would imply an extraordinary density for the star: 10^{16} times the density of water. An object the size of an amoeba made of such stuff would have a mass of 10 tons. Are such densities possible?

The answer is yes. Gravity alone creates a pressure inside any star or planet, and this pressure is greatest at its center. Stars don't usually collapse under their own gravitational self-attraction because they develop an opposing pressure from the high temperatures arising from nuclear fusion reactions taking place inside them (the same reactions convert hydrogen to helium and produce the light they radiate). Eventually, however, most of the hydrogen is used up. When this happens the star begins to collapse on itself, leading to a rise of temperature. This rise ignites a new nuclear reaction: the conversion of helium to heavier elements like carbon and oxygen. This new reaction produces a great expansion of the star into a "red giant." What happens next depends on the mass of the star. If it was a small star (our sun is small), it will collapse to form a "dwarf" star—dark, cold, and very dense, but not a black hole. But if it happens to have a mass of 2.5 or more times that of the sun—and there are many such stars—it has a different destiny. The nuclear reaction in which helium forms heavier elements goes faster and more intensely, until iron, the most stable of elements (see Figure 15.4), is formed. Again, however, the tendency to gravitational collapse when the reaction has nearly run out of fuel leads to still higher temperatures and an explosion to form a supernova. In this explosion, nuclei previously formed are torn apart by the high temperature, and elements heavier and less stable than iron are formed, including uranium.

When this explosion has done its work, part of the star is spread out in space as a rapidly expanding gas, and the rest forms a highly compact body. The extraordinary pressure of the star's own gravity overcomes the resistance of atoms to being pushed together, and the electrons and protons are combined into neutrons. The pressure, which depends on the mass of the star, may or may not be great enough to overcome the even greater resistance of nuclei to being pushed too close to each other. If it is not great enough, the star becomes a "neutron" star, whose properties will not concern us here. If, however, the star was massive enough, the nuclear resistance is overcome, and a black hole is formed.

Black holes have a size as well as a mass. Their radius is defined as the distance from the center of the black hole to the point at which the gravitational force is just great enough to prevent light from leaving. Nothing inside this radius, light or atoms, can get out.

While neither light nor atoms can escape, the gravitational attraction remains and can capture any matter in its vicinity: clouds of gas from nearby stars, nearby stars themselves, and even other black holes. The captured matter spiraling in should emit characteristic radiation before actually reaching the surface of no return. There are sources of radiation in space having the expected characteristics that suggest the presence of black holes, one of them near the center of our own galaxy.

Black holes are important to us in two ways: they have an interesting thermodynamics, and they have a bearing on the fate of the universe.

Thermodynamics of a Black Hole

A black hole can capture radiation just as it can capture atoms. The energy of the captured light increases the energy and therefore the mass of the black hole. This raises a question as to whether black holes can violate the second law. We know from the properties of bodies at temperature equilibrium with radiation (see the discussion of black-body radiation in Chapter 10) that even the coldest bodies, as long as they are not at absolute zero, radiate some energy away. If all bodies not at absolute zero lose energy to a black hole, and if the hole has the property of absorbing all matter and energy that falls on it and emitting none, then one must regard it as being at a temperature of absolute zero. But if it is at zero temperature, it should have zero entropy as well, according to the third law (Chapter 14). Now any matter and radiation it captures must have initially had some entropy. If the black hole swallows them up, their entropy disappears, leading to a net *decrease* of entropy, which is a violation of the second law.

We are not arguing that black holes can't really behave this way; maybe the second law can be violated under such extreme circumstances. We have already seen the breakdown of the first law in GR; the second law was originally justified as a consequence of Newton's mechanics, and later as a consequence of quantum mechanics also, but it is not clear yet that it is consistent with GR.

But in fact the second law seems to work even for black holes. Theoretical treatments combining quantum mechanics and the general theory of relativity, by Jacob Bekenstein and Stephen Hawking, have reached the following conclusions:

1. Black holes do not have zero entropy; in fact they have very large entropies, much greater than the entropy of the gaseous or stellar matter from which they are formed. They have 10^{20} times as much entropy as a quantity of an ideal gas of the same mass and filling the entire visible universe, 10^{10} light-years in diameter. (This enormous value of black hole entropy has been calculated from macroscopic rather than microscopic considerations. So far there is no clear understanding of how to calculate W, the number of possible microstates, for a black hole, and while the entropy values may well be correct, it has not yet been possible to give a microscopic justification for them.) So the formation and growth of a black hole, rather than posing a problem for the second law, serves to confirm it.

2. Black holes have an energy given by the usual $E = Mc^2$, where M

is the mass of the hole. Now bodies with both an energy and an entropy can be shown to have a temperature also, as a mathematical consequence of the two laws. The temperature is surprisingly low: about 10^{-7} kelvins for the smallest possible black hole that can be formed by stellar collapse, lower still for more massive ones.

3. Having a temperature, a black hole should radiate energy just as any other body warmer than absolute zero does, but how can it if its gravity is too great for radiation to escape? The remarkable answer, provided by Hawking, is that it does anyway. The radiant energy is not emitted from inside the hole but from just outside the surface, produced from empty space by the intense gravitational field and compensated for by a decrease in the mass (energy) of the black hole. The spectrum of this radiated energy can be calculated by combining general relativity and quantum mechanics; it is exactly the black-body spectrum for the temperature of the hole as calculated from the two laws of thermodynamics.

The temperature of space is about 3 kelvins. We know this because we can detect black-body radiation corresponding to that temperature coming from the sky. Space, then, is much warmer than any black holes (10^{-7} kelvins) that may be in it. This implies that black holes, if they are out there, are receiving more radiant energy than they are emitting. They cannot begin to lose energy until the universe itself is cooled below 10^{-7} kelvins. How long that may take will be discussed, but not answered, later.

Given that black holes can radiate, losing energy and mass accordingly, and should eventually disappear, how long will it take, if the universe ever cools enough to make it possible? The formula for the energy radiated by a black body given in Chapter 10 gives us an answer. A body of the minimum black hole mass (2.5 times the sun's mass) will disappear in 10^{66} years. The universe has lasted (since the big bang—see below) for 10^{10} years. So any black holes of this mass now existing will not disappear until the universe has lasted at least 10^{56} times its present age. How long is that? The time period doesn't even boggle the mind; it is so vast that it merely numbs us.

The Fate of the Universe

A number of possible universes are consistent with both GR and the experimental facts, and we don't know which of them, if any, is the one we actually have. It is usually assumed by astrophysicists that ours is (after averaging over large enough volumes) uniformly filled with energy (including, but not limited to, energy in the form of matter). There are two reasons for using this assumption: one is that there is no evidence to the contrary and considerable evidence in its favor, and the second is that the equations of GR can more readily be solved for this kind of model.

Even if uniform density is assumed, the possible universes are radically different depending on whether the curvature of space is positive, negative, or zero. We will not need to consider these possibilities, however; the main question for our purposes is whether the universe is finite or infinite in extent.

1. If the density of mass/energy is greater than a critical value that can be calculated from the theory, the universe is finite in extent and positively curved, just as the surface of the earth is finite in extent and positively curved.

2. If the density of mass/energy is less than this critical value, the universe is infinite in extent and therefore contains an infinite quantity of energy. Such a universe has a negative curvature.

3. There is a borderline case, with the energy density exactly equal to the critical value. Such a universe is infinite in extent and contains an infinite quantity of energy, but it is flat, not curved. Euclid's geometry would describe this kind of universe. It would seem highly unlikely that the energy density should be exactly right to make the universe flat, but there is some evidence that this may really be our universe, and we may need a theory to account for what would otherwise be an extraordinary coincidence.

All three of these possible universes have one feature in common: they are not static. As time goes on they do not remain the same. The finite-sized (positively curved) universe must first expand, reach a maximum size, and then contract as a result of the gravitational attraction of the matter in it. In the two infinite universes, the density of matter is not high enough to provide enough gravitational attraction to limit expansion, which therefore continues forever. The distance between any two stars or galaxies will continue to increase, the more so the greater their distance apart.

Which universe is ours?

All three of these universes have a definite beginning in time, but only the closed one has an end (of sorts: it may be a new beginning also). All three show expansion, the closed universe only in the first phase of its existence, the others throughout all time.

The observational evidence we currently have only shows that (1) there *was* a beginning and that (2) the universe is now expanding.

There are three primary pieces of evidence for a beginning to the universe and for the current state of expansion:

1. The stars and galaxies are moving away from us here on the earth, and those that are farther away are moving away faster. Their speeds of recession were measured by the American astronomer Edwin P. Hubble, who detected the "red shift" of the light they radiate to us. This recession could mean that the earth is the center of the universe, but it is more humble and more parsimonious to assume that the stars and galaxies throughout space are uniformly spreading apart, and that the distance between any two stars or any two galaxies increases more rapidly the further apart they are. There is no center of the universe, but from any observation point, not just our own, all other stars would appear to be receding.

2. Space is filled with radiation whose energy is distributed according to the black-body curve corresponding to a temperature of about 2.7 kelvins, very cold indeed. Either the universe has always been cold, or else it was once much hotter than now, and the wavelengths of the black-body radiation were stretched out by an expansion of the universe into what we find today. This stretching is consistent with the general theory of relativity.

3. The abundances of hydrogen and helium in the universe, estimated spectroscopically, fit very well the values that would be expected if the universe were once so hot that no helium nuclei but only the separated particles, neutrons and protons (hydrogen nuclei), of which they are composed could have existed. Subsequent gradual cooling would then have allowed these particles to coalesce into the relative amounts of helium and hydrogen we actually find.

Before these experimental observations were made, a few scientists (the Russian Alexander Friedmann and a Belgian priest, Georges Lemaître), exploring the consequences of GR, had realized that the universe must have had a history; GR did not permit a static universe. It is remarkable that Einstein was not one of the pioneers in this line of speculative research. When he first formulated GR he believed the universe must be static, as did most scientists at that time, and he introduced a modification of his theory to account for it, a step he later regretted.

We have said earlier that the presence of mass curves space-time in its vicinity. The models of the universe provided by relativity start with the reasonable assumption, not so far contradicted by any observation, that matter is spread fairly uniformly through space. "Uniform" here means uniform on a large enough scale. There is empty space between our sun and nearby stars, and emptier space between our galaxy and others. Matter can be said to be uniformly distributed only if we average over regions of space large enough to include large numbers of galaxies. There is some controversy at present as to how large the number must be.

If matter curves space in its vicinity, and matter is uniformly distributed through all of space, then all of space is curved. The general theory of relativity goes further: it predicts not only that space is curved but how the curvature changes with time. It predicts specifically that the universe could have once consisted of a highly curved space with a very high density of energy, which all at once began to expand at a definite moment in time. If it did, we would expect the three essential experimental observations we described above: the recession of the galaxies, the presence of black-body radiation at a low temperature (the stretched-out and cooled-down radiation of an earlier high temperature), and the observed abundance of hydrogen and helium. This picture of the universe is called the "big bang" theory.

Entropy in the Past

How has the entropy of the universe varied in the past? When we speak of the "entropy of the universe" we must distinguish between the finite entropy of a finite (positively cured) universe that is now expanding and will eventually collapse and the entropy of an infinite universe that will expand forever. As noted earlier, the entropy of an infinite universe is infinite, so we consider instead the entropy density: the average entropy of very large but specified volumes of space, with the specified volume expanding (or contracting) as the universe expands (or contracts). We will somewhat loosely talk about the "entropy" of either kind of universe, but it is important to keep the distinction always in mind.

For all processes that have gone on in the past, the entropy has either increased or remained the same. We will list here a number of those processes that have involved significant increases.

1. *The stabilization of matter.* At the extremely high temperatures of the early universe ($T = 10^{11}$ K); particles of matter can be formed from high-energy photons (electromagnetic radiation), and in turn particles of matter can collide and disappear to form high-energy photons.

Matter and radiation are mutually in balance at such temperatures. The rate of destruction of either equals its rate of formation. All that is required is that the energy of the photons (which equals $h \times f$: see Chapter 14) equals the energy of the particles (mc^2). Under such conditions of balance, matter can be thought of as a form of radiation, and the entropy density of such a universe remains unchanged by expansion. The fall of temperature, which would decrease the entropy of the radiation in a fixed volume, is compensated for by the increase of the volume. The total energy is essentially equally divided between matter and radiation.

As the universe continues to expand and the radiation and matter cools, a temperature is eventually reached at which the photons no longer have the energy to form particles, although the particles that are present continue to undergo collisions and annihilation to form photons. During this process the entropy rises. Most of the particles of matter present during this period are annihilated to form radiation. As they thin out in numbers, and as further cooling takes place, the few particles remaining cease to destroy each other, and the matter present at this stage is what we have now.

2. Gravitational collapse. When the temperature has fallen enough to allow electrons and protons to come together to form hydrogen atoms, the hydrogen is spread out in space as a gas. Such extended clouds of gas are unstable because the gravitational attraction between different portions of the extended cloud tends to pull them closer together. This tendency is resisted by the kinetic energy of the individual atoms. If, however, a large enough aggregate once forms, gravity will hold it together permanently to make one of those large, highly localized spheres of hydrogen we call "stars." The process is called "gravitational collapse."

It might be thought that gravitational collapse should cause an entropy decrease, not an increase. After all, a gas uniformly distributed in space has the largest possible entropy, and the entropy is less when the gas is all bunched up in one corner of its container. Gravity is a very weak force compared with the electrical forces that hold matter together, but it gets stronger and stronger the more mass there is and, as we might guess from the example of black holes, can overcome all other forces. A uniform cloud of gas of sufficiently large mass has a high gravitational potential energy, and when it "falls" in on itself, that potential energy is converted first to kinetic energy and, through frictional effects, to a heating of the gas. The effect of the rise in temperature on the entropy more than compensates for the compaction of the gas, and the total entropy increases during collapse.

3. *Nucleosynthesis in stars.* Once stars heated to high temperatures by gravitational collapse are formed, nucleosynthesis begins. The matter that collapses to form the stars was mostly hydrogen (one electron and one proton) and "heavy" hydrogen, or deuterium (one electron and a nucleus consisting of one proton and one neutron). Hydrogen, as can be seen from the curve of binding energy (see Figure 15.4), is not stable with respect to forming heavier elements. At the high temperatures of stars, the reaction of hydrogen and deuterium to form helium begins and increases the temperature still further. This reaction is the primary source of the energy that stars (including the sun) radiate, and there is a large entropy increase when it occurs. The most stable nuclei are those of iron and a few other elements of similar nuclear masses. These form more slowly, and they would be the ultimate end of all nuclear reactions if other things did not intervene.

4. *Black hole formation.* Stars burn out eventually. Those that are small end up as "dwarf" stars. Larger ones explode into supernovas and then may collapse into neutron stars or black holes. It is during this final supernova explosion that the elements heavier than iron are formed and distributed through space. Collapse of a star to a black hole involves an enormous entropy increase.

Black holes have the capacity to draw surrounding matter into themselves, including surrounding stars as well as other black holes. Stars are not randomly placed in space but are organized into galaxies of about 10^{11} stars, with nearly empty space between the galaxies. It is possible that galaxies can eventually collapse into a single black hole. There is evidence that a black hole with a mass of 10^6 stars is present near the center of our galaxy (the Milky Way) and growing at the expense of surrounding stars.

This much is past history, about which general relativity and astronomical observations are in agreement. What happens next?

An Ideal Universe

The expansion of the universe began with something like an explosion, but the spreading out of its matter into a larger and larger space is resisted by the gravitational attractive force tending to keep the matter together. We have described three basic scenarios for the future. Which will happen depends on how much matter there actually is in the universe, which is as yet uncertain.

1. If there is too little (and when we count up what we can see we find too little), its attractive force is too weak to turn things around, the expansion will go on forever, and space has a negative curvature, implying an infinite extent.

2. If there is too much (and there are some reasons to think we can't see all there is), the attractive force is enough to reverse the expansion eventually and produce collapse to a small size. This scenario corresponds to a positive curvature of space, making our three-dimensional space finite, something like the two-dimensional surface of the earth.

3. It is also possible that the density of matter is just on the borderline between these two alternatives, though it would seem a fortuitous accident if it were. In this case the universe expands forever, but at a steadily decreasing rate that approaches zero in sufficient time. Space would be flat, as Euclid and everybody else up to the modern era took for granted, and of infinite extent.

How does entropy fare in these scenarios? Let us deal with this question first as though there were only two: an indefinitely expanding, infinite universe and a finite one that collapses. Before we do so we would like to give an example of an "ideal" expanding universe in which the entropy density remains the same all the time, to serve as a kind of reference for the real universe. Our ideal universe will be one that contains radiation and essentially no matter, and since the entropy and energy of radiation is important for real universes as well as imaginary ones, we will review some of the thermodynamics of radiation first.

In Chapter 10 we discussed black-body radiation: radiation in temperature equilibrium with matter. Such radiation has a characteristic energy spectrum, which depends on the temperature but not on the specific properties of the kind of matter emitting and reabsorbing it. In Chapter 14 we described the quantum-mechanical view of electromagnetic radiation: although such radiation comes with various wavelengths and frequencies, its properties are not identical with those of "ordinary" waves. Instead it combines wave properties with properties resembling those of particles of matter traveling at high speeds. The particle-like entities of electromagnetic radiation, each of which carries a definite quantity of energy, are called "photons" and have some properties in common with gas molecules. We can therefore usefully compare and contrast collections of photons with collections of gas molecules, both when they are at equilibrium and when they are not.

Molecules vs. Photons

1. Molecules of the same kind can have different kinetic energies because they can travel at different speeds. In contrast, photons all travel at the same speed (c, the speed of light), but they can have different energies because they can have different frequencies (or, alternatively, different wavelengths).

2. In a collection of molecules at a given temperature, there is one particular distribution of speeds, the Maxwell-Boltzmann distribution, that has the largest entropy and is therefore overwhelmingly more probable than any other. This is the distribution when equilibrium is reached. All other distributions have less entropy. Likewise, in a collection of photons inside a furnace at a given temperature, there is one particular distribution of photon frequencies, the black-body distribution, that has the largest entropy and is therefore overwhelmingly more probable than any other. This is the distribution when equilibrium is reached. Again, all other distributions have less entropy.

3. If we have two separate containers of molecules, each at equilibrium but at two different temperatures, each will have the Maxwell-Boltzmann distribution for its temperature. If we place the two containers side by side, remove the walls between them, and let the gases mix, initially we will not have a Maxwell-Boltzmann distribution in the mixture: the sum of two Maxwell-Boltzmann distributions corresponding to different temperatures is not identical to a Maxwell-Boltzmann distribution for any single temperature. But as time goes on, collisions between the molecules will transfer and redistribute the energy to bring about an increase of entropy and a Maxwell-Boltzmann distribution with an intermediate temperature.

If we have two separate furnaces, each at equilibrium but at different temperatures, each will have a black-body distribution of photons for its temperature. If we place the two furnaces side by side, remove the walls between them, and let the photons mix, we will not have a black-body distribution in the mixture. As time goes on, absorption and re-emission of radiation by the walls will redistribute the radiant energy and bring about an increase of entropy and a black-body distribution with an intermediate temperature.

4. The average translational energy of a molecule in an equilibrium (Maxwell-Boltzmann) distribution is $\frac{3}{2}kT$. This is so *because* the equipartition theorem is an excellent approximation for the translational energy of gas molecules. The average energy of the photons in an equi-

librium (black-body) distribution is kT. This is *in spite of* the failure of the equipartition theorem to work for the higher-frequency photons.

5. The total energy of a collection of gas molecules depends on both the temperature and the number of molecules in the collection, but the total energy of the radiation in a furnace depends on the temperature *only*. We can change the number of molecules in a container by adding more. We can change the number of photons in a furnace only by changing the temperature.

Entropy in the "Ideal" Universe

Let us return now to our example of an "ideal" expanding universe and complete our argument that its entropy density would always remain the same. We have described the microwave radiation that we can detect coming from space having a black-body energy spectrum for a temperature of 2.7 K. This once-hot radiation, which was in thermal equilibrium with matter at an earlier, hotter stage of the universe, has had each wavelength increased in proportion to the expansion since that earlier stage, and thus has been cooled from 3,000 K to 2.7 K. This is not the only radiation in space: there is also radiation from the stars, whose surface temperatures are around 5,000 to 10,000 K, and infrared radiation from "dwarf" stars and planets like ours.

Suppose, however, that the only energy in the universe were just this 2.7 K black-body radiation: there would be no other radiation and no matter (except for a pair of small stars that we introduce to serve as markers for expansion). As this hypothetical universe expands, the temperature falls in proportion: the general theory of relativity shows that as the distance between the two reference stars doubles with the expansion, each wavelength in the black-body radiation doubles. The radiation keeps the *form* of black-body radiation, but its *temperature* falls to half. How does the entropy of the radiation change with the expansion? If we calculated the entropy in a fixed volume, say in a volume of 10^6 cubic light-years, it would *decrease*, as a result of the cooling. But if we take into account the expansion of the reference volume we should use for calculating an entropy density—and if the distance between two reference stars doubles, that volume must expand eight-fold (2^3)—the entropy in that expanded volume remains unchanged.

Such a universe, expanding forever, never reaches the kind of equilibrium state envisaged by Clausius. The temperature never stops falling, each doubling of the distance between two reference stars halving it, but at each stage the entropy density has the largest possible value at that moment, and nothing can be done to increase it. It is a "heat death," although not quite the one Clausius foretold.

It can be shown from the thermodynamics of black-body radiation that while the entropy density of such a universe remains the same on expansion, the energy density of the black-body radiation in it falls. This lost energy does not show up somewhere else: it exemplifies the breakdown of energy conservation in the general theory of relativity.

Expansion or Contraction?

We must now distinguish what can happen if the universe collapses in the future from what can happen if it continues to expand indefinitely.

If the mass/energy density is high enough to make the universe finite in size (positively curved), at some point in the future the universe will begin to contract. As it does, the wavelengths of the radiation will shorten and the temperature of the radiation will rise. The smaller the universe becomes, the hotter it will get. This description sounds like a return to the initial state of the big bang, and a return therefore to the low entropy the universe started with. The entropy, if this is to happen, would therefore have to decrease, which means the second law would break down. If the direction of time is determined by the increase of entropy, it seems that time would run backward, whatever that might mean. Indeed, some physicists have speculated on this as a possibility, but the consensus is that a decrease of entropy will not take place, that the entropy will continue to increase even during collapse. This implies that the universe does not retrace its steps exactly, even though it eventually becomes hotter and hotter as it gets smaller and smaller, and the final state reached on collapse would therefore be more disordered than the initial state from which the expansion began.

We have discussed the extraordinarily great entropy of a black hole. One speculative hypothesis, recently proposed, about a collapsing universe is that black holes formed during the era of expansion will continue to grow in size during collapse, thus increasing the total entropy of everything. The growth comes from several factors, among them the absorption of mass and radiant energy from outside and the coalescence of black holes to form larger ones. Eventually, according to this hypothesis, one giant black hole of very high entropy would be left, after which a new expansion of the universe might begin.

If the universe is to engage in a sequence of expansions, collapses, and re-expansions, with the entropy never decreasing, it would do so as a rubber ball bounces, the height of each successive bounce diminishing but the entropy never decreasing as the kinetic energy of the bouncing ball is converted by friction into a rise in temperature.

For an indefinitely *expanding* universe, we know of no process that can cause an entropy decrease. Questions may be raised, however, about the inevitability of a heat death.

Clausius's concept of the heat death was based on an analogy between the universe and a finite system, with a finite quantity of matter and energy. For such a system, there is no doubt that there is a finite maximum possible entropy when equilibrium is reached, and when this entropy is reached the capacity for further change is exhausted. No change of temperature, or pressure, or of any other property is possible.

An infinite and endlessly expanding universe differs from a closed finite system in important ways, however. First, there is no final state, no final temperature, no final pressure, no final equilibrium state at all. Change continues indefinitely. Second, the breakdown of energy conservation makes the applicability of the second law questionable.

The Real Expanding Universe

The real universe is not so neat and tidy as an "ideal" universe with only black-body radiation. There is matter as well as radiation, and the radiation is not black-body radiation corresponding to a single temperature but rather a mixture. There is the 2.7 K microwave radiation left over from the big bang, and then there is radiation from all the stars, which have various surface temperatures around 5,000 to 10,000 K, as well as dwarf stars and planets with lower temperatures radiating in the infrared range.

This, of course, is another way of saying that the universe is not at temperature equilibrium. Some of the matter in it is hot, some cold, and the radiation we detect reflects this nonequilibrium temperature distribution. Whereas we would expect a finite system eventually to reach equilibrium, an expanding universe is not likely to. As matter and energy become more and more dilute, the encounters between different molecules, or between molecules and photons, which in a finite system enable equilibrium to be reached, take place more and more rarely.

The 2.7 K black-body radiation illustrates this point. It is colder than the average temperature of matter, but over the lifetime of the universe it has not come to temperature equilibrium with matter. There is too little matter, and with too small an absorptive power, for the photons to encounter. Each photon of that radiation that we detect with our instruments here on earth has been traveling uninterrupted in space since the time when the temperature of the universe was 3,000 K, about 700,000 years after it began.

The Fate of Matter

There is strong evidence in favor of the existence of black holes in our universe. Let us assume, then, that we are now in an era in which black holes are forming, with consequent enormous entropy increases.

We have mentioned the possibility that there are black holes that may eventually grow to galactic size. We noted earlier that stars are not distributed at random in space but are organized into galaxies of 10^{11} stars. There is now evidence that galaxies themselves are not distributed at random either but are organized into *supergalaxies,* and it is at least possible that even these are parts of a larger whole. This suggests that we may legitimately distinguish alternative histories for galactic black holes.

Black holes radiate their energies away eventually, although at inconceivably slow rates. A stellar mass black hole, with a temperature of 10^{-7} K, has a radiative power of 10^{-24} watts and a lifetime of 10^{66} years. Galactic-size black holes, with masses 10^{11} times as great, would last 10^{33} times as long. On that enormous time scale, however, galactic black holes may drift together because of their gravitational attraction and coalesce to form supergalactic holes before they have had time to disappear, and the process of coalescence may continue beyond that stage, forming holes of ever-increasing masses. In a universe of infinite extent and of uniform density, there is an infinite quantity of matter, and if there are black holes at all, there are an infinite number of them. The universe of negative curvature expands forever. In such a universe objects sufficiently far apart can never meet, so even if the radiation of black holes did not put an end to them, the coalescence of black holes would end at some time. While the borderline universe, with zero curvature, also expands forever, the expansion rate slows as time goes on, and eventually objects in it are no longer receding from each other. Under such conditions, indefinitely continuing coalescence of black holes cannot be ruled out.

So we have two alternative possibilities:

1. Black holes continue to coalesce, faster than they disappear into radiant energy.
2. Black holes disappear into radiant energy, faster than they can coalesce.

The first alternative implies no heat death at all. The entropy of a black hole was shown by Bekenstein to be proportional to the *square* of the mass: when two holes coalesce, the total entropy is four times that of a single hole, rather than twice as great. A detailed mathematical analysis shows that the entropy density in a universe of indefinitely coalescing black holes increases without any limit. There would be no final maximum value for the entropy.

If we have the second alternative, and the black holes vanish into radiation, we have a universe something like the model universe we considered earlier, with one difference. Since the temperature of a black hole varies as the mass changes, the total radiation produced over their lifetimes by black holes is not black-body radiation: it does not correspond to a single temperature. The universe would not be at its maximum entropy density, but in the normal course of events it has no way of getting there.

Human Intervention

Life cannot continue unless it can feed on entropy-increasing processes, which are a necessary condition for life to continue, although not a sufficient one. There are too many scenarios—wars, plagues, and environmental disasters including but not limited to the capture of the earth by a black hole—that could wipe life out before a heat death ends it. But suppose we can avoid all these disasters by conscious choice and ingenious technology; what possibilities would be open to the human race if it survives until a late stage of the expansion of the universe? Can we find entropy-increasing processes to keep us going indefinitely? Could we, for example, collect black holes, and move them close enough together so as to be able to make use of the entropy increase as they coalesce?

We are now entering the realm of science fiction, but do not forget that once in a while science fiction turns into science fact. Physicists have speculated about possible solutions to the heat death. A recent review by the astrophysicist Steven Frautschi concludes that other entropy-increasing, deteriorative events will probably bring matter to an ultimate, unchangeable state, and thus life will come to an end eventually, no matter how ingenious our technology.

So it appears that in spite of the new insights of quantum mechanics and relativity, the grim inference for the future drawn from the second law in the nineteenth century is not far wrong. Whether this is a matter for despair or indifference is a question we leave for others to answer.

Afterword

<hr/>

We have tried to do four things in this book. We began by explaining
what the laws of thermodynamics say. We noted that the laws forbid
some things—anything that involves a disappearance or creation of
energy or a decrease of entropy—and permit others. We also pointed
out what the major limitation of thermodynamics is: it cannot tell
whether or when the things it permits will happen.

Second, we showed why these laws follow from the molecular properties of matter. The discovery of the conservation of energy followed
directly from the idea that matter is made up of molecules in motion.
Entropy, originally discovered as a macroscopic property of matter having no clear intuitive meaning, can be understood much more readily
from a molecular point of view as being directly related to the relative
probability of various states of a collection of molecules. The law of
increasing entropy becomes an almost self-evident statement: given a
chance, collections of molecules, like shuffled decks of cards, will be
found in more probable rather than less probable states. We also
showed how difficult it has been to justify completely the laws of thermodynamics from a kinetic-molecular theory. In the nineteenth century, when the laws were formulated, there was no accurate theory of
how molecules move, and even now, with what we are strongly convinced is a correct theory of molecules—quantum mechanics—there
are still unsolved problems and even paradoxes in the way of a satisfac-
tory justification.

Third, we have demonstrated the extraordinary range of applications
of the laws: to physics, chemistry, biology, and geology; to the operation

of refrigerators and automobile engines; to the extraction of iron from its ores and to the manufacture of diamonds; to the processes by which muscles and kidneys function; to the energy balance of the earth and things that might upset it; and to much else.

Fourth, we have discussed how twentieth-century science has changed both the laws of thermodynamics, made it easier to justify them on molecular grounds, and extended their applicability to astronomy and cosmology. Quantum mechanics not only provided a better molecular theory than Newton's laws; it also eliminated certain ambiguities in the molecular meaning of entropy and accounted for its disappearance at very low temperatures. Not long after the concept of energy and its conservation were enshrined as one of the most fundamental principles of science, the special theory of relativity changed its meaning. In the new theory energy was combined with other physical quantities previously considered independent of energy, such as mass and momentum, as one component of a mathematical entity called a tensor. The tensor is a way of describing the local concentrations of these physical quantities, and in the general theory of relativity it is responsible for what we would once have called gravitational forces but now call the curvature of space. So energy, in its new incarnation, remains of fundamental importance, but a cloud has been cast on its most significant property: its indestructibility. In problems in which the expansion of the universe needs to be taken into account, energy need not be conserved.

We can summarize by saying that thermodynamics has something to say about everything but does not tell us everything about anything.

The history of thermodynamics is a story of theories once regarded as pinnacles of scientific achievement discarded and replaced by others. It might make one wonder whether anything science has discovered will stand the test of time. We can't give guarantees to the contrary, but we can distinguish between those constructs of science like the phlogiston theory or the caloric theory (both discussed in Chapter 3) that have vanished, leaving only a trace or two in the vocabulary of the field, and those that have been superseded but not discarded. Among the latter we count Newton's mechanics, which is now known to be an approximation for medium-sized, slowly moving bodies but which it is still the approach of choice, and the first law in its nineteenth-century form, which still covers the great majority of applications of the energy concept. And when quantum mechanics, the general theory of relativity, and the theory of fundamental particles are combined into a single

grand unified theory, Newton's laws and nineteenth-century thermodynamics will almost certainly remain as useful as ever in their domain.

Philosophers of science have tried to characterize the difference
between the natural sciences and other fields of rational study by this
feature of the theories we have described: they don't get replaced by
new discoveries but rather get incorporated by them. Such sciences are
said to be "cumulative." Some take the view that this trait defines the
boundary between "science" and anything else.

We cannot guarantee that thermodynamics will never suffer the fate
of the caloric theory, but we will let Einstein have the last word:

> A theory is the more impressive, the greater the simplicity of its
> premises is, the more different kinds of things it relates, and the
> more extended is its area of applicability. Therefore the deep im-
> pression that classical thermodynamics made upon me. It is the
> only physical theory of universal content which I am convinced
> that, within the framework of applicability of its basic concepts, it
> will never be overthrown.

Appendix:
Math Tools

~~~~~~

Stephen Hawking, whose *Brief History of Time* was written for the non-specialist, reports that he was warned while writing it that each equation he introduced would cut his readership by half again. Ten equations, according to this advice, would mean 1/1,024 as many readers as he would have had without them. He therefore confined himself to $E = mc^2$, which he felt was so familiar that it hardly counts as an equation.

Roger Penrose, another scientist writing for the nonspecialist, repeats this story in *The Emperor's New Clothes*, but he bravely included many equations. His offered this advice to the nonmathematically trained reader: "If you are a reader who finds any formula intimidating (and most people do), then I recommend a procedure that I normally adopt myself when such an offending line presents itself. The procedure is, more or less, to ignore that line completely and to skip over to the next actual line of text!"

We will be as brave, or as foolhardy, as Penrose, in including a lot of equations, but more cowardly, in that they will only be equations of algebra. Most of our readers will have had algebra in high school, though many will not remember much of it. We feel we have no choice, given our goals. We offer this appendix as a review and a reminder of what was once known.

Our advice to the nonmathematical reader will, however, be different from Penrose's. Do not skip the equations, or stare at them incomprehendingly; translate them into English.

Translating Mathematics

Mathematicians are often told by new acquaintances on learning their profession, "I can't even balance my checkbook." This is guaranteed to annoy mathematicians, who are no better at balancing their checkbooks than anyone else. Mathematics is not arithmetic or computation but something else, a search for the most general and economical way to say things. This search unfortunately leads to the formal and abstract mode of saying those things that many find intimidating.

There is a famous story about J. Willard Gibbs, a physicist, physical chemist, and one of the few American scientists of international stature before the twentieth century. Gibbs was a quiet and self-effacing member of the Yale University faculty for thirty years; he is said to have made only one speech to a meeting of his colleagues in all that time. During the course of a heated discussion as to which discipline—classical languages, modern languages, or the sciences—trains the mind best, he got up, said, "Gentlemen, mathematics is a language," and sat down.

We will follow Gibbs's lead: not only is mathematics a language, but its statements are complete sentences, with nouns, modifiers, verbs, and other parts of speech. An equation is an example of a sentence: the verb is the equals sign. The nouns in equations are numbers, some of which may be known, like 7 or the square root of 2, and some not yet known (x, y) but soon, we may hope, will be. Just as a conjunction, like "and," joins two nouns, like "Jack" and "Jill," together to make a new noun "Jack and Jill," the arithmetical symbols $+$, $-$, $\times$, and $/$ (for division) make new numbers out of several others. We will usually follow the algebraic convention of not using the sign $\times$ for multiplication, except where its use prevents misunderstanding. When two numbers are written adjacent, as in xy or $9z$, multiplication is assumed. Division will be written as a fraction, using the $/$, as in $10/2 = 5$.

Another part of speech we will use is not quite an adjective, which designates a quality possessed by a noun (as in "black cat"), but rather a modifier that converts one number into another. Let us call such a part of speech an *operator*. Examples are "take the square root of," symbolized by $\sqrt{}$, "multiply the number y by itself," symbolized by y^2, and so on. A purely verbal example of an operator is the prefix *un-*, which converts an adjective like *distinguished* or *generous* into its opposite.

Once these equivalences between "English" and "mathematics" are understood, it is a simple matter to translate an equation with its symbols into a verbal statement. Each time an equation is encountered, the reader should attempt to "translate" it. Let us illustrate the method with some equations we use in this book.

$$\text{Kinetic energy} = \tfrac{1}{2}mv^2$$

This equation is made less intimidating by the use of words rather than symbols on the left side. It tells us that to calculate the "kinetic energy," whatever that is, square the speed v, multiply the result by the mass, and take one-half the product. For a speed of 3 meters per second and a mass of 8 kilograms, one calculates a kinetic energy of 36 joules. The units we used, meters, seconds, and joules, are not in the equation but must be supplied by the user.

$$P = \frac{NkT}{V}$$

This is as complicated as any of our equations get. It relates the pressure P of an "ideal" gas to four other quantities: the number of molecules N in the gas, the temperature T (what is called the absolute temperature), the volume V in which the gas is contained, and finally a constant k, a number that is the same for all gases but whose numerical value depends on the units in which P, V, and T are measured. In international scientific units, P is measured in pascals (see Chapter 4), V in cubic meters, and T in kelvins, and the constant k is 1.38×10^{-23} joules/kelvin. Knowing the value of k, we can calculate P if N, T, and V are given. Alternatively, we might know P, N, and T and calculate V. To show that V is now the "unknown" quantity, we may rewrite the equation according to the rules of algebra:

$$V = \frac{NkT}{P}$$

In any problem that requires calculating a numerical value, the user must know which units to use to get a sensible answer. One does not combine a distance in miles with a speed in meters per second. The choice of units requires care: pressure, for example, may be encountered in atmospheres, pounds per square inch, millimeters of mercury, or pascals.

Not only can we translate from math to English, we can also change English into math. Consider the following verbal problem. A car is traveling at 40 miles per hour. How far will it travel in 1 hour, or in 2.5, or in 7? The answers are easy: 40 miles, 100 miles, 280 miles. The equation is

$$d = 40t$$

in which d is the distance traveled, t the time elapsed.

A more general form of this equation is

$$d = vt$$

where v is the speed of the car. (It is not engraved in stone that the algebraic symbol for a quantity must be the initial letter of the word. The symbol v is customarily used for *velocity;* the distinction between speed and velocity is discussed in Chapter 15.) The terms of this equation, d, v, and t, take on different numerical values depending on the problem and the context. As the problem was originally worded, v stays the same as different values of t are considered, and different values of d are found accordingly. The speed v is said to be a *parameter:* it can have different values in different problems but will usually remain the same in a particular one. The distance d and the time t are called *variables;* time passes as the car moves, and so both d and t vary steadily from 0 to larger and larger values. The equation then describes a relation between the two variables d and t, which can be plotted as a graph (Figure A.1).

Proportionality

The relation between d and t for a car moving at a constant speed v is an example of proportionality. The distance traveled is *proportional* to the time elapsed, which means the distance doubles when the time doubles and increases five-fold if the time increases five-fold. This is true regardless of the value of the speed v, which is called the *constant of proportionality.* There are times when we know from measurement or by theory that one variable y is proportional to another x even when we don't yet know the constant of proportionality. We can write the relation

$$\text{either as} \quad y \propto x \quad \text{or} \quad y = cx$$

even when the value of c is not yet known, though we may hope to determine it eventually.

If y is proportional to x and z is proportional to y, we have

$$y = cx \quad \text{and} \quad z = by = b(cx) = dx$$

in which d, a new constant, is the product of the constants b and c.

Operators

The operators "take the square root of" and "square" are familiar. Squaring a number requires only multiplication, but "take the square root of" is harder. Both are equally easy with a handheld scientific calculator.

Operators convert one number to another, according to a particular computational procedure. We can usually undo the effect of any operator with a second operator. If we square the square root of a number we find ourselves back where we started: the square of the square root of x is just x again. The square root of the square of x is a little more complicated; numbers have two square roots having the same numerical value, one positive and the other negative, so the square root of x^2 may be either x or $-x$. If we ignore that complication and consider only the positive square root we have:

Square root of (x squared) = Square of (square root of x)

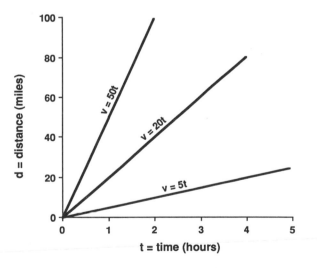

Figure A.1   Distance vs. Time
A graphical representation of the distance a car traveling at a given speed will cover over time. Three speeds are represented, 50 miles per hour, 20 miles per hour, and 5 miles per hour. The speed v is called a "parameter" here because it remains constant in a particular case but can have different values in other cases. Distance d and time t arc called "variables" because their values change during the discussion of a particular case.

Two operators either of which undoes the effect of the other are said to be *inverses*.

We will use only a limited number of mathematical operators in this book, squares and a few other "powers" of numbers, square roots, exponentials, logarithms, and factorials. The exponential and the logarithm are inverses, and we will explain them first.

Consider the relation $y = 2^x$. It looks not too different from $y = x^2$, but beware. We contrast the results in Table A.1. 2^x first increases sluggishly, barely keeping up with x^2, but it soon takes off and rapidly outstrips the latter value.

Many will be familiar with this behavior, but not under the name *exponential*, in compound interest. A bank deposit on which interest is compounded has a rate of increase proportional to its size. The bigger it gets, the faster it grows. Eventually, compounding leads to a faster rate of growth than any fixed but not-compounded rate of interest. Five percent interest compounded annually will eventually give us a larger balance than 10 percent simple interest based only on the initial deposit. The latter rate of growth, a steady $100 per year on a $1,000 initial deposit, is called "arith*metic*" (accent on the third syllable), while the compound interest rate is called *exponential*, a term often used inaccurately to mean "very, very fast growth," which it need not be initially. A

Table A.1
Comparison of x^2 and 2^x

x	x^2	2^x
0	0	1
1	1	2
2	4	4
3	9	8
4	16	16
5	25	32
6	36	64
7	49	128
8	64	256
9	81	512
10	100	1,024

note of financial advice if you own bonds: they pay simple interest, so if you want the advantage of compounding, reinvest the interest as you receive it.

One other well-known example of exponential growth is the assumption made by Thomas Malthus concerning human populations. In his view, the larger a population gets the more babies are born, so the population grows faster as it gets larger. As he further assumed that food supply grows only arithmetically—110 million more tons of wheat each year, each year 20 million more gallons of milk—it is no wonder that he predicted disaster. One hundred and fifty years after Malthus's death, there is still considerable argument as to whether his assumptions about either population or food supply are correct.

Exponential Notation

We will be dealing in this book with numbers enormously, inconceivably large and with tiny, tiny ones. Rather than write a 1 followed by 23 zeroes to represent the number of atoms is some moderate quantity of a substance (let's do it just once: 100,000,000,000,000,000,000,000) we will write 10^{23}; this is called *exponential notation,* and it is conventionally used in science. The number 23 in the preceding example is called the *exponent* or the *power* of 10.

We all know how to multiply a number by 100, just add two zeroes at the end of the number: 100 times 10,0000 is thus 1,000,000. In exponential notation we would write the multiplication as $10^2 \times 10^4 = 10^6$, suggesting that to *multiply* two numbers expressed in exponential notation we *add* the exponents.

We can use the same rule for division: $10^6/10^2 = 10^4$. When dividing a large number by a smaller one we subtract the exponent in the denominator from that in the numerator. This rule can be made convincing by working a few examples. But what if the number in the numerator is *smaller* than the one in the denominator? Subtracting exponents would give us a negative exponent: what could we possibly mean by ten multiplied by itself -3 times? We can find, by "experimenting" with dividing small powers of 10 by larger powers, that if we *interpret* or *define* 10^{-1} as $1/10$ and 10^{-4} as $1/10,000$, subtracting exponents gives the right answer in these problems also. So that is what we will do.

As 10^x gets very large very fast as x increases (exponential growth), it follows that $10^{-x} = 1/10^x$ gets very small very fast as x increases (exponential decay). Exponential decay describes the decrease with time of the quantity of radioactive substances like radium or radon, but the example of a negative exponential we will have most to do with is the Maxwell-Boltzmann distribution, which gives the relative numbers of molecules with different amounts of energy. That number depends exponentially on the negative of the ratio of the molecular energy to kT:

Relative number of molecules with energy $\varepsilon \propto 10^{-0.4343\varepsilon/kT}$

The constant 0.4343 arises from the use of the base 10 for our exponential representation; another choice of base, to be discussed below, eliminates this constant. The important point is that the number of molecules having a particular energy ε decreases very rapidly as the value of the energy ε increases. High-energy molecules are therefore rarer the higher the energy, but raising temperature increases their number.

As the above equation illustrates, exponents need not always be whole numbers. What meaning can be given to fractional exponents, such as 10 raised to the $\frac{1}{2}$ power? Our experience with giving meaning to negative exponents should suggest that if we can assign a meaning to such exponents that makes them useful, we should do it.

Let's begin with the requirement that whatever these exponents are going to mean, multiplication of two numbers will be accomplished by adding the exponents and division by subtracting the exponent in the denominator from that in the numerator. Let us try this with $10^{1/2} \times 10^{1/2}$. By the addition rule, the product should be 10^1, or just 10 itself. This implies that the number $10^{1/2}$ multiplied by itself is equal to 10. But that is just the definition of the square root of 10. So we have a meaning now for $10^{1/2}$, and in fact we also have a numerical value: the square root of 10 can be found on a hand calculator to be 3.162277 . . .

By an extension of this procedure, a numerical meaning can be assigned to any power of 10, positive or negative, whole number or fraction. In Table A.2 we list values of $y = 10^x$ for various values of x between -4 and 4. One can see the extremely rapid "exponential" increase of y as x increases, and the extremely rapid decrease of y as x becomes negative: this latter behavior illustrates the "negative exponential curve" we described in our discussion of the Maxwell-Boltzmann energy distribution in Chapters 4, 7, and 14.

Logarithms

So far we have in effect treated y as an unknown to be found when x, the exponent of 10, or the power 10 must be raised to to get y, is a known number. If we change our perspective and regard y as the known number, what is the exponent of 10, or the power 10 must be raised to, to get y? When we ask the question in this way it is customary to change language also; we call x the *logarithm of y to the base 10*. The symbol we use is "log" or, more specifically, "$\log_{10}$." The use of 10 is arbitrary; in the modern world we most commonly use a number system based on 10, but other number systems and therefore other bases for logarithms are used when they make more sense in other contexts.

So we can summarize what we have said about exponentials and logarithms by saying that the following two equations mean the same thing:

$$y = 10^x \quad \text{and} \quad x = \log_{10} y$$

We can see that the operators *exponential* and *logarithm* are inverses:

$$\log 10^x = x$$

Table A.2
The Exponential Function

| | $y = 10^x$ | |
x	Scientific notation	Decimal notation
-4	10^{-4}	0.0001
-3	10^{-3}	0.001
-2	10^{-2}	0.01
-1	10^{-1}	0.1
0	10^0	1
0.2	$10^{0.2}$	1.585
0.4	$10^{0.4}$	2.512
0.6	$10^{0.6}$	3.981
0.8	$10^{0.8}$	6.31
1	10^1	10
2	10^2	100
3	10^3	1,000
4	10^4	10,000

In Table A.3 we list values of $x = \log_{10}y$ for various values of y between 1 and 400. The inverse relation between the exponential and logarithmic functions is shown by comparing the values of y in this table and those in Table A.2 when $x = 0.0, 0.2, 0.4, 2,$ and 3. In contrast to the rapid increase of $y = 10^x$ as x increases, note that $x = \log_{10}y$ increases only very slowly as y increases: when y is multiplied by 10, x increases by only one unit.

Suppose we want to multiply two numbers, a and b, to obtain their product, c. Let us further suppose hand calculators have not yet been invented. Let us finally suppose also that we have a table available in

Table A.3
The Logarithmic Function

y	$x = \log_{10}y$
1	0.000
1.585	0.200
2	0.301
2.512	0.400
3	0.477
4	0.602
5	0.699
6	0.778
7	0.845
8	0.903
9	0.954
10	1.000
20	1.301
30	1.477
40	1.602
100	2.000
200	2.301
300	2.477
400	2.602

which the logarithms (in base 10) of numbers are listed. We look up the logarithms u and v of a and b ($u = \log_{10} a$, and $v = \log_{10} b$, or equally well, $a = 10^u$ and $b = 10^v$). In this problem, c is the desired unknown, and so its logarithm z is also unknown: $z = \log_{10} c$, and $c = 10^z$.

Obviously $c = a \times b = 10^u \times 10^v = 10^{u+v} = 10^z$. Hence z, the log of the desired unknown c, is the *sum* of the logs of the numbers a and b whose *product* is wanted. So we add u and v to obtain z, and then again use the table of logarithms to find the number c.

By the same reasoning, the number d, the result of dividing a by b, is given by

$$d = 10^w = a/b = 10^{x-y} \qquad \text{or} \qquad \log d = w = \log a - \log b$$

Here are some examples:

$$\log (2 \times 10^6) = \log 2 + \log 10^6 = 0.301 + 6 = 6.301$$

$$\log (7 \times 10^{-4}) = \log 7 + \log 10^{-4} = 0.845 + (-4) = -3.155$$

$$\log (6 \times 7) = \log 6 + \log 7 = 0.778 + 0.845 = 1.623$$

The logarithm of (6×7), of course, is the logarithm of 42. If we know this, we can calculate the logarithm of any number that is a product of 42 and a power of 10. For example, 4.2 is equal to 42×10^{-1}, and 42,000 is equal to 42×10^3, so:

$$\log 4.2 = \log (42 \times 10^{-1}) = \log 42 + \log 10^{-1}$$
$$= 1.623 + (-1) = 0.623$$

$$\log 42,000 = \log (42 \times 10^3) = \log 42 + \log 10^3$$
$$= 1.623 + 3 = 4.623$$

To summarize, the use of logarithms replaces multiplication or division by addition or subtraction. This was one of their main uses in the days before calculators and computers. It might seem unnecessarily complicated to use logs to multiply two numbers together, since we can always do it "long-hand," but when we want to multiply seven numbers, or multiply two and then divide the product by three others, the saving in time and effort is obvious. Printed tables of logarithms enabled one to look up the logarithms of any two or seven numbers, and once the logs were found the log of the number sought was obtained by adding (or subtracting) the logs of the numbers to be multiplied (or divided). Then the table was consulted again to convert the log of the number sought to the number itself. As we have seen, the addition or subtraction of logs is exactly the same rule as adding exponentials when you multiply and subtracting them when you divide. The old-fashioned slide rule does the same job faster, but with less accuracy.

One use of logarithms that may be familiar even to those without much mathematical training is the logarithmic graph. A logarithmic scale is used to show the changes in some quantity that increases or decreases enormously—for example, world population or the U.S. national debt over the last two hundred years. Very often, instead of plotting the quantity undergoing very large scale changes, we plot its logarithm. This has the effect of enabling us to show very large changes without using very large sheets of paper. The tell-tale sign of a logarithmic plot is that equal intervals on the graph correspond to multiplication by equal powers of 10 (see Figure A.2).

Other Bases

In information theory, discussed in Chapter 9, the information content (I) of a message written in binary digits is defined as the logarithm to the base 2 of the number of digits (N). Hence

$$I = \log_2 N$$

We use base 2 precisely because we are using the binary number system rather than the decimal (base-10) system. The same rules apply:

$$\text{If } I = \log_2 N, \text{ then } N = 2^I$$

$$\log_2(M \times N) = \log_2 M + \log_2 N$$

$$\log_2 2^{101} = 101$$

In higher mathematics and in physical science it is more convenient and simpler to use as base a number e, which like π is an unending decimal: $e = 2.7182818\ldots$

Where does the number e come from? We all appreciate the advantages of compound interest on a bank deposit, and banks first made this type of account even more appealing by compounding quarterly and then, with the availability of computers, by compounding daily. There is no reason, given computers, why they could not compound every second, or every thousandth of a second. If they did, how well would one do?

Let us imagine a bank paying interest at the rate of 100 percent a year. Then if the year is not divided up into quarters or seconds, one would earn one additional dollar for each dollar of the initial deposit in the first year. If the compounding is done quarterly, the bank pays 25 percent four times a year on the actual deposit, including principal and previously accrued interest. The result is that at the end of the year one has gained $1.4414 per dollar of the original deposit, a result calculated by multiplying the original deposit by 1.25 four times. Daily compounding pays 100 percent/365 or 0.2740 percent each day, giving at the end of a year a gain of $2.7146 per dollar deposited. How much better do we do with compounding 10,000 times a year? The improvement is not very great: $2.718146 per dollar. Obviously a principle of diminishing returns is operating.

The number e, 2.7182818 . . ., an unending decimal, is the limit

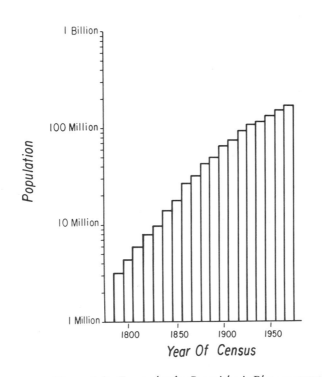

Figure A.2   Example of a Logarithmic Plot
U.S. population since 1790 is here plotted using a logarithmic scale for the vertical (population) axis. If one were to ignore the scale and judge only by the size of the bars indicating population size, one would think that population had merely quadrupled. In fact, though, population increased from about 2 million to about 200 million during the time period shown. It would be difficult to plot a range this large using a linear scale (one in which equal size intervals on the graph represent equal increments of the variable): either the initial 2 million would be barely visible, or the graph would have to be about a meter high.

reached when we divide up the interval of compounding indefinitely. Why it is useful in higher mathematics is another book.

Since we can multiply e by itself four times, divide by the square of e, or take its square root just as we can with 10 or 2, all the rules given for logarithms to those bases apply to base e also.

It is simple to switch from one base of logarithms to another: the conversion is made by multiplying any logarithm in one base by an appropriate constant determined by both bases, as the following equations show:

$$\log_{10}x = 0.3010\log_2x = 0.4343\log_ex$$

$$\log_2x = 3.322\log_{10}x = 1.4343\log_ex$$

$$\log_ex = 2.303\log_{10}x = 0.6931\log_2x$$

In the above formulas x is any positive number, greater or less than 1.

Scientific hand calculators have keys for calculating logarithms both to base 10 and base e, customarily symbolized by "log" and "ln," respectively.

When we use the base e for logarithms, the Maxwell-Boltzmann formula is simplified:

$$\text{Relative number of molecules with an energy } \varepsilon \propto e^{-\varepsilon/kT}$$

The simplification is one of the reasons the number e is useful. Another example of such a simplification is given in the next section.

Factorial Numbers

How many arrangements of a deck of 52 cards are there? The answer is easy to write down but hard to calculate. There are 52 different cards that could be in the top position in the deck. Once a card is selected for that position, there are only 51 remaining. There are therefore 52 × 51 ways of selecting the first two. Continuing on, there are 52 × 51 × 50 ways of selecting the first three cards and, just as clearly, when we have specified what the first 51 cards are, there is only one card left for the last, and no choice at all.

The number of arrangements of the deck is therefore the product of all whole numbers from 52 down to 1. This is a very large number indeed; our scientific hand calculator has a key for it marked "$x!$" that gives the value as 8.066×10^{67}. The number $x!$, called "factorial x," often comes up in problems where the number of arrangements of anything is needed. It is a number that increases very rapidly as x increases, even faster than 10^x. When $x = 10^{23}$, a number that arises in molecular arrangement problems, the magnitude of $x!$ is beyond conceiving. Taking the logarithm of $x!$ reduces this magnitude to more manageable proportions.

From the definition of $x!$ as a product of all whole numbers up to x, we can see that its logarithm would be the sum of the logs of all whole numbers up to x, which would be incredibly tedious to calculate when x is 10^{23}. Fortunately, an approximation enables the logarithm to be calculated very easily and with sufficient accuracy for the determination of entropies:

$$\log_e x! \approx x \log_e x - x$$

Using logs to the base 10, the formula would be:

$$\log_{10} x! = x \log_{10} x - 0.4343x$$

A Note on Energy Units

~~~~~~

The three most familiar units used for measuring energy are the kilowatt-hour, the Calorie (or kilocalorie), and the British Thermal Unit (BTU). The scientific unit, the joule, is less familiar. A steel ball of 1 kilogram mass (2.2 pounds) traveling at a speed of 1.41 meters per second (2.75 miles per hour) has a kinetic energy of 1 joule.

1 kilowatt-hour	=	3,600,000 joules
1 Calorie	=	4,184 joules
1 BTU	=	1,054 joules

As a rough approximation, therefore, there are one thousand Calories in a kilowatt-hour, and four thousand BTU in a kilowatt-hour.

Power is the *rate* of energy use. One watt equals one joule per second.

Glossary of Symbols

There are more quantities in science that need to be represented by symbols than there are letters in the Roman, Greek, and Hebrew alphabets combined. In addition, we have chosen as often as possible to use the conventional scientific symbols for the quantities discussed in this book. There is therefore some unavoidable duplication of symbols used. Usually the context will help make the meaning clear.

Subscripts are often added to symbols so that further distinctions of meaning may be made—for example, v_1 and v_2 for the speeds (v) of two separate bodies, or ε_{H_2O} for the energy of a water molecule. In the following list we give the page on which a symbol is first defined to direct the reader to a more detailed explanation of the term.

Page	Symbol	Definition
12	F	Force
12	d	Distance
21	w	Weight
21	g	Force of gravity per kilogram, at earth's surface
21	m	Mass
21	v	Speed
27	E	Energy
27	PE	Potential energy
27	KE	Kinetic energy
52	I	Electric current (amperes)
52	R	Electric resistance (ohms)

Page	Symbol	Definition
58	Q	Quantity of heat
58	W	Quantity of work
58	Δ	"Delta": a change in . . .
59	c	Speed of light in a vacuum (3.0×10^8 meters/second)
78	M	Mass of a macroscopic body
78	V	Speed of a macroscopic body
85	P	Pressure of a gas
85	A	Proportionality constant in the relation of gas pressure to temperature
85	t	Temperature (Celsius degrees)
86	T	Absolute temperature (kelvins)
89	ε	Average energy of a molecule
89	K	Proportionality constant in the relation of average energy to absolute temperature
91	k	Boltzmann's constant (1.4×10^{-23} joules/kelvin)
92	N	Number of molecules in a collection
94	m	Mass of a molecule
98	V	Volume in which a gas is confined
101	L	Number of molecules in one cubic centimeter of a gas
119	$Eff.$	Efficiency of a heat engine
119	Q_H	Heat taken in by a heat engine at the higher temperature of operation
119	Q_L	Heat given out by a heat engine at the lower temperature of operation
121	T_H	Higher temperature of operation of a heat engine
121	T_L	Lower temperature of operation of a heat engine
123	$C.O.P.$	Coefficient of performance of a refrigerator or heat pump
138	S	Entropy
154	W	Number of simple outcomes, microstates, or messages

Page	Symbol	Definition
158	$\gg$	Is much greater than
166	C	Number of "cells" of molecular size into which a large volume is divided
167	$\cong$	Is approximately equal to
168	log	Logarithm (see Appendix: Math Tools)
183	N_A	Avogadro's number, the number of hydrogen atoms (approximately) in one gram of hydrogen $(= 6.02 \times 10^{23})$
188	C	Proportionality constant in the equation relating entropy S to the logarithm of W
200	a, b, c, x	Numbers used in a program generating "random" numbers
201	π	Ratio of circumference of a circle to its diameter $(= 3.14159 \ldots)$
210	x_n	Number of rabbits in the nth year
210	A	Proportionality constant in the "logistic" equation
217	I	Information content of a message
217	$\log_2$	Logarithm to base 2
218	$\log_e$	Logarithm to base e
234	L	Wavelength of a wave
234	f	Frequency of a wave
234	v	Speed of a wave
234	A	Amplitude of a wave
248	s	Stefan-Boltzmann constant, the proportionality constant in the relation between the energy radiated per second per square meter of a black body and the fourth power of the absolute temperature $(= 5.7 \times 10^{-8}$ joules per second per square meter per kelvin$^4)$
249	l_{max}	Wavelength at which the maximum amount of energy is radiated by a black body
249	w	Wien's constant, the proportionality constant in the relation between the wavelength of maximum energy and $1/T$ $(= 2.9 \times 10^{-3}$ meter-kelvins)

Page	Symbol	Definition
284	W_{max}	Maximum work that can be performed by the combustion of a given quantity of fuel (or by any other chemical reaction)
285	ΔS_{net}	Change in net entropy during the combustion of a fuel (or by any other chemical reaction) carried out so as to do no useful work whatever
285	ΔG	Change in Gibbs free energy when a chemical reaction takes place
329	e	Energy in one photon of light
329	h	Planck's constant, the proportionality constant in the relation between the energy of a photon and the frequency of the light (= 6.63×10^{-34} joule-seconds)
352	t	Time interval read on a clock
352	β	Factor in special relativity giving the shrinkage of meter sticks, slowing of clocks, and increase of mass observed on a moving body
354	m	Apparent mass of a moving body
354	m_0	Mass of the body at rest
355	RKE	Relativistic kinetic energy
374	M	Mass of a black hole

References

General

The titles we list here cover the history of thermodynamics, the basic scientific background in physics and mathematics, and the technological background. Many of these works will be referred to again in the following sections on specific topics. Their levels of difficulty vary.

For the history we recommend D. S. L. Cardwell's fascinating *From Watt to Clausius: The Rise of Thermodynamics in the Early Industrial Age* (Ithaca, N.Y.: Cornell University Press, 1971; rpt., Iowa City: University of Iowa Press, 1989) and *The Kind of Motion We Call Heat: A History of the Kinetic Theory in the Nineteenth Century*, by Stephen G. Brush (Amsterdam: North Holland, 1976). Both are excellent, but the latter requires more in the way of scientific background. *The Dictionary of Scientific Biography*, edited by Charles C. Gillispie (New York: Scribner's, 1970), is not only a source of biographical material but also gives good summaries of the scientific work, and it is available on the reserve shelves of many public libraries. Its drawback: living scientists are excluded. We will refer to it in what follows as *DSB*.

For the laws of physics, you may best rely on your own high school physics text, if it is still at hand. If not, we suggest James S. Trefil, *Physics as a Liberal Art* (New York: Pergamon Press, 1978), a good reference on basic physics for the nonscientist. *The Character of Physical Law*, by Richard P. Feynman (London: Penguin, 1992), is also good. At a more advanced level, using calculus but often understandable without it, is the classic *Feynman Lectures on Physics: Mainly Mechanics, Radiation, and Heat*, by Richard B. Feynman, R. B. Leighton, and M. Sands (Reading, Mass.: Addison Wesley, 1963). A fully authoritative reference on thermodynamics, requiring calculus, is *Heat and Thermodynamics*, 6th ed., by M. W. Zemansky and R. H. Dittman (New York: McGraw-Hill, 1981). An exposition for nonscientists of Newton's mechanics, energy, entropy, relativity, and quantum mechanics is by Nathan Spielberg and Bryon D. Anderson, *Seven Ideas That Shook the Universe* (New York: Wiley, 1985).

For the mathematical background, use your old high school algebra text or another classic: *Mathematics for the Million,* by Lancelot Hogben (New York: Norton, 1983). It is one of the best introductions to many kinds of mathematics.

On the practical side of energy we suggest the *McGraw-Hill Encyclopedia of Science and Technology* (New York: McGraw-Hill, 1987), D. J. Rose's charmingly written and informative work, *Learning about Energy* (New York: Plenum, 1986), and the September 1990 issue of *Scientific American,* which is devoted entirely to energy.

The social, intellectual, and cultural consequences of the development of thermodynamics are described by Stephen G. Brush in *The Temperature of History: Phases of Science and Culture in the Nineteenth Century* (New York: Burt Franklin, 1979).

Work, Force, and Perpetual Motion

From the general references we suggest Trefil, *Physics as a Liberal Art,* either book by Feynman, and *DSB* on Galileo, Newton, and Leibniz. The early history of mechanics is given in *A History of Science: Ancient Science through the Golden Age of Greece,* by George Sarton (Cambridge, Mass.: Harvard University Press, 1952). Various perpetual-motion machines are described by A. W. J. G. Ord-Hume in *Perpetual Motion: The History of an Obsession* (New York: St. Martin's, 1977). Some articles on litigation between an inventor and the U.S. Patent Office include: R. Jeffrey Smith, "An Endless Siege of Implausible Inventions," *Science,* November 16, 1984, p. 817; Malcolm S. Browne, "Perpetual Motion," *New York Times,* June 4, 1985; and Boyce Rensberger, "Is It Perpetual Motion?" *Washington Post,* July 23, 1986.

Energy and the First Law

Most of the references cited in the "General" section above address the concept of energy, its history, and the law of its conservation. See especially the *DSB* articles on Lavoisier, Rumford, Kelvin, and Mayer, and Sanford C. Brown's biography, *Count Rumford, Physicist Extraordinary* (Garden City, N.Y.: Doubleday, 1962). It is fun to browse in *The Collected Works of Count Rumford,* 2 vols., edited by Sanford C. Brown (Cambridge, Mass.: Harvard University Press, 1968). Many of the original papers relevant to the discovery of the first law are reproduced in *Energy: Historical Development of the Concept,* edited by R. Bruce Lindsay (Stroudsburg, Pa.: Dowden, Hutchinson and Ross, 1975). Other historical essays include: T. S. Kuhn, "Energy Conservation as an Example of Simultaneous Discovery," in *Critical Problems in the History of Science* (Madison: University of Wisconsin Press, 1955); J. B. Conant, "The Overthrow of the Phlogiston Theory: The Chemical Revolution of 1775–1789," and D. Roller, "The Early Development of the Concepts of Temperature and Heat: The Rise and Decline of the

Caloric Theory," both in *Harvard Case Histories in Experimental Science,* volume 1, edited by J. B. Conant (Cambridge, Mass.: Harvard University Press, 1957); and Stephen G. Brush, "Should the History of Science Be Rated X?" *Science,* vol. 183 (1974), pp. 1164–1172.

Energy from the molecular standpoint is discussed in Trefil's book and in Brush, *The Kind of Motion We Call Heat.* The *DSB* articles on Maxwell and Boltzmann are worth reading.

Entropy and the Second Law

The history is covered by Cardwell, *From Watt to Clausius,* and in the *DSB* articles on Sadi Carnot, Lazare Carnot, Clausius, Kelvin, and James Thompson; the basic science by Feynman's lectures and Zemansky and Dittman, *Heat and Thermodynamics;* practical implications of the second law by Rose, *Learning about Energy.* A Dover reprint of Sadi Carnot's monograph, *Reflections on the Motive Power of Fire,* is out of print, but there is a new translation by Robert Fox (New York: Lilian Barber Press, 1986; Manchester: Manchester University Press, 1986). The *McGraw-Hill Encyclopedia* has articles on practical aspects of the heat pump (vol. 8, p. 360), the steam engine (vol. 17, p. 351), and refrigeration (vol. 15, p. 257). Martin H. P. Bott discusses the earth as a heat engine in *The Interior of the Earth: Its Structure, Constitution, and Evolution,* 2d edition (New York: Elsevier, 1982), pp. 354–358.

For the molecular interpretation of entropy see Brush's *The Kind of Motion We Call Heat,* Hogben on the theory of probability, and the *DSB* articles on Jean Perrin and Einstein. A more detailed biography of Perrin and a discussion of his scientific work is Mary Jo Nye's *Molecular Reality: A Perspective on the Scientific Work of Jean Perrin* (London: MacDonald, 1972).

Chaos and the Paradox of Irreversibility

James Gleick's deservedly popular *Chaos: Making a New Science* (New York: Viking, 1987) is the best place to begin. David Ruelle's *Chance and Chaos* (Princeton: Princeton University Press, 1991) gives the perspective of one of the discoverers of the phenomenon. *Does God Play Dice? The Mathematics of Chaos* (New York: Basil Blackwell, 1989), by Ian Stewart, is written at a popular level in spite of its subtitle. Edward Lorenz's classic paper is "Deterministic Non-periodic Flow," *Journal of Atmospheric Science,* vol. 20 (1963), pp. 130–141. Other articles include Robert Pool, "Is It Chaos, or Is It Just Noise?" *Science,* January 6, 1989, pp. 25–28; and Roderick V. Jensen, "Classical Chaos," *American Scientist,* vol. 75 (1987), pp. 168–181.

Roger Penrose offers his radical conjecture that a complete theory of physics, combining quantum mechanics and general relativity, will not be time-reversible in *The Emperor's New Mind: Concerning Computers, Minds, and the Laws of Physics* (Oxford: Oxford University Press, 1989). *The Arrow of Time: A Voyage through Science to Solve Time's Greatest Mystery,* by Peter Coveney and Roger Highfield

(New York: Ballantine Books, 1990), describes various attempts to solve the irreversibility paradox.

Information, Randomness, and Maxwell's Demon

Maxwell's Demon: Entropy, Information, and Computing, by H. S. Leff and A. F. Rex (Princeton: Princeton University Press, 1990), is a comprehensive history and discussion of Maxwell's demon. The short essay by Martin J. Klein, "Maxwell, His Demon, and the Second Law of Thermodynamics," *American Scientist,* vol. 58 (1970), pp. 84–97, is also informative, as is the *DSB* article on Smoluchowski. The entropy cost of making an observation, and its bearing on the paradox of the demon, is discussed in articles by Charles H. Bennet: "Demons, Engines, and the Second Law," *Scientific American,* vol. 244 (November 1987), pp. 108–116, and, with Rolf Landauer, "The Fundamental Physical Limits of Computation," *Scientific American,* vol. 242 (July 1985), pp. 48–56. The precise distinction between the terms *random* and *ordered* is clarified by Gregory J. Chaitin in "Randomness and Mathematical Proof," *Scientific American,* vol. 232 (1975), pp. 47–52. Brian Hayes reports a calculation in which pseudo-random numbers turned out surprisingly to be not random enough, in "The Wheel of Fortune," *American Scientist,* vol. 81 (1993), pp. 114–118.

Maxwell's description of his "demon" is from a letter to a scientific colleague and friend, P. G. Tait, and is quoted from a biography of Tait by C. G. Knott, *Life and Scientific Work of Peter Guthrie Tait* (Cambridge: Cambridge University Press, 1911).

Entropy and/or Information

The classic work on information is Claude H. Shannon and Warren Weaver's *Mathematical Theory of Communication* (Urbana: University of Illinois Press, 1963). K. G. Denbigh and J. S. Denbigh discuss the objectivity or subjectivity of entropy in *Entropy in Relation to Incomplete Knowledge* (Cambridge: Cambridge University Press, 1985). Richard P. Feynman, in one of his lectures (chapter 46, entitled "Ratchet and Pawl"), shows beautifully how Brownian movement leads to Clausius's macroscopic relation for the ideal reversible efficiency of a Carnot engine.

Radiation

Electromagnetic radiation in general, and black-body radiation in particular, are discussed in Feynman's lectures, Trefil, and Zemansky and Dittman. *Light,* by Mitchell I. Sobel (Chicago: University of Chicago Press, 1987), is excellent. The *DSB* has articles on Gustav Kirchhof, the discoverer of black-body radiation, and on Wilhelm Wien. The *McGraw-Hill Encyclopedia* has an article on heat radiation (vol. 8, p. 363).

On the greenhouse effect, we recommend John Gribbin's *Hothouse Earth: The Greenhouse Effect and Gaia* (New York: Grove, Weidenfeld, 1990); Stephen H.

Schneider's *Global Warming: Are We Entering the Greenhouse Century?* (San Francisco: Sierra Club Books, 1989); and, from the U.S. Environmental Protection Agency, "The Greenhouse Effect: How It Can Change Our Lives," *EPA Journal,* vol. 16, no. 1 (January/February 1989). By the time this book will have appeared, however, much more will be known about the effect; for the newest developments, check current issues of *Scientific American, Science,* and *American Scientist,* as well as daily newspapers.

Chemistry

Gilbert N. Lewis and Merle Randall's *Thermodynamics and the Free Energy of Chemical Substances* (New York: McGraw-Hill, 1923), is a classic for chemists, and though it is somewhat out of date much that it contains is still interesting. A more up-to-date introduction to chemistry is by John W. Hill, *Chemistry for Changing Times,* 4th ed. (Minneapolis: Burgess, 1984). But, as we noted in regard to other subjects, a high school text on chemistry will provide necessary background. The *McGraw-Hill Encyclopedia* has articles on diamonds (vol. 5, pp. 181–184), iron metallurgy (vol. 9, pp. 406–409), carbonated soft drinks (vol. 7, pp. 286–287), and hemoglobin (vol. 8, pp. 403–406). More advanced discussions of the hemoglobin-oxygen reaction are in I. Tinoco, K. Sauer, and J. C. Wang, *Physical Chemistry: Principles and Applications in Biological Sciences* (Englewood Cliffs, N.J.: Prentice-Hall, 1978), and Lubert Stryer, *Biochemistry,* 3d ed. (San Francisco: W. H. Freeman, 1988).

Biology

Basic discussions of energy in biology are in the following: Lubert Stryer, *Biochemistry,* 3d ed. (San Francisco: W. H. Freeman, 1988); Isaac Asimov, *Life and Energy* (New York: Doubleday, 1962; out of print, but worth hunting for); Franklin M. Harold, *The Vital Force: A Study of Bioenergetics* (San Francisco: W. H. Freeman, 1986); and two works by Harold J. Morowitz, *Foundations of Bioenergetics* (New York: Academic, 1978) and *Energy Flow in Biology: Biological Organization as a Problem in Thermal Physics* (Woodridge, Conn.: Ox Box Press, 1979). Energy use during exercise is discussed in R. J. Shephard, *Physiology and Biochemistry of Exercise* (New York: Praeger, 1982); G. A. Brooks and T. D. Fahey, *Exercise Physiology: Human Bioenergetics and Its Applications* (New York: Wiley, 1984); and Rodolfo Margaria, *Biomechanics and Energetics of Muscular Exercise* (Oxford: Clarendon, 1976).

Transport through cell membranes is discussed in the works by Stryer and by Harold cited above, and by Wilfred D. Stein in *Channels, Carriers, and Pumps: An Introduction to Membrane Transport* (New York: Academic Press, 1990).

Bernd Heinrich's *Bumblebee Economics* (Cambridge, Mass.: Harvard University Press, 1979) is not pertinent to our discussion of biology, but it is a fascinating application of the first law to the energy-gathering strategy of bees.

The following works state the creationist position: Robert E. Kofahl, *The*

Handy-Dandy Evolution Refuter (San Diego: Beta Books, 1977); Duane T. Gish, "The Origin of Biological Order and the Second Law" (pp. 67–90), and Emmet L. Williams, "Resistance of Living Organisms to the Second Law" (pp. 91–110), both in *Thermodynamics and the Development of Order*, edited by Emmet L. Williams (Kansas City, Mo.: Creation Research Society Books, 1981). References critical of the Creationist position include: Robert Shapiro, *Origins: A Sceptic's Guide to the Creation of Life on Earth* (New York: Summit Books, 1986); Robert W. Hanson, ed., *Science and Creation: Geological, Theological, and Educational Perspectives* (Washington, D.C.: AAAS/Macmillan, 1986); Laurie R. Godfrey, ed., *Scientists Confront Creationism* (New York: Norton, 1983), especially the essay by John W. Patterson, "Thermodynamics and Evolution" (pp. 99–116); and Ernan W. McMullin, ed., *Evolution and Creation* (Notre Dame, Ind.: University of Notre Dame Press, 1985). A somewhat deeper and more speculative discussion of the relation between thermodynamics and evolution is in B. H. Weber, D. J. Depew, and J. D. Smith, *Entropy, Information, and Evolution: New Perspectives on Physical and Biological Evolution* (Cambridge, Mass.: MIT Press, 1988).

The Age of the Earth

There are two fine studies of the controversy, one by Joe D. Burchfield, *Lord Kelvin and the Age of the Earth* (New York: Science History Publications, 1975), and one by A. Hallam, *Great Geological Controversies* (Oxford: Oxford University Press, 1983). The paper by Kelvin in which he seems willing to sacrifice the second law is reprinted as paper number 265 in volume 6 of his collected works, *Mathematical and Physical Papers*, edited by Sir Joseph Larmor (Cambridge: Cambridge University Press, 1911). It is titled "An Attempt to Explain the Radioactivity of Radium" and was originally published in the *Philosophical Magazine*, vol. 13 (March 1907), pp. 313–316.

Quantum Mechanics

Begin with either book by Feynman, or with Trefil or Spielberg. The *DSB* biographies of Bohr, Heisenberg, Planck, and Schrödinger are also interesting. More comprehensive treatments at a popular level are by Fred Alan Wolf, *Taking the Quantum Leap: The New Physics for Non-scientists* (San Francisco: Harper and Row, 1981), and Heinz R. Pagels, *The Cosmic Code: Quantum Physics as the Language of Nature* (New York: Simon and Schuster, 1982). The quantum dilemmas of the nature of reality are engagingly confronted in David Mermin's article, "Is the Moon There When Nobody Looks? Reality and the Quantum Theory," *Physics Today*, vol. 38 (1985), pp. 38–47.

Relativity

In addition to the Spielberg book and the *DSB* article on Einstein, we suggest Banesh Hoffmann's *Relativity and Its Roots* (San Francisco: W. H. Freeman,

Scientific American Books, 1983), which we consider one of the best popular books we have seen on this subject. Also good are Robert Geroch's *General Relativity from A to B* (Chicago: University of Chicago Press, 1978), and Delo E. Mook and Thomas Vargash's *Inside Relativity* (Princeton, N.J.: Princeton University Press, 1987).

Cosmology

The First Three Minutes: A Modern View of the Origin of the Universe, by Steven Weinberg (New York: Basic Books, 1977), is out of date on many points but it is still the best introduction to the subject. A more up-to-date source is by Robert M. Wald, *Space, Time, and Gravity: The Theory of the Big Bang and Black Holes*, 2d ed. (Chicago: University of Chicago Press, 1992). We personally have found Stephen Hawking's *A Brief History of Time: From the Big Bang to Black Holes* (New York, Bantam, 1988) very difficult to understand, but who are we to contradict the judgment of millions? Roger Penrose's *The Emperor's New Clothes: Concerning Computers, Minds, and the Laws of Physics* (Oxford: Oxford University Press, 1989) is somewhat idiosyncratic, but it covers just about everything in physics. Two articles specifically concerned with the future of entropy under various scenarios are by Steven Frautschi, "Entropy in an Expanding Universe," *Science,* vol. 217 (1982), pp. 593–599, and Martin Rees, "The Collapse of the Universe: An Eschatological Study," *The Observatory,* vol. 89 (1969), p. 193.

Black Holes

Evidence for the existence of black holes is discussed by Charles H. Townes and Reinhard Genzel in "What Is Happening at the Center of Our Galaxy?" *Scientific American,* vol. 247 (April 1990), pp. 46–55, and by Martin Rees in "Black Holes in Galactic Centers," *Scientific American,* vol. 263 (November 1990), pp. 56–66. Speculations about the role they may play in a contracting universe are reported by Frederic Golden in "Theory and Whimsy Take Physicists on Tour through a Black Hole," *New York Times,* June 9, 1992, pp. C-1 and C-9.

Cosmology is a field in which startling discoveries are being made almost every week, and these are reported in the daily newspaper no less than in journals like *Scientific American, Science,* and *American Scientist.*

Einstein's statement quoted in the last chapter of this book is from "Autobiographical Notes," in *Albert Einstein: Philosopher-Scientist,* edited by P. A. Schilpp (Evanston, Ill.: Library of Living Philosophers, 1949). It was also quoted in an article worth reading in its own right, M. J. Klein's "Thermodynamics in Einstein's Thought," *Science,* vol. 157 (August 4, 1967), pp. 509–516.

Credits

Page 114, quote from Sadi Carnot: From *Reflections on the Motive Power of Fire by Sadi Carnot and Other Papers on the Second Law of Thermodynamics by E. Clapeyron and R. Clausius*, ed. E. Mendoza (New York: Dover, 1960).

Page 221, quote from James Clerk Maxwell: From Cargill Gilston Knott, *Life and Scientific Work of Peter Guthrie Tait* (Cambridge: Cambridge University Press, 1911).

Page 240, Figure 10.6 (top): Reprinted from M. P. Thekaekara, "Solar Radiation Measurement," *Solar Energy* (1976), pp. 309–325, with permission from Pergamon Press Ltd, Headington Hill Hall, Oxford OX3 0BW, UK.

Page 249, Figure 10.8: From M. W. Zemansky and R. H. Dittman, *Heat and Thermodynamics*, 6th ed. (New York: McGraw-Hill, 1981); reproduced with permission from McGraw-Hill, Inc.

Page 271, Figure 11.3: From I. Tinoco, Jr., K. Sauer, J. C. Wang, *Physical Chemistry: Principles and Applications in Biological Sciences*, 2d ed. (Englewood Cliffs, N.J.: Prentice-Hall, 1985); © 1985, reprinted by permission of Prentice-Hall, Englewood Cliffs, New Jersey.

Pages 306–307, quote from John Playfair; pages 308 and 311, quotes from Lord Kelvin; page 313, quotes from Charles Darwin and T. H. Huxley; page 315, quote from Ernest Rutherford: From Joe D. Burchfield, *Lord Kelvin and the Age of the Earth* (New York: Science History Publications, 1975).

Page 360, Figure 15.4: From William W. Porterfield, *Inorganic Chemistry: A Unified Approach* (Reading, Mass.: Addison-Wesley, 1983), reprinted with permission from William W. Porterfield and from Academic Press, publisher of the forthcoming new edition.

Page 391, quote from Albert Einstein: From M. J. Klein, "Thermodynamics in Einstein's Thought," *Science*, vol. 157 (August 4, 1967), pp. 409–516.

Index